At an elementary level, this volume ties together the episodes in the life history of a star, be it a modest one like the sun or a massive object destined to form a supernova. It explains how we find the chemical composition of a star, and how elements are manufactured in the dense central cores late in a star's life. The role of the interstellar gas and dust whence new stars are formed is described in detail. An account is given of how mysteries of high-energy particles like cosmic rays are explored – the new frontiers of astrophysics.

Atoms, stars, and nebulae

The nebula 30 Doradus. (Mount Stromlo Observatory.)

Atoms, stars, and nebulae

Lawrence H. Aller

Third edition

The right of the
University of Cambridge
to print and sell
all manner of books
was granted by
Henry VIII in 1534.
The University has printed
and published continuously
since 1584.

CAMBRIDGE UNIVERSITY PRESS

Cambridge

New York   Port Chester   Melbourne   Sydney

Published by the Press Syndicate of the University of Cambridge
The Pitt Building, Trumpington Street, Cambridge CB2 1RP
40 West 20th Street, New York, NY 10011-4211, USA
10 Stamford Road, Oakleigh, Melbourne 3166, Australia

First edition published by The Blakiston Co, Philadelphia, PA, USA 1943
Second edition published by Harvard University Press 1971
Reprinted 1975
Third edition published by Cambridge University Press 1991

Printed in Great Britain at the University Press, Cambridge

British Library cataloguing in publication data

Aller, Lawrence H. (Lawrence Hugh)
Atoms, stars, and nebulae.–3rd. ed.
1. Astrophysics
I. Title
523.01

Library of Congress cataloguing in publication data available

ISBN 0 521 32512 9 hardback
ISBN 0 521 31040 7 paperback

SE

To Hugh and Margo

Contents

Preface   xi

Acknowledgements   xiii

1   Introducing the stars and nebulae   1

2   Stellar rainbows   15

3   Atoms and molecules – stellar building blocks   35

4   The climate in a stellar atmosphere   57

5   Analyzing the stars   72

6   Dwarfs, giants, and supergiants   104

7   What makes the stars shine?   127

8   The youth and middle age of a common star   142

9   Wind, dust, and pulsations: a star's last Hurrah!   167

10   The interstellar medium and gaseous nebulae   204

11   Uncommon stars and their sometimes violent behavior   255

12   High-energy astronomy   281

Epilogue   314

Appendix A   Designations of stars and nebulae   317

Appendix B   Some physical quantities and relations useful in astronomy   319

Appendix C   The ionization and excitation formulas, and curves of growth   324

Appendix D   Astronomical constants   330

Appendix E   Stellar magnitudes and colors   331

Appendix F   Interstellar molecules   347

Appendix G   The determination of stellar masses   350

A selected bibliography   351

Name index   356

Object index   360

Subject index   362

Preface

Half a century has elapsed since Leo Goldberg and I set out to write the first edition of *Atoms, Stars, and Nebulae*. It was an exciting epoch when great progress was being made. Quantitative chemical analyses of stars were being successfully carried out. Subatomic sources of stellar energy had been specifically identified and there was under way the first groping toward a comprehensive picture of the life history of a star. We were just beginning to recognize the interstellar medium as the material from which new stars and possibly solar systems were emerging – perhaps under our very eyes.

Subsequent developments have greatly extended these early efforts. A vast range of astronomical phenomena – from chemical composition differences between stars, to details of the spectrum-luminosity diagram and stellar variability – have been embraced in a ubiquitous theory of element building, stellar structure, and evolution. Bits and pieces of seemingly unrelated data now often emerge as essential connecting links in a steadily sharpening picture.

All the physical processes involved in our interpretation of observed features of stars and nebulae then had one thing in common: everywhere matter shone because it was hot, reflected and scattered light, or fluoresced. Then, in the late 1940s, as the new technique of radio astronomy was developed, a brand new window was opened on the universe. Through this window the outer world looked strangely different. Copious amounts of power were emitted by streams of charged particles moving with nearly the velocity of light in vast magnetized clouds in the deep recesses of space. Additional windows are now available. The infrared, the domain of heat radiation where we could see but darkly, is intensively being explored – thanks to great technological advances. Observations with satellites flown above the earth's atmosphere have wonderfully expanded our horizons. The International Ultra-violet Explorer, IRAS, and Einstein are but three examples of instruments that have revolutionized our understanding of the ultraviolet, the infrared, and the X-ray regions. Ground-based radio observations, together with X-ray and gamma-ray detectors flown in satellites, have established the active field of high-energy astrophysics. The mysterious cosmic rays, long a province worked by a small band of devoted physicists, were shown to be an integral part of the expanding scene. Radio galaxies and quasars revealed powerhouses of unbelievably high wattage radiating in remote space, while pulsars made sense only in terms of incredibly dense cores of defunct stars, where the very nuclei of the atoms, themselves, were simply squeezed beyond redemption. In some instances, matter was even further crushed into black holes from which nothing, neither particle nor radiation, can ever escape.

In the following pages, I have attempted not only to summarize the state of the art in what might now be called 'classical astrophysics' but have also tried to explore that mysterious magnetized realm of high-energy particles and bizarre power sources, where we are yet hampered by the fragmentary character of much of our information.

At the end of the 1930s, some astronomers felt that we already stood at the threshold of an understanding of the essential processes of stellar origin, development, and expiration. In spite of what appears to have been steady progress towards understanding these particular processes, the wealth of data supplied through the new radio, infrared, ultraviolet, X-ray, and gamma-ray windows has presented us with a greater array of puzzles than ever before. They might be likened to a creator's Zen koans, loaded with apparent contradictions, but carrying in their resolution a fuller understanding and enlightenment as to the nature of the enchanting universe in which we live.

Possibly the single most tantalizing feature is the omnipresence of magnetic fields. Detectable fields a hundred thousand times weaker than the earth's field are found in interstellar space. At the other extreme, the feebly glowing dense cores of some defunct stars have magnetic fields twenty million times stronger than that of the earth!

Composing even a quick, broad-brush sketch of contemporary astrophysics has been a frustrating assignment. The explosive expansion of the field has constituted only one problem. So many facets of new knowledge are disclosed that it is very difficult to select or emphasize those that may lead to the most exciting future progress.

Lawrence H. Aller

Acknowledgements

First of all, I must express my gratitude to my former colleague, Leo Goldberg, with whom I collaborated in the first edition of *Atoms, Stars, and Nebulae*. His example of lucid writing, and his good advice have served as an inspiration to me for many years.

Numerous people have helped in the preparation of this third edition by reviewing the manuscript and supplying data and illustrations. Mirek Plavec read the entire text and furnished valuable suggestions throughout. Benjamin Zuckerman critically reviewed Chapters 7, 8, 9, and 10 for which he made very helpful comments. D.M. Popper supplied up-to-date information on fundamental properties of stars; i.e. temperatures, radii, luminosities, and masses. I express my thanks to Nancy Houk for evaluating the section on spectrum and luminosity classes and for supplying illustrations. Lee Anne Willson critically assessed the material on pulsating stars, particularly Miras. Mike Jura, Mark Morris, and Bruce Balick supplied helpful comments and illustrations for the chapter on gaseous nebulae and the interstellar medium, Chapter 10. S.E. Woosley provided illuminating insights on the section on supernovae; Sumner Starrfield supplied valuable data pertaining to ordinary novae. Ned Wright read Chapter 12 on high-energy astrophysics, while Hugh Aller made many much appreciated suggestions for this chapter.

As for the numerous illustrations, it seemed best to give the credits in the captions, but I would particularly like to express my gratitude to Bruce Balick for loaning me his collection of CCD image-prints for planetary nebulae, and to C.L. Carilli for the remarkable VLA images of Cygnus A. For introducing me to some of the word-processing techniques, I am thankful to C.D. Keyes.

1

Introducing the stars and nebulae

To people of ancient times the universe was a stable, if not always secure, place, created, so it seemed, for the sole convenience of humanity. That man's abode, the earth, should occupy the dominant central position could scarcely be doubted, while the sun's justification for existence was to provide mankind with light and life-sustaining energy. The gleaming stars, fixed in the revolving celestial sphere, were regarded as bits of a cosmic mural designed to beautify the night.

It was only natural, too, that the details of the celestial scenery should have become identified with heroes and objects of mythology, identifications that remain in current use as names of star groups, or constellations. Thus the unexcelled constellation of the winter sky is Orion, the mighty hunter whose club is upraised against the charging bull Taurus (Fig. 1.1). Three bright equally spaced stars represent his belt; a misty group of stars forms his sword. Behind Orion his two dogs pursue Lepus, the fleeing hare. Marking the eye of the greater dog is sparkling Sirius, the Dog Star. To the ancient Egyptians Sirius was the popular Nile star, whose rising just before the sun foretold the impending flooding of the Nile. Sirius was distasteful to the Greeks, however, for they believed that the blending of its rays with those of the August sun produced hot summer weather. In Persian mythology, Sirius was Tishtrya, the Great Rain Star, who battled Apaosha, the demon of drought. Immortalized in the constellations are Hercules, the Nemean Lion, Hydra, Perseus and Andromeda, and the equipment of gods and heroes – Jason's ship *Argo*, the Harp of Orpheus, and the arrow from the bow of Chiron.

With the passage of time, these legends, which represented people's earliest attempts to relate themselves to their surroundings, became replaced by objective studies of the stars. The astronomical explorer has found the universe a treasure-house of existing discoveries wherein, to add zest to the chase, each great addition to knowledge has brought forth scores of fresh unsolved problems. Mysteries will continue to appear as long as there are people to ponder them.

In this book we embark on a journey of astronomical exploration in which the reader may sample a little of the thrill of discovery. During the course of our journey we shall probe the seething atmospheres of the stars and even dig into the interiors themselves. We shall encounter all kinds of curious objects, not only single stars, multiple stars, dwarf stars, giant stars, pulsating stars, stars with fantastic magnetic

Fig. 1.1. Mythological map of the sky in the neighborhood of Orion.

fields, and some whose surface layers are occasionally torn off in cataclysmic stellar explosions, but also clouds of gas and smog, the bizarre pulsars, sources of X-rays and gamma rays, and mysterious emitters of huge amounts of energy, whose nature is not understood at the present time.

Our course among the stars has already been charted, for, in broad outlines at least, the geography of the local regions of the universe is known. The earth is but one of a family of planets, satellites, planetoids, and meteoric particles that revolve periodically about the sun. In its turn the sun is but one of a vast host of stars, about two hundred thousand million, which are grouped together in the form of a thin lens-like system. This stellar system, which contains all naked-eye stars as well as many millions too faint to appear visually, is known as the Galaxy or the Milky Way System. The sun's position is at a point approximately two-thirds of the way from the center to the circumference. The broad outlines of galactic structure are now pretty well established, thanks largely to new observational techniques described in

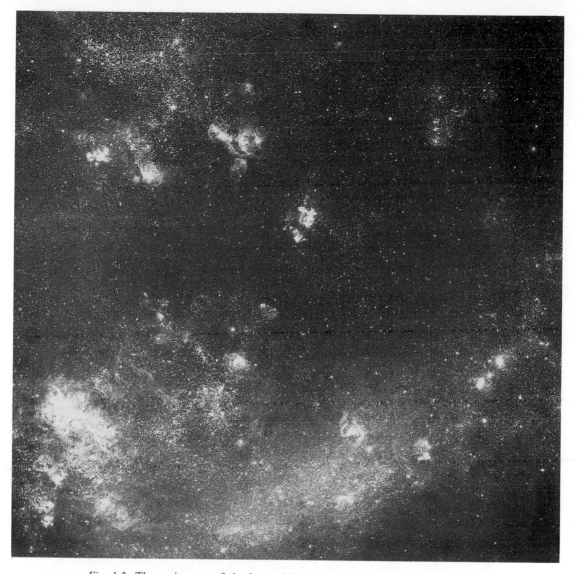

Fig. 1.2. The main part of the Large Magellanic Cloud. This galaxy has a large number of nebulosities or glowing clouds of gas as well as many stars and star clusters. The most spectacular nebula is 30 Doradus, at the centre of the nebulosities lower left (see frontispiece). (Photographed with the Uppsala Schmidt Camera at Mount Stromlo Observatory, courtesy Bengt Westerlund.)

later chapters, but many details still elude us, since we are within the system. Our Galaxy, however, is but one of thousands of millions of far-flung systems that lie in the observable universe. The nearest external galaxy is the Large Magellanic Cloud in the southern hemisphere (See Fig. 1.2). Another example is the Triangulum Spiral, M33 (i.e. no. 33 in the catalogue of nebulae and clusters compiled by Charles Messier). In structure our Galaxy more closely resembles (Fig. 1.3) than the

Fig. 1.3. The Triangulum Spiral, Messier 33. This spiral galaxy, which is much smaller than the Andromeda Spiral, M31, or our own Galaxy, is seen nearly in plan. It has well-defined spiral arms but no prominent central bulge. There are large numbers of gaseous nebulae and luminous stars, similar to those found in our own Galaxy or in the Magellanic Clouds. (Lick Observatory, University of California.)

irregular Magellanic Clouds, but it is much larger than M33. Studies of the Andromeda Spiral, M31, have revealed its similarity in size, form, and stellar content to our Milky Way. In both of these galaxies there are pronounced bulges and many of the stars in the main disks are arranged in well-defined spiral arms. Our emphasis is going to be on the gaseous nebular, dusty, and stellar contents of our own Galaxy, but we shall also make reference to other systems.

Despite the space-penetrating powers of large optical telescopes as well as powerful radio telescopes there is no indication that we have reached the boundaries of the universe, if any indeed exist. Most of our tour of exploration will be within the confines of our own Galaxy or the local group of galaxies, but we have reason to believe that our sample is more or less typical of the universe as a whole.

The local group of galaxies – Andromeda (M31) and its companions M32 and NGC 205, the Triangulum Spiral (M33), the two Magellanic Clouds, and several fainter objects such as NGC 6822 and IC 10, and of course our own Galaxy – contains samples of most of the principal types of stellar systems. Their relative proximity and the advent of new, large telescopes equipped with efficient instrumentation make them attractive objects for studies of gaseous nebulae, clusters, and individual stars.

The voyages between stars are likely to be smoggy, for interstellar space is strewn with great clouds of gas and solid grains of dust that dim and redden the light from the stars beyond. Like powerful searchlights, bright stars illuminate many of these clouds, revealing them to the astronomical explorer as bright nebulae. This interstellar matter is spread so thinly that, by comparison, the density of gas present in the best laboratory vacuum seems enormous. Yet, despite its extreme tenuity, enough dust is scattered between the stars to hide from view distant regions of our Galaxy. The interstellar gas emits radiation in the optical, infrared (heat), ultraviolet, and radio-frequency ranges, and much progress has been made by studying the interstellar matter particularly with radio telescopes and infrared detectors.

In addition to the above-mentioned radiations (all of which are examples of what are called electromagnetic radiation (see Chapter 2), fast-moving, stripped nuclei of atoms of familiar elements, mostly hydrogen, are found in space and impinge upon the earth. These are called cosmic rays. They are actually particles, and their properties are discussed in Chapter 12.

One of the features of our tour of discovery is that it can be made without the usual perils of exploration. In fact, owing to the magical powers of light rays, X-rays, gamma rays, and radio waves, we can explore far corners of the universe without leaving the comfort and security of the earth. Radiations that are absorbed in the earth's atmosphere (most of the infrared or heat radiation, some radio waves, and all X-rays and gamma rays) can be studied by rockets and satellites flown above the earth's atmosphere.

The astronomer of a century ago mapped the positions of stars upon the sky and designated their locations much as a geographer maps the earth from accurate measurements of latitude and longitude upon its surface. The positions of the stars are found from the directions of the light rays they emit, but direction is only one characteristic of light rays. Starlight also carries a message about the physical nature of the stars, their masses, brightnesses, chemical compositions, surface temperatures, and even the nature of their internal structure. Radio waves from clouds of interstellar gas tell something about their temperature, density, and chemical composition, and reveal the presence of large-scale magnetic fields. Only relatively recently have we learned to read these hieroglyphic messages from the stars and nebulae. Modern physics, which describes how atoms behave and how they can radiate light, has made this analysis possible. The story of the interpretation of stars and nebulae, so highly dependent on the findings of modern physics, is one emphasized in this book.

We shall start by describing the most obvious properties of stars and nebulae as revealed by their optical radiation. Historically, of course, no other option was possible. Restriction to the optical region led to the conception of an essentially 'thermal' universe. Radiation was emitted by hot bodies and subsequently scattered and absorbed by cooler objects. The development of radio astronomy, ultraviolet, and X-ray techniques, showed this view to be much too simplistic and opened the door to the exciting vision of a more complex universe.

Stellar distances and brightnesses

Four obvious questions will occur immediately to anyone interested in probing the physical nature of the stars, namely, how distant, how bright, how big, and how heavy they are. To answer these questions we must employ measuring rods and scales that can be applied over large distances. An astronomer uses the same principle to measure the distance of a star that a surveyor uses to measure the distance across a lake. Fig. 1.4a illustrates the surveyor's problem; Fig. 1.4b the astronomer's. The former measures the length of the line AB and the angles ABC and CAB. The determination of two angles and an included side serves to fix the dimensions of the triangle ABC and side AC or BC may be computed. Analogously, the astronomer uses as his baseline AB the diameter of the earth's orbit around the sun. When the earth is at A the star lies in the direction AC; six months later the earth is at B and direction of the star is now BC. One-half of the angle of displacement, i.e. the angle BCD or ACD, is called the parallax of the star. The amount of the shift clearly depends on the proximity of the star, the more distant ones being the least affected. (Actually, the star is not at rest but is moving in a straight line with respect to the sun, and additional observations must be secured to obtain both the parallax and the motion of the star across the line of sight. The motion across the line of sight is called the 'proper motion' and is measured in angular units.)

The unit of stellar parallax is the second of arc (or arcsec or ″); it is 1/3600 of a degree, i.e. it is about the angle subtended by a small coin at a distance of 4 kilometers (2.5 miles). So small an angle cannot be distinguished with the unaided eye, but modern telescopes permit parallaxes of 0.01 arcsecs to be measured with fair accuracy. An astrometric satellite to replace the ill-fated Hipparchos, would enable the measurement of the parallaxes of about 100 000 stars to an accuracy of 0.002

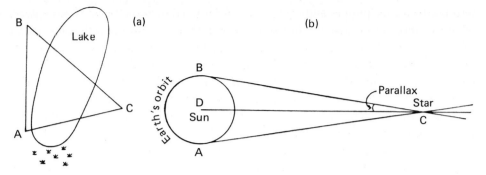

Fig. 1.4. Measurement of distances by triangulation. (a) The surveyor measures the distance A to B, plus the angles at A and B, from which the triangle ABC can be constructed and the distances across the lake obtained. (b) The astronomer uses the diameter of the earth's orbit as the baseline and measures the parallax angle of the star.

arcsecs, that is about five times better than the best ground-based instrument. This accuracy of angular measurement corresponds roughly to the angle subtended by a man standing on the moon. In the radio range, even smaller angles can be measured with the Very Long Baseline Intercontinental (VLBI) radio array.

The parallax of α Centauri, the nearest star, is 0.752 arcsecs, corresponding to a distance of about forty million million kilometers or twenty-five million million miles. To express distances such as this in kilometers or miles is more awkward than giving the distance from London to New York in millimeters; hence stellar distances are often expressed in light-years, at least in popular writing. One light-year, the distance traversed in one year by a ray of light traveling at 299 793 kilometers (186 000 miles) per second, is nearly 9.5 million million kilometers (6 million million miles). The nearest star is 4.33 light-years away; Sirius, which appears as the brightest star in the sky, is at a distance of 8.7 light-years, and our entire system of stars, the Milky Way, probably measures 100 000 light-years across. On such a scale our solar system seems tiny indeed: if we represent the distance from the earth to the sun by 15 millimeters, 1 light-year corresponds very nearly to 1 kilometer. (If the distance to the sun is represented by one inch a light-year is about 1 mile.)

Two other units of distance, the astronomical unit and the parsec, are also useful in astronomy. For expressing distances intermediate between the kilometer (or mile) and the light-year, the radius of the earth's orbit, which is called the astronomical unit (abbreviated AU), is commonly used. (This unit should not be confused with the ångstrom unit, abbreviated Å, used to express the wavelength of light.) The parsec is the distance of a star whose parallax is 1 arcsec; it is equal to 206 265 AU or 3.26 light-years. Since parallax is inversely proportional to distance, the distance in parsecs is simply the reciprocal of the parallax in arcsecs. Thus a star 10 parsecs or 32.6 light-years away has a parallax of 0.1 arcsec, a star 100 parsecs or 326 light-years away has a parallax of 0.01 arcsec and so on.

The surveyor's method of measuring parallax is inadequate for any but the very nearest stars; angles smaller than 0.01 arcsec cannot be measured accurately by ground-based techniques and even our best-designed astrometric satellites will be able to supply parallaxes for stars only within about 500 parsecs, that is only to about 1/15th the distance to the center of the Galaxy. Fortunately, astronomers have devised ways of estimating the distances of remoter stars. One can take advantage of the fact that the stars are in motion, both in regard to one another and with respect to the sun. The actual velocity of motion along the line of sight can be measured by the Doppler effect (see Chapter 2). Then, by measuring the apparent angular motions of selected stars across the sky in different parts of the celestial sphere, one can obtain average or statistical distances much as one could estimate the distance of a lighted speedboat seen on a harbor at night if one knew its actual speed in the water. Other methods, which are described later, are based on the principle that we measure accurately the intrinsic luminosities of certain kinds of stars that we can recognize in distant parts of the Galaxy or even in other stellar systems. Then, from the apparent brightness of the star and its known intrinsic brightness, we can get its distance, since the brightness of a point source of light diminishes as the square of its distance. If α Centauri were 8.66 instead of 4.33 light-years away it would appear one-fourth as bright.

Conversely, if the distance of a star has been found, we may, knowing its apparent brightness, establish its true brightness. Our current practice of expressing the apparent brightness of a star as seen in the sky in terms of magnitudes was initiated 2000 years ago, when ancient astronomers graded the stars from the first (brightest) to the sixth magnitude, the latter being just barely visible to the naked eye. For more than a century, the magnitude scale has been so adjusted that a star of the first magnitude is exactly 100 times as bright as one of the sixth magnitude. The scale goes as a geometrical progression, that is, the brightness ratio corresponding to a one-magnitude step is constant. Thus, a first-magnitude star is 2.512 times as bright as a second-magnitude star, which in turn is 2.512 times as bright as a third-magnitude star, and so on. The original scale of six magnitudes has been extended to include the very faint as well as the very bright stars. Stars as faint as the 23rd or 24th magnitude can be detected with the aid of photoelectric cells or with what are called charge-coupled devices (CCDs) with large telescopes. The brighter stars in the sky, like Aldebaran and Altair, are of the first magnitude. The two very brightest stars in the sky, however, have negative magnitudes; thus the magnitude of Canopus is -0.7, and that of Sirius is -1.6. On the same scale the apparent magnitude of the full moon is -12.7 and that of the sun is -26.8.

Stellar magnitudes may be measured with the eye or with other light-sensitive devices such as the photographic plate, with the photoelectric cell, or with charge-coupled devices, with the aid of appropriate filters (see Chapter 6). By using different filters the color of a star may be measured. The visual magnitudes measured by early observers have been replaced by photoelectric magnitudes measured with a yellow filter – the so called V-magnitudes. If we wish to express the apparent brightness of a star taking into account all the radiation it emits – infrared, red, green, blue, violet, and ultraviolet (Chapter 2) – we use the bolometric magnitude. The bolometric magnitude is a quantity derived from the observations and the temperature of the star (see Chapter 4); it will be an observed quantity only when stellar brightnesses can be measured from above the earth's atmosphere. Both very cool and very hot stars are very much brighter bolometrically than visually, since most of their energy is emitted as radiation to which our eyes are not sensitive.

Were all stars equally distant from us, their apparent magnitudes would represent their true relative brightnesses. In practice, we define the intrinsic luminosity of a star by its so-called absolute magnitude, which is the apparent magnitude it would have at a standard distance of 10 parsecs = 32.6 light-years (see Appendix E). The bolometric absolute magnitude of the sun is +4.69. This is the quantity that is important when we want to actually compare the energy outputs of stars. The absolute 'photoelectric visual' magnitude of the sun is +4.83 (according to Popper; see Appendix D), which means that at a distance of 10 parsecs it would be comfortably visible on a clear moonless night. Arcturus, whose distance is about 33 light-years, would appear at about its present brightness. Sirius would be about 1/14th as bright as at present and no longer conspicuous. Rigel, in the constellation Orion, which is 50 000 times brighter than the sun, would outshine any object in the present night sky save the moon.

Until recently, all of what we knew of the universe had been discovered by the detection and measurement of radiation by optical methods, that is with devices employing ordinary lenses and mirrors. A great technological breakthrough was provided by radio astronomy. It was found that stars, gas clouds, and galaxies also emit radio waves in addition to light and heat waves. Many radio telescopes are parabolic reflectors, thus resembling instruments of the optical astronomer, but even the largest of these give limited angular resolution. Modern developments entail the use of arrays of radio reflectors (or 'dishes' as they are commonly called). The individual dishes can be moved relative to one another along tracks and placed in different positions and configurations. With such a device, which is called an interferometer, it is possible to obtain observations of very high angular resolution. The largest and most successful example of this type of telescope is the Very Large Array (VLA) in New Mexico in the USA. It is also possible to link radio telescopes in different hemispheres. Then, measurements of a thousandth of an arcsec are possible.

As seen through the eye of a radio telescope, the sky has a totally different 'appearance' from that in visible light. Most of the radio radiation comes from gas clouds rather than from individual stars; hence the familiar constellations are not seen in the radio telescope, but are replaced by a variety of radio sources that have quite a different arrangement in the sky.

Weighing the stars

The motion of the earth about the sun makes possible the determination of stellar distances. Curiously enough, the circling of one star about another permits the determination of stellar masses. Like all planets, and stars too for that matter, the earth is imbued with a wanderlust. Were the restraining influence of the sun's gravitational attraction suddenly to be removed, the earth would fly off in a straight line and eventually lose itself in interstellar space. Just as the earth is kept in its path by the gravitational attraction of the sun, so also are a large number of stars denied a carefree existence by the gravitational attractions of companion stars. Stars so inhibited pursue circular or ellipitical orbits about each other. The more massive the two stars, the faster will they move about each other, which we may see from a simple analogy.

Suppose we were in a spaceship in interstellar space, where there was no gravitational attraction, so that we floated freely about, and suppose further that it was necessary to measure the mass of a small solid object. Since gravity would not exist inside the spaceship, we could not just put the object on a scale and weigh it; some other technique would have to be used. If a spring scale were available, the unknown mass could be found by attaching the object to the scale and swinging them both in a circle at the end of a string. The spring scale would measure the tension in the string, which would depend on the speed of revolution and the mass of the object. The tension would be greater, the greater the mass or the greater the speed of revolution. From the measured tension and the speed of whirling, we could find the mass of the object.

By an analogous procedure the astromomer weighs the stars. The rate of motion of two stars in a double-star system about each other depends on the gravitational force between them. This attractive force, analogous to the tension in the string, is proportional to the masses of the stars (and also to the inverse square of the distance between them), according to Newton's law of gravitation. By observing the time required for the two stars to circle each other (the period) and measuring the distance between them, we find the restraining force and hence the masses.

Double-star, or binary, systems are common among the stars. In fact, groups have been found in which three, four, five, and even six stars revolve about one another. Some of these multiple systems merit a brief description.

α Centauri consists of two stars that revolve about each other in 80 years in rather elongated ellipitical orbits, so that at times they approach as near as 11 astronomical units (a little more than the distance of Saturn from the sun) and sometimes they recede to 35 astronomical units (nearly the distance of Pluto from the sun). The brighter component is almost a duplicate of the sun, save that it is a little brighter and perhaps a little heavier and a little hotter. The fainter component is cooler and less massive. In 1915, R.T.A. Innes discovered a faint red star 2 degrees away that shares the same motion through space as α Centauri but is 15 000 times fainter than the sun. It is at least 10 000 or 12 000 astronomical units from the brighter pair and must take about a million years to complete its orbit.

Of particular interest is the lesser Dog Star, Procyon, a binary with a period of 40.65 years and a mean separation of 4.55 arcsecs, which corresponds to a distance of 15.8 astronomical units, somewhat less than the separation of the sun and Uranus. The brighter (V-magnitude of 0.35) star has a mass about 1.75 times that of the sun. The companion is a very faint star (with a magnitude of 10.8). It is an aged, superdense star, commonly called a white dwarf (see Chapter 9). Now the orbit of the bright star is known in terms of shape, orientation, and diameter (in arcsecs). Also the motion of the bright star in the line of sight can be measured from its spectrum (see Chapter 2) and, since the orbit is known, the orbital speed can be found. Then from the period one obtains the actual diameter of the orbit in kilometers or miles, which can be compared with the diameter of the orbit obtained from the separation of the stars in arcsecs and parallax. The relation is

$$\text{Diameter of orbit} = \frac{\text{Diameter (arcsec)}}{\text{Parallax (arcsec)}} \times 149\,600\,000 \text{ km.}$$

In this way Kai Strand obtained an independent check on the stellar parallax, which he found to be in good agreement with the trigonometric value.

Among multiple stars ζ Cancri and Castor are worth mentioning. The brighter component of ζ Cancri is itself a binary consisting of two nearly equally bright components revolving about each other with a period of 59.7 years. The fainter component also comprises two stars, one of which can be detected only by its gravitational effects on the other. They revolve about each other with a period of 17.5 years. This fainter pair revolves about the brighter one with a period of 1150 years. The masses of all four stars are comparable with that of the sun.

As seen in a telescope, Castor is a double star, whose components, separated by a distance of about 80 astronomical units, move about each other with a period of 340 years. Spectroscopic observations (see Chapter 2) show that both stars are actually double, with periods of about 9 and 3 days, respectively. An even more interesting result is that more than 1 minute of arc in the sky away from Castor is a faint star, Castor C, associated physically with the brighter pair. This object itself consists of two faint red stars, smaller and less massive than the sun, separated by about 2 700 000 kilometers (1 700 000 miles) and revolving about each other in less than a day. Thus Castor is a sextuple star whose components appear in pairs.

To obtain reliable masses from measurements of visible double stars, accurate parallaxes are required. A 10 percent error in the parallax will give about a 30 percent error in the masses. Hence we see the importance of satellite studies. Fortunately, as Henry Norris Russell and Einar Hertzsprung independently showed, it is possible to get good average values of the masses of stars whose parallaxes are known, even if we observe the motion of a star over only a part of its orbit.

The multiple systems are composed of all kinds of stars, large and small, cool and hot; consequently we are able to determine masses for most varieties of stars. When the weighing operation is completed, we find that the most massive stars are between 50 and 100 times heavier than the sun; the lightest stars have probably between one-fifth and one-tenth the solar mass, with the majority weighing a bit less than the sun.

In many instances a star trapped in a double-star system is forced to reveal not only its mass but also its size. Double-star components are frequently so close together that even the most powerful telescope fails to reveal them separately. If, however, the plane of the orbit is so tilted as to appear edgewise in the sky, the

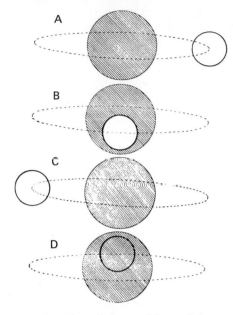

Fig. 1.5. The relative positions of the stars in an eclipsing binary system during three-quarters of a period. A small bright star revolves around a larger one of much lower surface brightness. At B the bright star is in front of the dim one, whereas at D it is behind it.

passage of one star in front of the other will produce a periodic eclipse, not unlike an eclipse of the sun by the moon. Such double stars are known as eclipsing binary stars. In general, each star masks the other once during a revolution, thus producing two eclipses per cycle. If the two stars are of equal size and brightness, the amount of light received on the earth will be cut in half twice during a revolution. Usually, however, the components of known eclipsing stars are of unequal brightness and size; the pairing of a large, faint star with one that is small and bright is a frequent occurrence. Such a pair of stars is shown diagrammatically in Fig. 1.5. The passage of the bright star across the faint one produces a partial eclipse of the latter and a resultant dimming of the total light. Half a revolution later, the relative positions of the stars are reversed and, since the bright star is now obscured, the loss of light is much greater. If the observed brightness of the eclipsing binary is plotted against time, we find a periodic variation in the light as shown in Fig. 1.6. When the bright star is at positions in the orbit corresponding to A and C in Fig. 1.5, the light is undimmed. In D, the brighter of the two stars is obscured and the star is said to be at primary miminimum. In B, the faint star is partially obscured, only a small fraction of the light is lost, and the star is said to be at secondary minimum. It is clear that the duration of each eclipse, which may be learned from the light curve, depends both upon the diameters of the stars and upon the speed of revolution. Since, as we shall see in the next chapter, we may frequently determine the latter by means of the spectroscope, we can find the stellar diameters.

If, as is usually true, the plane of the orbit does not quite pass through the earth

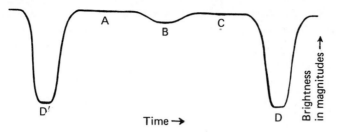

Fig. 1.6. The light curve of the eclipsing binary system shown in Fig. 1.5. The letters correspond to the positions indicated there. Note that the primary eclipse always occurs when the star of lower surface brightness is in front.

(i.e. the inclination is not exactly 90°), the situation will be as depicted in Fig. 1.5. From an accurately determined light curve, one can determine this tilt or inclination of the orbit, the sizes of the two stars in terms of the diameter of the orbit, and the ratios of the surface brightnesses of the stars (which depends on their surface temperatures).

We can do even more. If we know the orbital speeds in kilometers (or miles) per second and the period, we can find the size of the orbit. From the period and orbital radius we can find the masses of the stars, in terms of that of the sun (see Appendix G). Since the stellar sizes relative to their orbits are known from the light curve, and since the orbital size is found from the spectroscopic measurements, the diameters of the stars are known in kilometers (or miles). With both its diameter and its mass known, the density of star may be found. In some instances, where the light curve has been followed for several decades, we can even ascertain something about the rate at which the density increases toward the center in the heavier star of the eclipsing system. These studies show that the stars are not homogenous; rather the density increases markedly toward the center (see Chapter 8).

Tables 6.2 and 6.3 list masses, sizes, periods, and densities for some well-known eclipsing binaries for which data have been obtained. Perhaps the best known of these systems is Algol, the second brightest star in the constellation Perseus, which at intervals of 2.87 days suddenly fades to about one-third its usual brightness. The brighter component has about three times the diameter of the sun, while the larger but much fainter component measures 3.2 solar diameters. Further discussion of these systems is deferred to Chapter 6.

One further point must be mentioned. Although we can learn much from eclipsing binaries, we must recognize that many of the stars found in eclipsing systems are abnormal objects in the sense that similar stars are not found singly, nor in very wide binaries. The reason is that the evolution of a star in a close binary system is severely modified from the course that evolution would take if the stars were alone. A bright star accompanied by a large dim companion is the most likely type of eclipsing binary to be detected. This large dim companion has suffered some severe evolutionary effects (see Chapter 8). This situation offers some interesting possibilities for understanding certain remarkable variable stars.

Although the separation of stars in binaries ranges from a few times their diameters in systems of the W Ursae Majoris type (where two solar-sized stars swing about each other almost in contact) to thousands of astronomical units, most double-star systems seem to be built on a scale not greatly different from that of the solar system. Frequently it has been suggested that the formation of a solar system and that of a binary system are different aspects of the same fundamental process. Usually the primordial material is assumed to be collected in two or more large masses, forming a binary or multiple system, but sometimes much of it may be lost and a star may be surrounded by a system of planets.

There exists a lower limit to the mass that a body may have and yet shine as a star (see Chapter 8). This borderline mass is about 0.08 the mass of the sun. Less massive objects, often called brown dwarfs, may exist in great numbers but are difficult to detect. The best example is the companion to a white star called G29–38 or ZZ Piscium, discovered by E. Becklin and B. Zuckerman and further analysed by Jesse Greenstein. It is a small, cool, faint object, possibly somewhat resembling the planet Jupiter. Its surface is near 1000 °C (about 1270° Kelvin), its luminosity is about 0.0004 that of the sun, and its radius is about 0.15 that of the sun. It moves in a small orbit close to the white dwarf star.

Stars as luminous as ten million suns and a hundred times the solar mass exist. Such objects rapidly lose mass and energy and are short-lived. The best known examples are found in multiple systems which might be likened to miniclusters. One example is the object R136, a hot source that powers the great 30 Doradus Nebula in the Large Magellanic Cloud. (The 30 Doradus Nebula is shown in the frontispiece of this book.) R136 consists of eight bright stars compacted into a volume subtending about one arcsec squared upon the plane of the sky, that is an actual volume about 0.75 of a cubic light-year at the distance of the Large Cloud. An even more remarkable object is the 'minicluster' in NGC 3603 where stars of similar luminosity are contained in an even smaller volume.

Mention must be made of remarkable objects of extreme density and low luminosity. Besides the white dwarf stars, with densities about 50 000 times that of water (which are the final stages of stars like the sun), there exist neutron stars and pulsars of even higher density (see Chapters 11 and 12). Black holes represent the ultimate extreme. No light can escape from them and all matter is irredeemably crushed. These objects are revealed by their gravitational effects and the fact that in their vicinity are produced high-energy particles and radiation (gamma rays).

Dwarf stars, normal stars, giants, supergiants, clouds of dust and gas all go to make up the Milky Way. But the fundamental building blocks for all material structures are tiny atoms, a few million-millionths of a centimeter in diameter. From atoms and molecules come the light rays that enable us to see and study the stars and nebulae. It is our good fortune that the kind of light emitted by atoms is controlled by their physical environment. Thus the light rays from galactic space carry with them vivid code messages of the climatic conditions in the stars and nebulae. We now turn to the story of how the message of starlight is decoded.

2

Stellar rainbows

The spectroscope

The fact that sunlight is composed of a mixture of colors was discovered in 1666 by Sir Isaac Newton. He admitted sunlight into a darkened room through 'a small Hole in [his] Window-shuts' and then allowed the light to pass through a triangular glass prism and to fall on the opposite wall of the room (Fig. 2.1). The original spot of white light was replaced by a brilliant rainbow or spectrum of colors, arranged in a band with violet at one end and changing slowly to blue, green, yellow, orange, and finally red at the other. By placing a second prism in reverse orientation behind the first, Newton demonstrated that they could be recombined and white light would be restored. Thus 'white' sunlight was proved to be actually a mixture of all the colors of the rainbow.

The glass prism sorts out the separate colored rays by changing their directions by amounts that depend on the color of the light. When a light ray passes from one medium to another its direction usually changes (Fig. 2.2). (Physically this bending, or refraction, of a ray of light arises from the fact that the velocity of light in the denser medium, such as glass, is lower than in air.) Were all light rays deviated by the the same amount in passing through a prism, the emergent light beam would be uncolored. However, the violet rays are bent more than the blue rays, the blue rays more than the green, the green more than the yellow, with the result that the original white light is spread, or dispersed, into its component colors. Similarly, droplets of water in the earth's atmosphere act like tiny prisms and disperse the sunlight to produce the rainbow.

The prism spectroscope (Fig. 2.3) is essentially patterned after Newton's experimental arrangement. To prevent overlapping of the separate colors, the light source is first focussed on a narrow slit, perhaps 0.005 centimeter wide. After passing through the slit, the diverging beam is collimated, or made parallel, by a lens at C, and then directed through a glass prism, D. The lens T then brings the rays to a focus along the line PP′. The spectrum at PP′ consists of a series of 'lines', which are images of the slit, each of a different color; we may examine it with an eyepiece, photograph it upon a plate or film, or use some other type of detector. In an alternative form of the spectroscope, the prism is replaced by a so-called diffraction grating. In its most

Fig. 2.1. Newton's experiment on spectra. (*Elémens de la philosophie de Newton,* Voltaire, 1738, Amsterdam.)

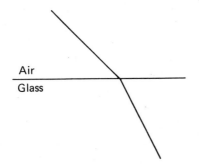

Air

Glass

Fig. 2.2. Refraction of light at a surface between two media. In glass, light travels at a slower speed than in air; violet light is slowed down more than red light.

widely used form the grating consists of a flat, reflecting surface upon which a series of exceedingly fine parallel grooves are ruled with a microscopically sharp diamond point. The grooves are uniformly and closely spaced, up to 1200 per millimeter or 30 000 per inch. When parallel white light falls upon the grooved surface, the component colors are reflected at different angles and are thus dispersed into a spectrum.

In conjunction with a large telescope, the spectroscope in its various forms is the single most important observational device used in astrophysics, and the results obtained with it will be the subject of much of the remainder of this book. Before discussing this instrument further, however, we shall first comment on the physical meaning of color.

The meaning of color

Just what is meant by the color of a light ray? The sensation of color is purely subjective, resulting from the response of the retina in the eye to some physical property of light. Laboratory experiments have shown roughly that light is propogated in the form of waves, at a speed (in vacuum) of 300 000 kilometers

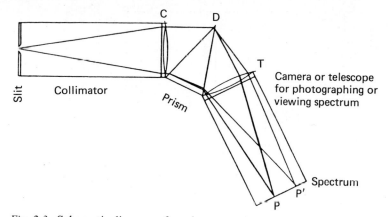

Fig. 2.3. Schematic diagram of a prism spectroscope.

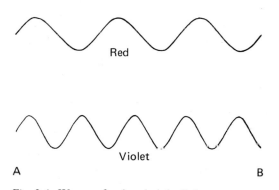
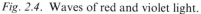

Fig. 2.4. Waves of red and violet light.

(186 000 miles) per second. The distance between successive crests or troughs in the waves is known as the wavelength. The interesting property of light waves is that the phenomenon of color, which is a physiological sensation, is directly related to the wavelength of the light; red light waves are the longest waves visible, the yellow ones are shorter, and the waves of violet light are the shortest that can be seen. The wavelength of red light, for example, is about 1/1500 of a millimeter or 25/1 000 000 of an inch, whereas the wavelength of violet light is only about 1/2500 mm or 16/ 1 000 000 of an inch. Two different waves of light, one red and the other violet, are shown schematically in Fig. 2.4. Both waves move from A to B in the same time, since the velocity of all light is the same in a vacuum. Since the violet ray has a shorter wavelength than the red ray, it undergoes a greater number of vibrations over the same distance. The number of such vibrations per second, or the frequency of the light wave, is equal to the velocity of light divided by the wavelength. Thus the frequency of short-wavelength violet light is 750 million million vibrations per second. (This number is usually abbreviated to read 7.5×10^{14}, where 10^{14} signifies the number 1 followed by 14 zeroes; similarly, the reciprocal of 10^{14} is written

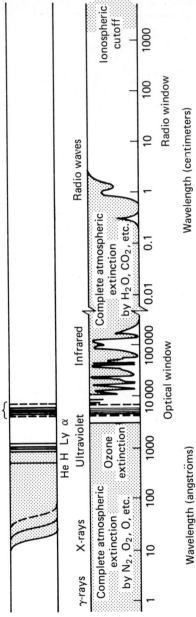

Fig. 2.5. The spectrum of electromagnetic waves from gamma rays to long radio waves. In the lower strip, the optical and radio 'windows' are indicated by white areas and the regions of atmospheric extinction by shaded areas. The upper strip shows the range of radiations accessible to detectors flown in a satellite or rocket above the earth's atmosphere. Absorption by hydrogen, and to some extent by helium, cuts off the light of distant stars in the ultraviolet far beyond the cutoff of the earth's atmosphere. Note, however, that γ-rays, most X-rays, and other regions of the spectrum can be observed without much interference. The narrow range of wavelengths to which the eye is sensitive is indicated. (Adapted from *Astrophysics: The Atmospheres of the Sun and Stars* L.H. Aller, 1963, 2nd edition, Ronald Press, NY, p. 4.)

10^{-14}.) The frequency of violet light is roughly 2500/1500 or $1\frac{2}{3}$ times that of the longer-wavelength red light.

Wavelengths are often expressed in ångström units, named in honor of the Swedish physicist, A.J. Ångstrom. One ångström (abbreviated Å) is equal to one hundred millionth (1×10^{-8}) of a centimeter or about four thousand millionths of an inch. It is now fashionable to express wavelengths in nanometers (1×10^{-9} meter); thus one nanometer (nm) equals 10 Å. The strong red radiation of atomic hydrogen has a wavelength of 6563 Å = 656.3 nm. We frequently express wavelengths in micrometers or in microns (one micron [μm] is 10 000 Å, 1000 nm, or one thousandth of a millimeter). Thus the above-mentioned red hydrogen radiation has a wavelength of 0.6563 μm.

The limited color sensitivity of the human eye confines the visible portion of the spectrum to a strip extending from 4000 Å in the violet region to about 7000 Å in the red. However, various devices for detecting radiant energy, such as the photographic plate, the photoelectric cell, and the thermocouple, show that the radiation spectrum extends far on either side of the visible region. Immediately shortward of the violet region lies the ultraviolet, which can be detected by photographic plates or photoelectric detectors. Solar ultraviolet radiation produces sunburn. Yet farther lies the region of the soft X-rays (10 Å to 100 Å approximately.) High-frequency or hard X-rays fall in the neighborhood of 1 or 2 Å; still higher frequencies are represented by gamma (γ) rays, which are emitted by elements like radium or artificially produced radioactive substances.

Longward of the red lies the infrared, which merges continuously through heat rays into the region of the microwaves, 'short' radio waves, and ultimately broadcast radio waves hundreds of meters long. Up to 1.0 μm or even 1.2 μm, the near infrared (i.e. that nearest to visible light in wavelength) can be studied by reticons, charge-coupled devices (CCDs) or even photographic plates. At yet longer wavelengths, we use detectors such as lead sulfide cells, lead telluride cells, Golay cells, and bolometers. Millimeter and centimeter waves can be detected by ultra-high-frequency radio receivers, while meter waves can be measured by more conventional receivers.

Although stars radiate energy at all wavelengths, most spectral regions cannot be observed from the ground, owing to absorption by atoms and molecules in the earth's atmosphere. All parts of the spectrum are at least partially extinguished by gases of the atmosphere, but in certain wavelength regions the absorption is so thorough that no radiation at all can penetrate the atmosphere, even when observations are made from mountain tops. The shaded areas in Fig. 2.5 show spectral regions suffering inroads of severe atmospheric extinction. No radiation of wavelength shorter than about 3000 Å can reach the surface of the earth. The shortest wavelength radiation, up to about 1000 Å, is absorbed by oxygen and nitrogen atoms at heights of more than 100 kilometers above the ground. From 1000 to 2300 Å, the radiation is blocked by molecules of oxygen and nitrogen, while between 2300 and 3000 Å, the absorbing agent is ozone. The spectral regions visible to the human eye are relatively free of obstruction, but at longer wavelengths large sections of the infrared spectrum are hid, mostly by molecules of water vapor and carbon dioxide. Finally, our atmosphere becomes totally opaque and remains so until the millimeter region is reached, when it again becomes transparent, this time to radio waves up to about 20 meters long, beyond which the ionosphere cuts out all radiation.

As we shall see, particularly in Chapters 10 and 12, the radio 'window' furnishes an amazing new view of the universe which emphasizes the role of high-energy particles and magnetic fields in interstellar space and in compact objects like pulsars. Investigation of spectral regions outside the optical and radio 'windows' requires telescopes and detectors flown above the earth's atmosphere in satellites and rockets. Solar and stellar ultraviolet spectra reveal extremely important clues to the structure of stellar atmospheres. X-rays are emitted by numerous celestial sources including catastrophic variable stars such as novae, supernovae, and sundry binaries containing white dwarfs, neutron stars, and black holes. Cygnus X-3, a binary hidden by dust clouds, does even better. Not only is it a strong X-ray source, but it also spews out gamma rays produced by the acceleration of charged particles in a dense neutron star.

In order to study the ultraviolet region, astronomers have had to devise spectrographs with special optical materials. In the X-ray region 'grazing incidence' optics are used. The incoming radiation is directed in such a way as to always make a very small angle with respect to the surface, instead of angles typically like 70°–90°. Crystals serve as the equivalent of diffraction gratings in optical spectroscopy.

Atomic thumbprints

Newton's discovery that a light source like the sun radiates a brilliant spectrum of color, although artistically appealing, is hardly as significant as the fact that different kinds of light sources are distinguished by different types of spectra. What are generally called 'Kirchhoff's laws' describe some elementary but basic principles of spectroscopy. Suppose that, simulating the experiments of G.R. Kirchhoff and R. Bunsen, we were to place the glowing white hot tungsten filament of an incandescent lamp before the slit of a spectroscope. We would find that the spectrum consists of a bright, continuous band of colors, very similar, in fact, to the rainbow. A piece of iron, or any other solid, heated to red or white heat, but not vaporized, likewise displays a continuous spectrum. But now if we employ as our light source a glass tube filled with rarefied hydrogen that is carrying an electric current and therefore is luminescent, we observe a spectrum radically different from that of a shining solid. In place of a brilliant continuum there are four bright, colored lines, or slit images, red, blue, blue–violet and violet, the last near the limit of visibility at 4102 Å. We note that the spaces between the lines appear black, and also that there is a remarkable regularity in the positions of the lines, with the separations between successive bright images decreasing steadily from red to violet. On the photographic plate, the series continues into the ultraviolet, with the lines crowding together until they terminate near 3650 Å; see Fig. 2.6, which shows lines of hydrogen and other elements emitted by the Orion Nebula. The spectrum of heated sodium vapor likewise shows discrete

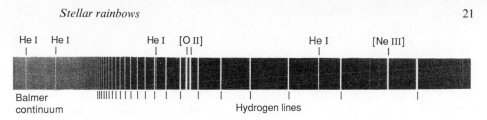

Fig. 2.6. The emission spectrum of the nebula in Orion, showing the Balmer series of hydrogen. This spectrum of a gaseous nebula shows how the hydrogen lines of the Balmer series converge to a limit, which is followed by a continuum. Note that lines of helium, indicated by He I, the so-called forbidden lines of oxygen [O II], and those of neon [Ne III] are also present. (Lick Observatory, University of California, 120-inch telescope with coude spectrograph.)

bright lines, notably a pair close together in the yellow and a series in the ultraviolet. Other glowing gases and vapors radiate bright-line spectra, too, but each element, be it hydrogen, helium, sodium, calcium, iron, lead, or radium, is marked by a different set of radiations, which the spectroscope sorts out as bright lines. Because no two elements display identical spectra, we see that nature has provided us with the means for fingerprinting every element. Once the spectra of the known elements have been recorded in the laboratory, the composition of any mixture may be ascertained, regardless of whether the sample to be analyzed is located on the earth or in a distant star or nebula.

If we now interpose cool sodium vapor between a hot tungsten filament and the slit of the spectroscope, we obtain still a third type of spectrum. In the visible part of the spectrum the brilliant continuum of color from the incandescent lamp appears unchanged except for two dark lines at precisely the same wavelengths at which the bright sodium lines had been seen before. The cooler sodium vapor has evidently absorbed light from the bright background, but only in those wavelengths it is capable of emitting. Similar results are obtained for other vaporized elements; their characteristic spectra appear as dark rather than as bright lines.

Experiments of the sort we have described led Kirchhoff to his three laws of spectroscopy; (1) an incandescent, that is glowing, solid or liquid or very dense gas radiates a continuous spectrum; (2) a rarefied glowing gas emits a characteristic bright-line spectrum; (3) the spectrum of a gas placed in front of a hotter source of continuous radiation consists of dark absorption lines at just those wavelengths that the gas emits when it is heated. Although the second and third laws are correct, the first law must be restated more carefully since it represents a sufficient but not a necessary condition for the production of a continuous spectrum. As we shall see in the next chapter, gases at low density can produce continuous spectra, so a continuous spectrum does not necessarily imply an incandescent solid or liquid, but a bright-line spectrum certainly implies a heated gas. In a very rarefied gas, the underlying continuous spectrum is often faint and far less conspicuous than the bright emission lines. A dark-line spectrum always implies a hotter, underlying source of continuous spectra, which may also be (as for the sun and stars) a heated gas.

In 1802, W. Wollaston, repeating Newton's experiment, found four dark lines in

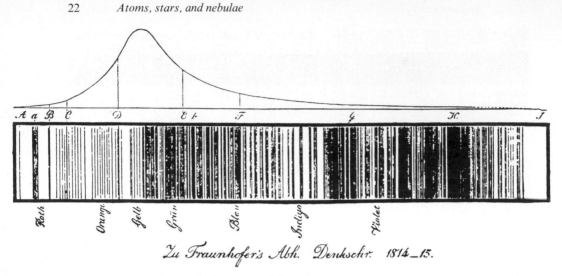

Fig. 2.7. Fraunhofer's map of the solar spectrum.

the spectrum of the sun and interpreted the lines as divisions separating the colors of white light: red, yellow–green, blue, and violet. Spectra of the sun obtained through liquid prisms containing nitric acid, oil of turpentine, oil of sassafras, and Canada balsam were similar, showing that the spectra did not depend on the dispersing medium. About 1815, J. Fraunhofer mapped 574 lines in the spectrum of the sun; a section of his map is shown in Fig. 2.7. The suggestion that these lines were caused by absorption in the atmosphere of the earth was disproved when Fraunhofer found that the spectra of several bright stars were quite unlike that of the sun (see Fig. 2.12). He also noticed agreements between the positions of lines of terrestrial elements and the dark lines of the solar and stellar spectra, but unfortunately attached no significance to the coincidences. In Fig. 2.8 we reproduce a portion of the solar spectrum in the near ultraviolet (i.e. the ultraviolet nearest to visible light).

Kirchhoff concluded from his observations that the sun and stars must be incandescent hot bodies surrounded by relatively cool thin atmospheres (Fig. 2.9). He proposed a simple model in which the elements comprising the gaseous atmosphere, or reversing layer, of a star absorb intense, continuous radiation emerging from the lower surface, or photosphere, and thus imprint their dark lines upon the spectrum. Kirchhoff's model is useful for visualizing the process of formation of spectral lines, and we shall employ it in some of our later discussions. In 1931, however, D.H. Menzel and others showed that this highly stratified model was an oversimplification; both the solar continuous radiation and the absorption lines originate in the same region of the atmosphere, which is still called the photosphere. It is true that on the average the continuous radiation comes from deeper atmospheric layers than do the absorption lines, but there is no sharp demarcation between the photosphere and the reversing layer. These developments revealed each star as a gigantic laboratory furnace in which matter often could be studied under extreme physical conditions not attainable on earth.

Sir William Huggins, an English amateur astronomer, and independently Sir

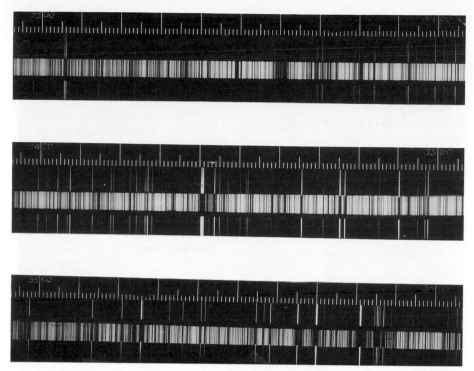

Fig. 2.8. A portion of the solar spectrum (center of each bar) with lines of the iron arc above and below it in the wavelength region 3300 3600 Å. The numerous coincidences between lines of the two spectra reveal the presence of iron in the sun. (Mount Wilson Observatory.)

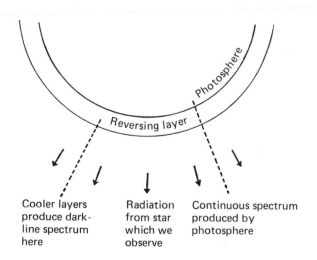

Fig. 2.9. Kirchhoff's model for the formation of the continuous and dark-line spectra of a star.

Fig. 2.10. The central region of the Orion Nebula. Note the star cluster and surrounding nebulosity which is caused to fluoresce by radiation from hot imbedded stars (see Chapter 10). The plate was obtained on 19 December 1957 by George H. Herbig at the prime focus of the Shane 3^m telescope on an IN emulsion which is sensitive to deep red and near infrared radiation. In conventional photographs and in visual observations, the cluster stars are overwhelmed by the bright nebulosity (see Fig. 10–21). (Lick Observatory, University of California photograph.)

Norman Lockyer, examined the spectra of a large number of stars, comparing the positions of the dark lines with those of bright lines emitted by elements in the laboratory. They found many coincidences and concluded that matter everywhere in the universe must be alike. The great Orion Nebula (Fig. 2.10), believed by many astronomers to be an aggregation of stars too far away and too close to one another to be resolved by existing telescopes, was expected to show a continuous spectrum. Huggins found instead, to his astonishment, that the spectrum (Fig. 2.11) consisted

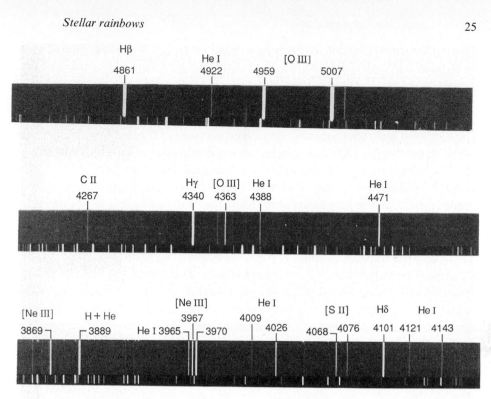

Fig. 2.11. The spectrum of the great nebula in Orion. A large number of emission lines of permanent gases, notably hydrogen, helium, oxygen, and neon, together with two lines due to sulfur, are indicated (wavelengths in ångströms). Many of these lines, in particular those of oxygen, neon, and sulfur noted here, are of the so-called forbidden type (see Chapter 10). (Lick Observatory, University of California, 120-inch telescope with coude spectrograph.)

entirely of a few bright lines, some of which could be identified with hydrogen and helium, although certain strong lines defied interpretation. These lines were originally attributed to a hypothetical element called 'nebulium', but were later identified as arising from doubly ionized oxygen (see Chapter 10). Later studies also showed a weak continuous spectrum to be present. This spectrum was actually produced by the rarefied hydrogen in the nebula. The Orion Nebula was thus found to be a cloud of low-density gas rather than a cluster of stars.

Sorting the stars

While Huggins was interested in the chemical composition of the stars, Father A. Secchi, at Rome, was attracted by the diversity in appearance among stellar spectra. Some, like the sun, featured large numbers of lines of metallic elements, notably calcium, sodium, and iron. Others showed only broad lines of hydrogen, while still others, the red stars, exhibited a wealth of complex detail, characterized by dark fluted bands. Secchi found that he could arrange the vast majority of stellar spectra into four distinct types, with all the stars in each group sharing roughly the same spectral features. This contribution was very important, for if the spectrum of a star were related to its physical characteristics, and if all the stars fell into one of four classes of spectra, the detailed study of one star might reveal the characteristics of many more. Secchi found that stars whose brightnesses fluctuated irregularly belonged to the class showing fluted spectra. Stars of Type 1, the blue and white stars, showed some tendency to collect in certain parts of the sky. For example, five of the stars in the Big Dipper, which form a physical cluster of stars moving through space in the same direction and with the same speed, are of this type.

Secchi's achievement was remarkable, considering that his observations were made visually, during long hours at the telescope. With the advent of photography, E.C. Pickering, director of the Harvard College Observatory, embarked on a huge program of spectral classification, with the collaboration of Mrs Williamina P.S. Fleming, Miss Antonia C. Maury, and Miss Annie J. Cannon. Pickering placed a large glass prism in front of the telescope objective, and used the lens to focus the spectra on the photographic plate. The advantage of the objective-prism technique is that a great many spectra may be photographed on a single plate, whereas the slit spectrograph records only one spectrum at a time.

The aim of the Harvard classification was to group the stars in such a way that the spectral features of one group merged as smoothly as possible into those of the next adjacent group. As the dark lines of hydrogen seemed to be common to all stellar spectra, the original plan called for labeling as Class A the stars with the most intense hydrogen lines, those with the next strongest hydrogen lines Class B, and so on down to Classes M and N, where the hydrogen lines are very weak. This scheme had to be modified for a number of reasons. Some of the classes, for example C, D, and H, had been derived from out-of-focus photographs and were spurious. Also the arrangement in order of decreasing hydrogen-line intensities produced discontinuities in the trends of other spectral lines. Class O, discovered later, was found to belong at the beginning of the sequence. As finally adopted, the classes follow the order O, B, A, F, G, K, and M. In addition, a few stars classified as R, N, and S appear to represent side branches jutting off from the main sequence near Class K. (If the reader finds it difficult to remember this peculiar arrangement of letters the sentence: 'Oh, Be A Fine Girl, Kiss Me Right Now, Sweet!' may help.)

The photographic plate shows such a wealth of detail that it has been necessary to divide each of the Harvard classes into subdivisions by affixing a number from 0 to 9 to each letter; thus the dark-line pattern of spectral Class A5 lies midway between those of A0 and F0.

According to this system, the sun was classified as G0 in the Henry Draper catalogue (see Appendix A), but modern work gives its classification as G2. In Fig. 2.12 the spectral classes, B, A, F, G, K, and M, as obtained by Nancy Houk and Michael Newberry with the Schmidt telescope of the University of Michigan, are shown. Notice how the spectra increase in complexity from Class B to Class M. The hydrogen lines grow steadily through Class B, reach their intensity maximum in Class A0, and then decline. Class B bears the imprint of helium, which is absent from the spectra of later types. In stars of Class O (not shown) helium is even more

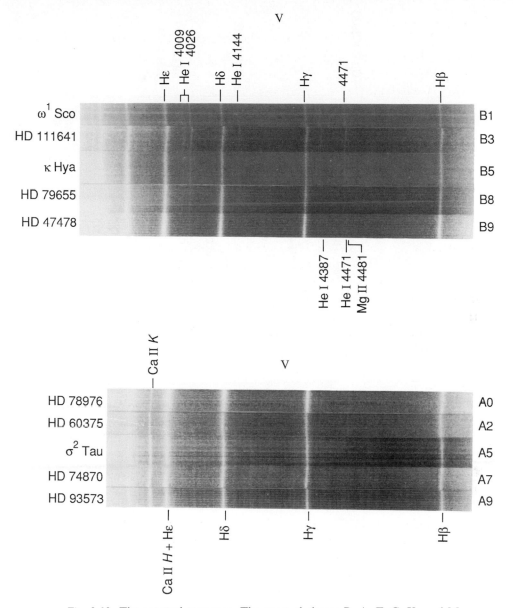

Fig. 2.12. The spectral sequence. The spectral classes B, A, F, G, K, and M are grouped in separate blocks and characteristic lines used for spectral classification are indicated. For Classes B, A, F, and G spectra of dwarf or 'main sequence' stars similar to the sun are shown. These are called luminosity class V stars (see Chapter 6). There were few dwarf K and no suitable M stars available on the Michigan objective prism survey plates so for these classes spectra of giant or luminosity class III stars are reproduced. (From *A Second Atlas of Objective Prism Spectra*, Nancy Houk and Michael V. Newberry, 1984, University of Michigan.)

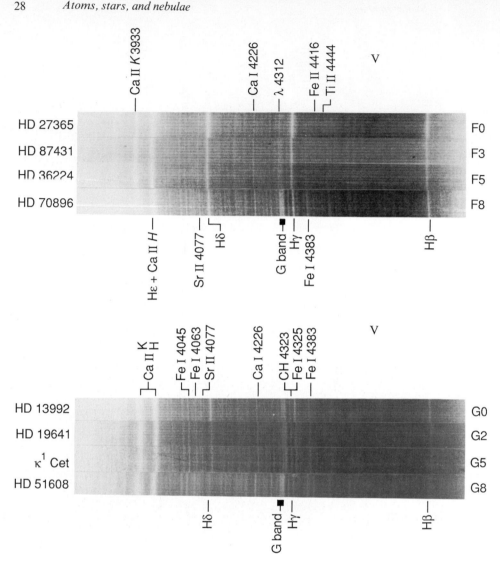

Fig. 2.12 (*cont.*)

prominent than in Class B, hydrogen is less prominent, and there occasionally appear bright emission lines. The lines of metals like calcium, sodium, nickel, and iron are first noticeable in Class A and rapidly grow in strength and numbers through Classes F, G, and K. The broad bands of molecular compounds creep into the picture in Classes G and K, becoming the outstanding landmarks in the spectroscopic maps in Classes M, R, N, and S. (See also Fig. 6.6, p. 115, which shows spectra of some other Class M2 stars.)

A very significant aspect of the spectral classification is that it also segregates the stars according to color. Furthermore, the colors along the sequence are arranged somewhat like those in a spectrum, the blue stars occurring at the beginning and the red stars at the end of the sequence. Thus the bright blue stars in the constellation of

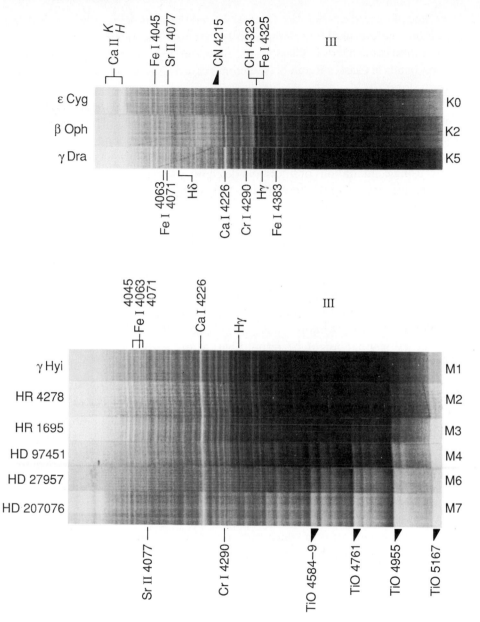

Fig. 2.12 (*cont.*)

Orion are of Class B. Sirius, a whitish star, is of Class A0, while the southern beauty, Canopus, is of class F0. Capella, brightest star north of the celestial equator and yellow like the sun, is of Class G0; Arcturus, the bright orange star of spring and summer, is of Class K0; and Betelgeuse and Antares, red stars of Orion and the Scorpion, respectively, are Class M stars.

The actual determination of the spectral class of a star depends upon the relative intensities of certain lines. The helium lines (in the hotter stars), the hydrogen lines, the K line of ionized calcium (see Chapter 4), and the 4226-Å line of neutral calcium are among the lines used for this purpose. In the cool stars one employs intensities of the titanium oxide bands in Class M, the zirconium oxide bands in Class S, and the carbon bands in Classes R and N (sometimes called Class C because they represent carbon stars).

These spectral differences are apparent even when the scale of the spectrum is very small, as it usually is with objective-prism plates or with plates secured with spectrographs intended for observations of very faint stars and distant galaxies. Spectral differences can also be recognized by accurate measurements of the color of a star, using combinations of filters and detectors to evalute the star's brightness in three or more, preferably narrow, intervals of the spectrum. Classifications of spectra can also be made by use of a spectrum scanner, in which the photographic plate is replaced by a photocell that is moved relative to the spectrum.

Another method involves an image tube scanner. A spectrum is formed on a phosphor, whose glow is monitored by an electronic sensor that sweeps along the spectrum and displays it on a television screen. Alternatively, the spectrum is recorded on a charge-coupled device (see Chapter 6).

It should be emphasized that this classification of stellar spectra was carried out solely on the basis of the appearance of the spectra themselves, without regard to physical causes that might be responsible. Many early workers believed these differences to be due to variations in chemical composition. Were the blackness of a spectral line dependent only on the abundance of the atom responsible for it, the stars could easily be arranged in order of steadily changing hydrogen abundance. It would be a most remarkable coincidence if stars arranged on this system also showed smoothly varying abundances of all other elements and if the hydrogen stars were always blue and the metallic stars red. In Chapter 4 it is shown that the variations in stellar spectra are due in most instances not to gross changes in chemical composition but to changes in temperature and density. Such variations in chemical composition as actually exist are usually relatively minor compared with effects of temperature or pressure. Exceptions are provided by highly evolved objects, such as cool stars in which carbon is more abundant than oxygen, by hot 'Wolf-Rayet' stars that show strong emission lines, and by aged superdense white dwarfs.

Stellar spectra are sometimes referred to as 'early' or 'late'. Thus Class B is 'earlier' than Class A, A is earlier than F, etc. This terminology is the relic of a long-defunct hypothesis of stellar evolution in which stars were presumed to start their lives as bluish or white objects, and gradually cool and become redder as they aged.

The spectroscope as a speedometer

The spectroscope reveals not only the compositions of stars, but also their speeds toward or away from the observer. To understand how the spectroscope can act as a speedometer, the reader should recall the high-pitched whistle that heralds the approach of a speeding train, and the sudden transition to a long-drawn-out wail that accompanies its passing and recession. The whistle emits sound waves of a definite frequency and wavelength, and the number of waves per second that strike the ear determines the pitch of the sound. When the train is in rapid motion toward the listener, the individual waves tend to crowd up on each other, and a greater number fall upon the ear every second. The increase in the number of vibrations per second is interpreted by the ear as a rise in pitch. Conversely, when the train is receding, the sound waves are drawn out and fewer of them per second strike the ear, which perceives that the pitch has fallen.

If light is propagated as a wave motion, a similar effect should operate, as was pointed out by Christian Doppler in 1842. Suppose that a source emits light of a certain frequency, which passes through the spectroscope and appears as a spectral line. The wavelength determines the position of the line, but when the light source is racing toward the observer, the light waves reach the spectograph more frequently and the wavelength seems shorter. Consequently, the spectral line is shifted from its normal position toward the violet. And when the light source is receding, the line moves over toward the red. The magnitude of the shift, which is known as the Doppler shift, is related to the speed of the light source by the equation

$$\frac{\text{Change of wavelength}}{\text{Normal wavelength}} = \frac{\text{Speed of source}}{\text{Speed of light}}.$$

(It makes no difference whether the light source or the observer is in motion; the important thing is the rate at which the two are approaching or receding from one another.) Thus, for example, the speed of light is 300 000 kilometers (186 000 miles) per second and if the light source is receding at 30 kilometers (18.6 miles) per second the position of a line at 5000 Å is shifted by 0.5 Å, an amount easily detected.

To measure the speed of a star, the spectral lines of a laboratory source – for example, iron, titanium, or thorium – are impressed on a photographic plate or charge-coupled device detector on either side of the stellar spectrum to serve as reference marks for measuring the position of stellar lines. By measuring the displacement of the stellar lines with respect to the comparison lines, the astronomer gets the radial velocity of the star, that is, its motion along the line of sight. The progressive displacement of a star's position on the celestial sphere, its 'proper motion', measures the angular speed at right angles to the line of sight. This angular speed can be converted to speed in astronomical units per year or kilometers per second provided we know the distance of the star. Then we can get the actual space motion of the star with respect to the solar system. Since we make the observations from the earth, we must correct for the effects of the earth's motion in its orbit. Figure 2.13 shows how the radial velocity of a star causes a shift in its spectral lines.

Very accurate methods of measuring radial velocities have been developed; some use patterns or 'stencils' of the whole spectrum to monitor simultaneously the displacements of many lines. When individual spectral lines can be identified in the radio range, precise radial velocity measurements are possible, even for objects hidden behind clouds of obscuring material.

Unless we are looking right down onto its pole of rotation, the Doppler effect

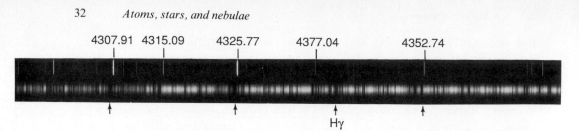

4307.91 4315.09   4325.77   4377.04   4352.74

Hγ

Fig. 2.13. A portion of the extremely complex spectrum of Arcturus, from which the radial velocity of the star can be determined. Lines of a comparison spectrum are shown at the top; corresponding lines in the stellar spectrum, several of which are indicated by arrows, are seen to be shifted toward the left, that is, toward shorter wavelengths, indicating that the distance between Arcturus and the earth is decreasing. In this cool star, the Hγ line of hydrogen at 4340 Å is no longer conspicuous. (Lick Observatory, University of California, 120-inch telescope with coude spectrograph.)

tends to blur out the spectral lines of a rapidly spinning star. One side of the star is approaching, the other side is receding, so that although a small patch of surface may emit a relatively sharp line, the net result is a broadened one. Altair (α Aquilae) and Sirius are both of similar spectral class but all the spectral lines in Altair are widened by the swift rotation of this star.

The spectroscope as a speedometer has also had important application in studies of the orbital motion of double stars. The components of many double stars are so close together that they cannot be separated by direct observation. However, when the plane of the orbit is tilted even slightly in the direction of the line of sight, each star appears now approaching, now receding, as it whirls about its companion. If the two stars are almost equally bright, the spectrum will exhibit a periodic doubling of the lines, when one star is approaching and the other is receding. Usually, however, one star is so much brighter than the other that only a single spectrum is seen and the lines of this spectrum oscillate to and fro as the velocity of the star with respect to the observer changes. A star whose duplicity is recognized from its spectrum is known as a spectroscopic binary. Mizar, the star at the bend of the handle of the Big Dipper, was the first star of this class to be detected, by E.C. Pickering in 1889. Many hundreds more have since been discovered. A catalogue by J.H. Moore and F.J. Neubauer of Lick Observatory gave orbits of over 500 spectroscopic binaries. A.H. Batten's catalogue (of 1978) lists 978 binaries and additional objects have been discovered more recently.

Polarized light

Infrared, visible, and ultraviolet light, X-rays, gamma rays, and radio waves are all electromagnetic waves, that is, waves in combined electric and magnetic fields (Fig. 2.14). Imagine a small, free body carrying an electric charge placed in the path of such a wave: it would be accelerated up and down as the electric field pointed in first one direction and then the other. The oscillations of the fields take place perpendicular to the direction of propagation of the wave; we speak of such waves as transverse

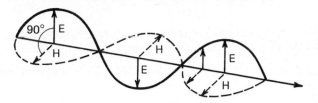

Fig. 2.14. An electromagnetic wave, consisting of a combination of an alternating electric field E and an alternating magnetic field H at right angles to each other.

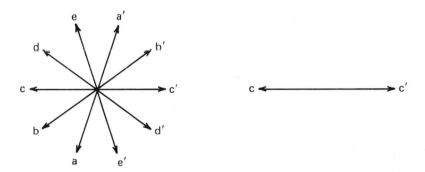

Fig. 2.15. Polarization of light: (left) orientations of the electric field at different instants; (right) all but one direction of vibration has been stopped by a polarizing filter.

waves, as distinct from longitudinal waves, such as sound waves or compressional waves in a solid or fluid.

An important property of transverse waves is that they can be polarized. Suppose we are looking along the direction in which a light wave is traveling (Fig. 2.15). At one instant the electric field will be along the direction aa′, an instant later it may be along the direction cc′, then along dd′, changing randomly at a great rate. If a piece of Polaroid is placed in the beam, all directions except, say, cc′ will be suppressed and the light ray is said to be plane polarized. By rotating the Polaroid, different directions of oscillation may be selected. If the light is initially polarized, the intensity of the transmitted light will be a maximum at some position of the Polaroid, zero at right angles thereto. If the light is partly polarized, there will be a variation of intensity as the Polaroid is rotated.

Another type of polarization is important. Suppose we looked along the direction of the beam and saw that the direction of the electric field rotated uniformly and with a frequency equal to the frequency of the light. The light would be said to be circularly polarized. A useful way to describe circularly polarized light is to think of two equally intense plane polarized pencils (mutually polarized at right angles to one another) but with the phase of one of the beams displaced $\pi/2$ or 90° with respect to the other. They are added together. A test of this beam with Polaroid would suggest that it was unpolarized. Suppose, now, that we inserted into the beam a slab of optically transparent material that had different properties in different directions

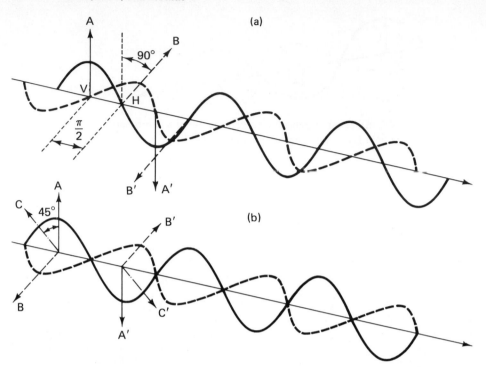

Fig. 2.16. Circularly polarized light.

(a) We can regard circularly polarized light as made up by the addition of two components of equal intensity, polarized perpendicular to one another, but with their wave patterns shifted by $\pi/2$ or 90°. The two directions of vibration are denoted by A and B. HB lies in the 'horizontal' plane (dashed line); VA lies in the vertical plane (solid line).

(b) With the quarter-wave plate introduced in the beam, A and B are now shifted so as to be in phase. the vertical vibration A is added to the horizontal vibration B to form a resultant vibration vector C that is inclined at an angle of 45°. We now have plane polarized light whose plane of vibration is tilted half-way between the original two out-of-phase sets of vibrations. Circularly polarized light has been converted to plane polarized light.

such that the component oscillating in the horizontal plane, *HB*, travels more slowly than the one in the vertical plane, *VA*. By carefully selecting the thickness of the slab, we can put the two waves back in phase such that the two maxima and minima occur at the same place with the angle of vibration making an angle of 45° with the horizontal direction, *HB* (see Fig. 2.16). Such a device is known as a quarter-wave plate. Since circularly polarized light has now been converted to plane polarized light, it can easily be cut out by placing a polaroid in the beam behind the quarter-wave plate.

Polarized light has been used to infer properties of surface rocks on the moon, on asteroids, and interstellar dust, but the most significant application of polarized radiation has been the identification of and measurement of celestial magnetic fields which play so important a role in modern astrophysics (see Chapters 3 and 12).

3

Atoms and molecules – stellar building blocks

Atoms and radiation

Where does light originate? When we press a switch at home, electrical energy flows through a wire and is somehow converted into light radiating from a tungsten filament. In some way, atoms, which are the tiny building blocks of all forms of matter, generate light of various colors or wavelengths when fed with fuel in the form of chemical or electrical energy. By what operation inside the atom is that light generated and why do different kinds of atoms radiate energy in different wavelengths?

Atoms are much too small to be seen; hence, experiments to find out their structure and behavior have to be conducted with large numbers of them. From the results of these experiments we may attempt to construct a hypothetical model of an atom that behaves like the true atom. Many such atomic models have been proposed in the past and have had varying degrees of success in reproducing the observed features of spectra. But all of them, at one time or another, have been contradicted by experiment. These failures have led to the conclusion that no purely mechanical model of the atom is entirely satisfactory; the laws of mechanics that govern the operations of large bodies break down when applied to ultramicroscopic particles. Entirely new laws of mechanics have had to be devised to cope with the behavior of atoms. These laws are embodied in the so-called wave mechanics, or quantum mechanics, which has thus far given a completely successful account of atomic behavior. The operation of these laws, although perfectly straightforward mathematically, is somewhat difficult to visualize. For this reason, even the scientist who makes his or her calculations according to the mathematical laws of quantum mechanics frequently thinks of the atom in terms of some simple mechanical model.

It may be worth while to recall here the differences between atoms and molecules. The chemists have shown that the many gases, liquids, and solids which make up the world are composed of the pure forms or combinations of fundamental substances, called elements, which may combine to form compounds. Thus water is composed of hydrogen (two parts by volume) and oxygen (one part by volume). The smallest particle of an element is an atom; the smallest particle of a compound is a molecule. The molecule of water consists of two hydrogen atoms bound to one oxygen atom, thus, HOH. We must distinguish between mixtures or alloys, such as brass, in which the atoms are loosely mixed with one another, and compounds, where the individual atoms that make up a molecule are tightly bound together.

What atoms are made of

Experiments in the laboratory have shown that the chief materials of atomic construction are three extremely small particles, which have been labeled electrons, neutrons, and protons. The electron, which carries a negative electric charge, is the least massive of the three. It would take 311×10^{26} (that is, 311 followed by 26 zeros) of them to weigh 1 ounce. Expressed in grams, the mass of an electron is 9.11×10^{-28} (1 gram $= 3.5 \times 10^{-2}$ ounce). The neutron and the proton are nearly equal in mass, weighing 1836 times as much as an electron, or 1.66×10^{-24} gram. A piece of dust 0.025 millimeter or 0.001 inch in diameter would still weigh a thousand million million times as much as a proton. The electric charge associated with atomic particles is conveniently expressed in terms of the charge of the electron, which is taken as -1. In the electrostatic system of units this charge is 4.803×10^{-10} esu; in the practical or SI system of units it is 1.602×10^{-19} coulomb. The proton carries a positive electric charge, equal in numerical magnitude to that of the electron, or $+1$, whereas the neutron, as its name implies, is electrically neutral.

Electrons, mesons (evanescent particles produced in cosmic-ray showers, see Chapter 12), and neutrinos (see Chapter 7) are examples of what are called leptons. Protons and neutrons are not believed to be the most fundamental of particles. Each is presumed to be made of yet smaller entities called quarks, which have some remarkable properties. Several varieties of quarks have been postulated to account for results from high-energy experiments, but here we need to be concerned with only two types: 'up' (u) and 'down' (d) quarks. They are characterized by spin, charge, and 'color'. Color is used in a symbolic sense, having nothing to do with our senses, since a quark is thousands of times smaller than the wavelength of light. Each type of quark may come in any one of three colors: yellow, red, and blue. The u and d quarks have charges of $+2/3$ and $-1/3$ of the charge of an electron, respectively. A proton consists of two u and one d quarks so the net charge is $+1$, while a neutron has two d and one u quark so its net charge is zero. The colors in a neutron or proton add up to yellow plus red plus blue, or no net color in each instance. Any free particle has no net 'color'. A free neutron is unstable. Left to itself, a neutron will decay into a proton plus an electron and a neutrino when a d quark turns into a u quark. Within the proton or neutron, the quarks are bound to one another by strong short-ranged forces. In fact, these forces are so powerful that it is impossible to pull out a free quark. However, in nearly all of our discussions we will not need to concern ourselves further with these sub-protonic particles or quarks.

In every atom, protons and neutrons, often in nearly equal numbers, are tightly bound together to form a closely packed nucleus, which is surrounded by one or more outer electrons. A small amount of atomic matter occupies a relatively enormous volume of space, for the electrons are probably separated from the nucleus by distances of the order of thousands of times the diameter of the nucleus. The 'cement' that binds this outer atomic structure together is the force of electrical attraction between the positive and negative charges. It is this attractive force that keeps atoms electrically neutral. Strip an atom of its electrons and the nucleus continually strives to capture others until the electrical balance is restored.

The number of protons and neutrons that constitute any nucleus, say that of an iron atom, may be learned from two observable quantities, namely, the mass of the atom and the number of outer electrons. Since the proton and the neutron weigh so much more than the electron, the total number of them in a nucleus determines the mass of the atom. Of this total, for a neutral atom, enough must be protons to equal the number of outer electrons and thereby provide electrical neutrality. The lightest of all elements is ordinary hydrogen, with a nucleus composed of a single proton, and with one outer electron; it has no neutron. The hydrogen atom weighs 1.673×10^{-24} gram, which is a bit more than the mass of one proton. A helium atom weighs approximately four times as much as hydrogen, and, with two outer electrons, must contain two protons and two neutrons within its compact nucleus (often called an alpha particle). Oxygen atoms are 16 times as massive as hydrogen atoms and have eight electrons; their nuclei consist of eight protons and eight neutrons.

The spectrum and the chemical properties of an atom depend essentially only on the number of its outer electrons. The differences in chemical properties between potassium, which has 19 outer electrons, and calcium, which has 20, are well known. Likewise the spectra emitted by calcium and potassium are entirely different. Disturbances of an atom's outer electrons by means of collisions with other atoms, or with a stream of electrons in an electric arc, produce the spectral lines we observe in a flame or an arc. To disturb the nucleus we must resort to far more drastic measures (see Chapter 7).

More than a hundred separate elements are known (Table 3.1). Of these, 88 appear on the earth as stable elements, radioactive elements such as thorium or uranium, or decay products of such elements. The remainder are unstable, but their nuclei have been created in the laboratory (see Chapter 7). Each atom has been given a number corresponding to the number of its electrons; thus, the atomic number of hydrogen is 1, that of helium is 2, of oxygen 8, and of uranium 92. The masses of atoms are usually expressed on a relative scale, which is based (by chemists) upon an adopted atomic weight of 16 for oxygen. Since oxygen contains 16 protons and neutrons, the mass of each of these particles must be unity. Why is it, then, if atoms are made up of integral numbers, 1, 2, 3, . . ., of the fundamental particles, that the atomic weights listed in Table 3.1 are not integers? Even the helium atom weighs slightly less (by about 0.7 percent) than four hydrogen atoms. The reason (see Chapter 7) is that when helium is formed from hydrogen in the stars, part of the mass disappears as energy. But this mass deficiency (which is often expressed in energy units) is a relatively small fraction of the mass. How, then, can we explain the atomic weight of chlorine, which is 35.457, or of zinc, which is 65.38?

It so happens that two or more electrically neutral atoms may be of the same

Table 3.1 *The chemical elements*

Element	Symbol	Atomic number	Atomic weight	Element	Symbol	Atomic number	Atomic weight
Hydrogen	H	1	1.008	Cadmium	Cd	48	112.41
Helium	He	2	4.003	Indium	In	49	114.82
Lithium	Li	3	6.94	Tin	Sn	50	118.70
Beryllium	Be	4	9.01	Antimony	Sb	51	121.76
Boron	B	5	10.81	Tellurium	Te	52	127.61
Carbon	C	6	12.01	Iodine	I	53	126.91
Nitrogen	N	7	14.01	Xenon	Xe	54	131.30
Oxygen	O	8	16.00	Caesium	Cs	55	132.91
Fluorine	F	9	19.00	Barium	Ba	56	137.36
Neon	Ne	10	20.18	Lanthanum	La	57	138.92
Sodium	Na	11	22.99	Cerium	Ce	58	140.13
Magnesium	Mg	12	24.31	Praseodymium	Pr	59	140.92
Aluminum	Al	13	26.98	Neodymium	Nd	60	144.25
Silicon	Si	14	28.09	Promethium	Pm	61[a]	147
Phosphorus	P	15	30.975	Samarium	Sm	62	150.36
Sulfur	S	16	32.066	Europium	Eu	63	151.96
Chlorine	Cl	17	35.457	Gadolinium	Gd	64	157.26
Argon	Ar	18	39.944	Terbium	Tb	65	158.93
Potassium	K	19	39.100	Dysprosium	Dy	66	162.51
Calcium	Ca	20	40.08	Holmium	Ho	67	164.94
Scandium	Sc	21	44.96	Erbium	Er	68	167.27
Titanium	Ti	22	47.90	Thulium	Tm	69	168.94
Vanadium	V	23	50.95	Ytterbium	Yb	70	173.04
Chromium	Cr	24	52.00	Lutecium	Lu	71	174.98
Manganese	Mn	25	54.94	Hafnium	Hf	72	178.50
Iron	Fe	26	55.85	Tantalum	Ta	73	180.95
Cobalt	Co	27	58.94	Tungsten	W	74	183.86
Nickel	Ni	28	58.71	Rhenium	Re	75	186.22
Copper	Cu	29	63.54	Osmium	Os	76	190.2
Zinc	Zn	30	65.38	Iridium	Ir	77	192.2
Gallium	Ga	31	69.72	Platinum	Pt	78	195.07
Germanium	Ge	32	72.60	Gold	Au	79	197.00
Arsenic	As	33	74.91	Mercury	Hg	80	200.60
Selenium	Se	34	78.96	Thallium	Tl	81	204.39
Bromine	Br	35	79.916	Lead	Pb	82	207.20
Krypton	Kr	36	83.80	Bismuth	Bi	83	209.00
Rubidium	Rb	37	85.48	Polonium	Po	84[b]	209
Strontium	Sr	38	87.63	Astatine	At	85[a]	210
Yttrium	Y	39	88.92	Radon	Rn	86[b]	222
Zirconium	Zr	40	91.22	Francium	Fr	87[a]	223
Niobium[c]	Nb	41	92.91	Radium	Ra	88[b]	226.05
Molybdenum	Mo	42	95.95	Actinium	Ac	89[b]	227
Technetium	Tc	43[a]	99	Thorium	Th	90[b]	232.12
Ruthenium	Ru	44	101.10	Protoactinium	Pa	91[b]	231
Rhodium	Rh	45	102.91	Uranium	U	92[b]	238.07
Palladium	Pd	46	106.4	Neptunium	Np	93[a]	237
Silver	Ag	47	107.88	Plutonium	Pu	94[a]	239

Table 3.1 (*cont.*)

Element	Symbol	Atomic number	Atomic weight	Element	Symbol	Atomic number	Atomic weight
Americium	Am	95[a]	243	Fermium	Fm	100[a]	253
Curium	Cm	96[a]	248	Mendelevium	Md	101[a]	256
Berkelium	Bk	97[a]	247	Nobelium	No	102[a]	253
Californium	Cf	98[a]	251	Lawrencium	Lw	103[a]	
Einsteinium	Es	99[a]	254	[d]			

Notes:
[a] Artificial elements, not found on the earth, but created in the laboratory.
[b] Naturally occurring unstable elements.
[c] Niobium was formerly called Columbium, Cb.
[d] Highly unstable elements from atomic number 104 to 109 are given in some tables. They are omitted here.

atomic number, and yet have different masses because they have different numbers of neutrons in the nucleus. Such atoms are said to be isotopes of the same element. The atomic weight of each isotope is nearly an integer, but, since each element may contain a mixture of stable isotopes, its average atomic weight need not necessarily be a whole number. Practically all elements have isotopes. Carbon, for example, has two stable ones, each containing six protons, but one with six neutrons and the other with seven neutrons; the atomic weights are 12.004 and 13.008, respectively. By far the most abundant carbon isotope is of atomic weight 12, hence the average atomic weight of ordinary carbon is 12.01. Carbon-13 (^{13}C) as it is called, is scarcely more than a trace of adulteration (1 percent) in the predominant carbon-12 (^{12}C). Since the spectra of atoms depend essentially on the numbers of their outer electrons, the spectra of different isotopes of the element are nearly identical. Hydrogen has two stable isotopes, ordinary hydrogen (previously described) and deuterium or heavy hydrogen. Deuterium's nucleus consists of a proton and a neutron, and thus its atomic weight is two. Since hydrogen of atomic weight 1.0 is overwhelmingly the most abundant, the final result is a mean atomic weight for hydrogen of just over one.

Some isotope nuclei and even nuclei of unique elements can be produced by bombardment by high-energy particles such as occur naturally in cosmic rays or are produced by accelerators (see Chapter 7). Often these nuclei are unstable, decaying to other nuclei in times ranging from a fraction of a second to many years. One of the best known of such isotopes is ^{14}C (which contains six protons and eight neutrons). It decays to ^{14}N with the emission of an electron from the nucleus. Cosmic-ray bombardment produces this ^{14}C on the earth. Since it obeys the chemistry of ordinary carbon, ^{14}C becomes involved in living material such as trees and bones, and steadily decays after the organism dies. W.F. Libby showed how the ^{14}C content of old organic remains could be used to date artifacts of old civilizations and primitive man.

To explain how atoms radiate light, we shall rely on an atomic model that has

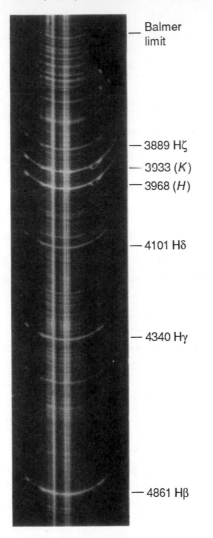

Fig. 3.1. The Balmer series of hydrogen in the solar chromosphere. The spectrogram was taken just as the moon cut off the bright disk or photosphere of the sun. It shows images of the limb of the sun in each spectral line emitted. The arcs are broken because of the irregularity of the lunar limb (in deep valleys some photospheric light emerges). Thus, we observe the rarefied, intrinsically much fainter overlying stratum of the solar atmosphere called the chromosphere. This is the buffer zone between the photosphere, whose temperature is about 6000 K, and the very hot corona, whose temperature is about 2 000 000 K.

served physicists for many years in the visualization of the behavior of the electrons within an atom. In the representation described, the electrons are pictured as revolving about the nucleus in much the same fashion as the planets revolve about the sun. But whereas the planets are prevented from escaping into space by the sun's gravitational attraction, the electrons are held within the atom by the force of electrical attraction between the positively charged nucleus and the negatively charged electron. Such a model closely reproduces the behavior of the hydrogen atom but must be modified to explain, even qualitatively, the behavior of more complex atoms.

When we speak of atomic behavior, we refer to the fact that each atom emits and absorbs light of certain wavelengths. Consider, for example, the hydrogen atom. The spectrum of this element in the discharge tube and in the stars is characterized by a precise regularity. The strongest line is the red line at 6563 Å, often labeled Hα, followed by the blue line Hβ at 4861 Å, the violet line Hγ at 4340 Å, and a sequence of others Hδ, Hε, . . . gradually drawing closer together until they merge near 3650 Å. J.J. Balmer, in 1885, showed that the wavelengths, λ, of this series of hydrogen lines could be represented accurately by the simple formula

$$\frac{1}{\lambda} = R\left(\frac{1}{2^2} - \frac{1}{n^2}\right),$$

where R is a constant. The wavelengths of the successive members of the series, beginning with the red line, are computed from the formula by setting $n = 3, 4, 5, 6,$ . . . Over 30 members of this series, known as the Balmer series, have been observed in the spectra of certain stars, of the sun's outer atmosphere, the chromosphere (Fig. 3.1), and of many gaseous nebulae (see Fig. 2.6).

T. Lyman found another series of hydrogen lines, in the far ultraviolet beginning at 1216 Å and ending at 912 Å, whose wavelengths could be represented by the formula $1/\lambda = R(1/1^2 - 1/n^2)$, with R as the same constant as before and $n = 2, 3, 4, 5,$ . . . F. Paschen discovered a series of infrared lines which could be represented by $1/\lambda = R(1/3^2 - 1/n^2)$, with $n = 4, 5, 6, 7,$ . . ., and F.S. Brackett discovered a far-infrared series that followed the formula $1/\lambda = R(1/4^2 - 1/n^2)$, with $n = 5, 6, 7, 8,$ . . . Two additional series, still farther in the infrared, have been observed by A.H. Pfund and by C.J. Humphreys, respectively.

Bohr's model of the atom

In 1913, Niels Bohr successfully explained the various hydrogen series by suggesting an atomic model in which the electron travels in a circular orbit about the proton. In his picture of the hydrogen atom, the motion of the electron is subject to very specific traffic rules, for only a restricted set of orbits is allowed – those whose radii are proportional to the squares of integers from 1 to infinity, that is, to 1, 4, 9, 16, . . . (Fig. 3.2).

In each of these orbits the energy of motion, or kinetic energy, of the electron is just balanced by the force of attraction that is exerted by the nucleus and prevents the electron from escaping. If an electron traveling in a particular orbit is to be made to travel in an orbit more distant from the nucleus, it must be supplied with energy from some outside source, because work must be done to pull it away from the nucleus that attracts it. A jostling encounter with another atom or the seizure of a passing light pulse may suffice to do the trick. But atoms are fussy; the electron will not change orbits unless it takes up precisely the required amount of energy, no more

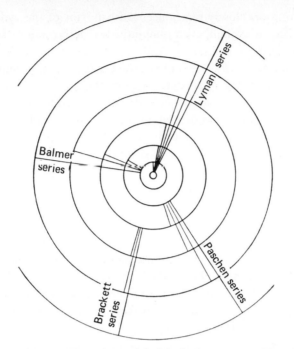

Fig. 3.2. The Bohr model of the hydrogen atom. Transitions depicting the first four spectral series are shown. The radii of successive orbits are proportional to the squares of consecutive integers, that is to 1, 4, 9, 16, 25 . . . The radius of the smallest orbit is 0.528 Å or 0.528×10^{-8} cm. In the low-density gases of the interstellar medium, transitions involving levels as high as $n = 600$ have been observed by radio-frequency methods. In such an orbit a hydrogen atom would have a radius of about a fiftieth of a millimetre. A particle that size would be visible in a small hand-held microscope.

and no less, to remove it to one of the other 'allowed' orbits. Bohr showed that if the amount of energy required to pull the electron from the ground, or lowest, orbit entirely free of the nucleus is represented by W, the amount of energy required to pull the electron out of the second orbit is $W/4$, out of the third $W/9$, and so on. In other words, the energies required are proportional to W/n^2, where $n = 1, 2, 3, \ldots$ If, for convenience, we call the energy zero when the electron is completely removed from the atom and at rest, the energy when the electron is in the lowest orbit is $-W$ (minus because work must be done to remove it from this level). The energy when in the second orbit is $-W/4$, and so on. Hence we speak of the energies of the allowed orbits as being equal to $-W/n^2$, where $n = 1, 2, 3, \ldots$ The quantity W, which depends upon the charge and mass of the electron and other constants, may be computed from Bohr's theory. A positive value of the energy indicates that not only has the electron been removed from the atom but it is flying away in space with a velocity of its own. One important point is that, although the negative energies are restricted by the condition $E = W/n^2$, the positive energies are not restricted at all. This means, of course, that the electrons flying freely about in space are not constrained to move with special speeds but may travel about with random speeds and directions.

When an electron is removed from an atom, the atom is said to be ionized. The energy necessary to tear an electron from the orbit of least energy entirely away from the atom is called the ionization potential, which is measured in electron volts. The ionization potential of hydrogen is 13.60 electron volts (eV), which means that if an electron is accelerated across a potential drop of 13.60 volts it will possess just enough energy, if it collides with a hydrogen electron, to detach it completely from its orbit of least energy.

Bohr postulated further that an electron may switch from an orbit of higher energy to one of lower energy. Since the transfer involves a loss of energy, Bohr supposed that the atom simultaneously releases a pulse or quantum of light and that the frequency, and therefore the wavelength and color, of the emitted radiation must be related to the difference in energy between the two orbits, or

$$E_a - E_b = h\nu,$$

where E_a is the energy of the electron when it is in the larger orbit, E_b is its energy in the smaller orbit, h is a numerical constant, called Planck's constant, and ν is the frequency of the emitted radiation.

From these postulates Bohr was able to calculate the wavelength of the radiation resulting from any jump the electron might perform. We have seen (Chapter 2) that the relation between the frequency of light and its wavelength is

$$\text{Frequency} = \frac{\text{Velocity}}{\text{Wavelength}}, \text{ or } \nu = \frac{c}{\lambda},$$

and that by the Bohr theory the energy in the second orbit, for example, is $-W/2^2$ and in some higher orbit, say the fourth, the energy is $-W/4^2 = -W16$. The reciprocal of the wavelength emitted by an atom when the electron 'jumps' from the fourth to the second orbit should be given by

$$\frac{1}{\lambda} = \frac{h\nu}{hc} = \frac{W}{hc}\left(\frac{1}{2^2} - \frac{1}{4^2}\right) = R\left(\frac{1}{2^2} - \frac{1}{4^2}\right),$$

where we write R for the constant W/hc. When the value of W as calculated from the Bohr theory, along with those of h and c, is inserted, we obtain $\lambda = 4861$ Å, which is the wavelength of the blue hydrogen line! If we write down the formula for jumps or transitions from any orbit, say the nth, to the second, we obtain

$$\frac{1}{\lambda} = R\left(\frac{1}{2^2} - \frac{1}{n^2}\right),$$

where $n = 3, 4, 5, \ldots$, which is just the empirical formula for the wavelengths of the Balmer series, including the numerical value of R. Similarly, all electron jumps terminating in the lowest orbit produce a series of lines in the far ultraviolet, the Lyman series, whose wavelengths may be obtained from

$$\frac{1}{\lambda} = R\left(\frac{1}{1^2} - \frac{1}{n^2}\right),$$

where $n = 2, 3, 4, \ldots$, and the constant R is exactly the same as before. The transitions ending in the third orbit give the infrared Paschen series and those ending on the fourth level the far-infrared Brackett series.

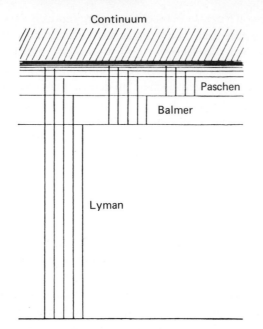

Fig. 3.3. The energy-level diagram for hydrogen. In this figure energies are plotted instead of orbits. Each level corresponds to an orbit in the Bohr model; the continuum above the uppermost level represents the energy that the electron may have when it has been completely detached from the proton forming the nucleus of the atom and is moving freely in space. Note that the greatest interlevel energy difference is that between the two lowest levels, corresponding to the two innermost Bohr orbits.

We may conveniently represent the energies of the Bohr orbits by plotting them as horizontal lines, or energy levels, as shown in Fig. 3.3 Transitions between the various levels are indicated in the figure by vertical lines. In a neutral hydrogen atom, the electron spends the vast majority of its time in the lowest orbit. In this condition, of course, the atom cannot radiate. The electron may be driven into one of the outer orbits either by a collision with a rapidly moving atom or free electron, or by absorbing a quantum of light whose wavelength coincides with one of the lines of the Lyman series. When the electron is in one of the outer orbits, the atom is said to be excited.

Once an electron arrives in a higher orbit, say the fifth, it may decide to jump to any one of the four lower orbits. But the decision must be made rapidly, for the electron lingers in an excited state only about a hundred-millionth of a second. A return to the lowest orbit will be accompanied by the emission of the fourth line of the Lyman series in the invisible ultraviolet. If the electron chooses to stop at the second level, the third line of the Balmer series, at 4340 Å, will be emitted and we shall have a minute flash of violet light. Similarly, jumps from the fifth level to the third and fourth result in the second line of the Paschen series and the first member of the Brackett series, respectively, both of which are invisible infrared radiations.

Fig. 2.6 shows that the lines of the Balmer series crowd ever closer together toward the violet end of the spectrum until they terminate at the series limit, which is followed by a continuous spectrum. Fig. 3.3 serves to explain the coalescing of the lines near the series limit. As we go to larger and larger orbits in the Bohr model, the difference in energy between proton and electron diminishes, until ultimately a minute amount of energy is sufficient to detach the electron completely. In our earlier discussion, we chose for the zero of energy the value that corresponds to the top of the series of horizontal lines in Fig. 3.3 The shaded region above represents a positive energy, the energy of the proton and electron after the electron has been torn away. There are no longer any restrictions on the electron's speed; it may fly about in a carefree fashion, although excessively high velocities are improbable.

We have seen that to produce a transition between any two of the Bohr orbits, or, in terms of Fig. 3.3, between the corresponding two levels (below the shaded region), a discrete amount of energy must be emitted or absorbed. This explains why the hydrogen lines appear only at certain wavelengths and no others. But the electron may escape from the atom provided it absorbs any amount of energy above the minimum required for ionization. The excess energy is used up in imparting a velocity to the free electron. The upper portion of the shaded region in Fig. 3.3 thus represents free electrons with high velocities, the lower portion those with low velocities. Consequently, the ionization of hydrogen atoms will produce a continuous absorption spectrum. Conversely, the capture of free electrons by protons produces a continuous emission spectrum at the violet end of the limit of each series. Figure 5.5 shows the continuous absorption at the limit of the Balmer series in the star ξ_2 Ceti and Fig. 2.6 shows the continuous emission at the Balmer limit in a gaseous nebula.

Complex atoms

We have discussed the spectrum of the simplest of all atoms, hydrogen. If we consider atoms with more than one electron, the problem becomes more involved. Each electron is free to travel in any number of allowed orbits, as before. But the energy of the atom depends upon the particular combination of orbits that are occupied by its electrons. The greater the number of electrons, the more numerous will be the possible combinations of orbits, and therefore the greater the number of spectral lines. The spectrum of iron (Fig. 3.4) is a good illustration of the intricacies of a complex atom. We find that a modified Bohr model is able to predict the exact number, but not the wavelengths, of the spectral lines that are observed for each atom.

The intricacy of an atom's spectrum, however, is not always in direct ratio to the number of its electrons. The reasons stem from the fact that there are limitations on the number of electrons that are allowed to move in orbits at the same distance from the nucleus. In the hydrogen atom, the electron normally moves in the smallest orbit. In helium, the two electrons revolve about the nucleus in orbits of the same size. Lithium has three electrons, two of which travel in identical orbits close to the

Fig. 3.4. The spectrum of iron compared with that of a hot star. A small portion of the emission-line spectrum of iron is compared with that of the corresponding region of the hot star, τ Scorpii. In each strip the bright, sharp, iron arc lines are on top. The lower spectrum is that of the star, which has a continuous background upon which dark lines are superposed. Notice the complexity of the iron spectrum. None of these arc lines of iron appear in the star's spectrum, because the stellar temperature is about 29 800 K, while that of the arc is only about 4000 K. In the stellar atmosphere the iron atoms have lost two or more electrons; hence there are no iron absorption lines corresponding to the emission lines in the arc, which are produced by neutral iron atoms. (Courtesy Mount Wilson Observatory.)

nucleus, while the third moves in a larger, outer orbit. Beryllium has four electrons, two in the inner and two in the more distant orbit. As the number of electrons is increased through the elements boron, carbon, nitrogen, oxygen, fluorine, and neon, the additional electrons are all found in the second orbit. No atom ever has more than two electrons in the first orbit. Similarly, no atom can have more than eight electrons in the second orbit. We may think of each set of orbits as a shell; the electrons tend to arrange themselves in shells about the nucleus. The first shell is completed at helium with two electrons; the second contains eight, and is filled at neon with $2+8$ electrons. If n denotes the number of a shell, $2n^2$ represents the number of electrons it may contain. It develops that the electrons in a closed shell are very tightly bound to the nucleus, and are excited to higher orbits only at the expense of a considerable amount of energy. When all but one of the electrons in an atom are in closed shells, it is the motion of this outside electron that is responsible for the spectrum. In this event the spectrum is roughly similar to that of hydrogen.

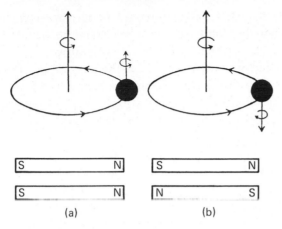

Fig. 3.5. The magnetic effects of spinning electrons. The curved arrows indicate the directions of revolution of the electron in its orbit and of its rotation about its axis. The straight arrows indicate angular motion in accordance with the right-hand rule: the arrow points in the direction of the thumb of the right hand when the fingers are curled in the direction of rotation.

The sodium atom, with its eleven electrons, falls into this hydrogen-like category. The outstanding feature of the sodium spectrum is a pair of strong lines in the yellow, the famous *D* lines in Fraunhofer's map of the solar spectrum. If we regard the two lines as a unit, the *D* lines form the first component of a series. Except for the doubling of the lines, this series closely resembles the Lyman series of hydrogen; the lines crowd closer and closer together and eventually approach a limit in the ultraviolet.

The doubling of each sodium line may be traced to the fact that the electron spins like a top at the same time that it revolves about the nucleus (Fig. 3.5). An electron, which is an electric charge, in motion is equivalent to a tiny electric current, which, as it flows in its closed circuit, generates a magnetic field. The revolution of the electron in its orbit about the nucleus generates one magnetic field, the spinning of the electron generates another. The energy of the atom depends upon the direction in which the electron is spinning. The energy is greater if, like the earth, the electron spins in the same sense as it revolves than if the spin is in the opposite sense. Both directions of spin occur in sodium atoms. As a simple, although not rigorous, analogy we may compare the behavior of the spinning and revolving electron with two bar magnets (Fig. 3.5). In case a, when the directions of spin and revolution are alike, the bar magnets are laid parallel, with their north poles side by side; in case b the magnets are antiparallel, with the north pole of one next to the south pole of the other. To shift the magnets from position b to position a, we must supply energy in order to overcome the mutual attractions of the opposite poles and the repulsions of like poles. The two positions therefore represent different energies. In the same fashion, each orbit in the sodium atom has two energies, corresponding to the fact that there are two directions in which the electron may spin. Consequently, each line appears doubled.

Now since the lines of sodium and similar atoms are doubled because of electron spin, and since the electron in the hydrogen atom is also spinning, we may ask why the hydrogen lines are not also split into components. The answer is that they do show a minute splitting, which has been detected in the physical laboratory, but which is too small to be observed in astronomical spectra. It will soon become evident, however, that the electron spin in hydrogen has an importance in astronomy which is quite out of proportion to the smallness of its effect on the visible spectrum.

There are two ways in which the spin of the electron may be deduced from observations of the spectrum. The first of these has already been described, namely, by the splitting of the lines, which comes about because, on the average, half of the atoms in a cloud of gas will at any instant have their electron spins oriented in one direction and half in the opposite direction. It can also happen that within a given atom an electron spinning in the same direction as it revolves about the nucleus may suddenly reverse the direction in which it spins. When this occurs, a quantum of radiation is released. Since the energy difference corresponding to the two orientations of the electron spin is very small, the wavelength of the quantum is very large, and may in fact be of the order of centimeters or meters, which is in the region of radio waves. When an electron is in an excited orbit like $n = 8$, the probability is tremendously greater that it will jump to a lower orbit than that it will reverse its spin. Hence it is much more likely to emit light waves than radio waves. In the hydrogen atom, the interaction between electron spin and orbital motion splits all of the excited energy levels ($n = 2, 3, 4, \ldots$) but does not affect the lowest energy level and therefore is a very weak if not completely negligible source of radio waves.

A much more powerful mechanism for the emission of radio waves by hydrogen atoms results from the fact that the nucleus of the atom also spins about its axis. The spinning of the nucleus generates a tiny magnetic field which interacts with the field of the spinning electron and produces an additional slight splitting of the energy levels. The special significance of the nuclear-spin–electron-spin interaction is that it also splits the lowest energy level of hydrogen, unlike the orbital-motion–electron-spin interaction, which affects only the excited levels. The difference in energy between the two possible orientations of the electron spin in the first orbit of hydrogen is such that when the electron spin reverses its direction a quantum of radiation of wavelength 21.1 centimeters is emitted. Since hydrogen is by far the most abundant element in the universe, and since in the gas clouds of interstellar space almost all of the hydrogen atoms are in the state of lowest energy, the 21-centimeter radiation is intense enough to be easily observed with radio telescopes, even though once any given atom is excited to the upper level it remains there on average for eleven million years before jumping to the lower level. In Chapter 10, we shall describe how observations of this line have resulted in important discoveries bearing on the structure of the Galaxy and on the physical state of the interstellar gas clouds.

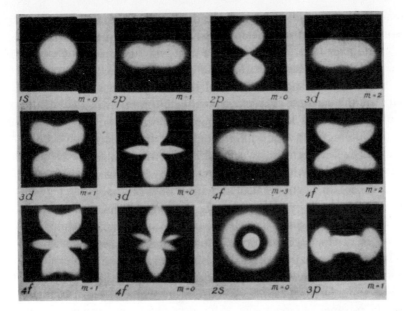

Fig. 3.6. The wave model of the atom. These photograhs, made by H.E. White with the aid of a mechanical model, suggest the pictures that might be obtained if the electron could be seen as a point of light and were photographed with a time exposure.

The wave atom

Since about 1925 the simple Bohr model of the atom has been replaced by a mathematical theory, which does not lend itself to pictorial visualization of the atom. The theory of the atom based on the laws of quantum mechanics shows that we cannot treat the electron as a point charge whose position in the atom at any instant may be strictly stated. Instead we may merely specify the likelihood or probability of finding the electron in any specified position. On this view the electron behaves for many purposes like a hazy cloud of electricity, as illustrated in Fig. 3.6, where the photographs are the sort we might expect to obtain of hydrogen if the electron could be seen as a point of light and were photographed with a time exposure, that is, each of these pictures corresponds to a time exposure of a hydrogen electron in one of its allowed states. There are many points of correspondence between the Bohr model and the wave model of quantum mechanics, one of them being that the electron is most likely to be found at the same distance from the nucleus as the Bohr theory predicts. Nevertheless, the chance of finding the electron at some other distance from the nucleus may also be very good. The quantum-mechanical theory of the atom has enjoyed many successes, and has become well established.

In any event, whatever model is used, we may always think of the atom as possessing a number of discrete states or levels of energy. The transfer of an atom from one state to another, by the absorption or release of energy, gives rise to either an absorption line or an emission line, as the case may be.

Ionized atoms

In Chapter 2 the statement was made that each of the atoms known in nature was distinguished by a unique and characteristic set of spectral lines. The statement is not strictly true, however, for by losing one of its electrons an atom effectively disguises its identity and radiates a completely new spectrum. E.C. Pickering, for example, examining the spectrum of ζ Puppis in 1896, found a series of unidentified lines, something like the Balmer series of hydrogen, at wavelengths of 3814, 3858, 3923, 4026, 4200, and 4542 Å, and concluded that they were 'due to some element not yet found on other stars or on the earth'.

The problem was clarified in 1913 by Bohr, who showed, in connection with his theory of the hydrogen atom, that the spectrum emitted by ionized-helium atoms would closely resemble that of hydrogen, with the important difference that the ionized-helium lines in the visible portion of the spectrum correspond in origin to some of the infrared hydrogen lines. The energy of an electron in its orbit depends not only upon the number of the orbit but also upon the square of the nuclear charge. The charge of the helium nucleus is twice that of hydrogen. Consequently, each spectral series of hydrogen has its analogue in ionized helium, except that the wavelength of each helium line is a quarter that of the corresponding hydrogen line. The lines that Pickering found correspond to the long-wavelength infrared Brackett series of hydrogen, consisting of electron jumps terminating in the fourth orbit. By an interesting coincidence, alternate members of the Pickering series fall within 2 Å of the Balmer lines and were therefore missed by Pickering. In 1922 H.H. Plaskett reported the discovery of these helium lines in the spectra of three Class O stars. The prediction of Bohr, therefore, was brilliantly confirmed.

The similarity between the spectra of neutral hydrogen and ionized helium is one example of a general rule: the spectrum of an ionized atom is qualitatively similar to that of the neutral atom with the same number of electrons, but corresponding lines are displaced toward the ultraviolet. The analogue of the *H* and *K* lines of ionized calcium is a pair of lines of neutral potassium in the red region of the spectrum. Similarly, the close pair of ionized-magnesium lines observed in the far ultraviolet spectrum of the sun near 2800 Å correspond to the yellow sodium *D* lines.

When several stages of ionization are involved, Roman numerals I, II, III, . . . are used to denote the spectrum of the neutral atom, the first ionized stage, the second ionized stage, and so on. Thus, Fe I refers to neutral iron, Fe II to singly ionized iron, Fe III to doubly ionized iron, and so forth.

The effects of magnetic fields upon the radiation of atoms

Soon after the experiments of Bunsen and Kirchhoff had established spectroscopy as an important new tool in physics and chemistry, the great experimental physicist,

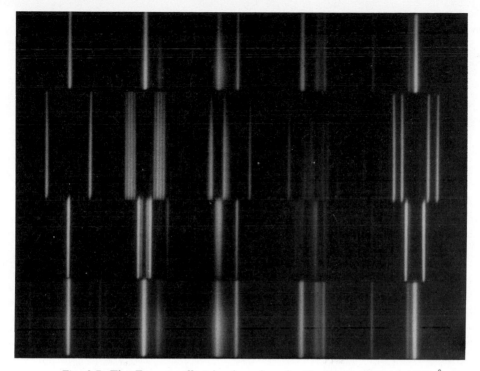

Fig. 3.7. The Zeeman effect in chromium for lines from 4613 to 4626 Å. In a magnetic field of 31 700 gauss the lines are split and the components are polarized, as shown in the two middle strips. The top and bottom strips shows the normal spectrum. In the second strip from the top we view the gas along the magnetic field. The line at the left shows what is called the simple Zeeman effect. Notice that there is no undisplaced component. The two displaced components are circularly polarized in opposite directions of rotation. In the third strip from the top we are viewing the radiation in a direction perpendicular to the magnetic field. The radiation is now plane polarized in the direction of the field. The next line shows a complex pattern; each component of the simple pattern is now replaced by several. (Courtesy Mount Wilson Observatory.)

Michael Faraday, looked for effects of magnetic fields on spectral lines. He found none, spectroscopes being too primitive, magnets too weak. It was not until 1896 that P. Zeeman was able to show that spectral lines could be widened and split by magnetic fields.

We have seen that when an atom undergoes a transition from one energy level to another a single spectral line is normally emitted. If, however, the atom is placed in a magnetic field, its energy levels become split into a number of sublevels or 'Zeeman states'. Each line is now divided into several components, the degree of their mutual separation depending on the field strength.

Typical observed patterns are shown in Fig. 3.7, where a magnetic field about sixty thousand times as strong as that of the earth (0.5 gauss) is employed. The simplest type of Zeeman patterns (Fig. 3.8) are shown by lines involving transitions between energy levels wherein the magnetic effects of electron spin are mutually

(a)	No field		Unpolarized
(b)	Parallel to field		Circularly polarized
(c)	Perpendicular to field		Plane polarized

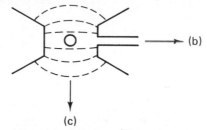

(c)

Fig. 3.8. The simple Zeeman effect. An emitting source is placed in a magnetic field. We observe a spectral line of a type called a 'singlet'. In strip (a) there is no magnetic field. When the magnetic field is turned on (b), the line is split into two circularly polarized components, if we observe it parallel to the field, that is, through a hole in the magnet. If we view the radiation from an angle perpendicular to the field, we see an unshifted, but plane polarized line (c). In the lower sketch the magnetic field directions are indicated by the dashed lines.

cancelled (singlets). When the spectrum line is viewed perpendicular to the field, it is plane polarized. When the source is observed parallel to the magnetic field, the central component is missing and the displaced components are circularly polarized. For most spectral lines the patterns are more complex (Fig. 3.7). Components viewed at right angles to a magnetic field are plane polarized, but along the magnetic field, the central components disappear and the outer components are circularly polarized. The important point is that the polarization (which can be measured by Polaroid and quarter-wave plates as indicated in Chapter 2) tells the direction of the magnetic field and the amount of the splitting tells the strength of the field. Thus the light emitted by an incandescent source can indicate whether or not a magnetic field is present.

The first celestial magnetic fields were found in sunspots. George Ellery Hale had invented the spectroheliograph, a device whereby photographs of the sun could be taken with very narrow band-passes in wavelength. Thus he could take pictures of the sun in the centers of spectral lines such as Hα 6563 Å or the H and K lines of ionized calcium. Near some sunspots he noted complex structures that reminded him of the pattern exhibited by iron filings scattered on a card held above a horseshoe magnet. He knew of Zeeman's then fairly recent discovery, so he searched for polarization of spectral lines of iron and other metals in the sunspots and demonstrated the existence of magnetic fields of as much as 3000 gauss – fields six thousand times stronger than that of the earth! The discovery was sensational. How are the fields produced? Eighty years later we yet cannot answer that question.

Hale's discovery illustrates an important scientific principle. A theory can be wrong but, if it suggests a good experiment, still make a valuable contribution. A static pattern of magnetized material analogous to iron filings is certainly incorrect. What we actually observe are cyclonic-like motions. Had Hale been an expert in fluid motion theory, he might never have thought of looking for magnetic fields. Actually, these strong magnetic fields are probably ultimately responsible for the cyclonic motions but the detailed story is not understood.

In addition to strong sunspot fields, the sun has a 'general' or 'dipole' field, akin to that of a magnetized steel ball or of the earth, but this field is very weak. If we analyzed the spectrum of sunlight from its whole disk, we would be unlikely to find any field at all. Sunspots are usually of what is called the bipolar type. That is, they appear in pairs with opposite polarities as though produced by a giant horseshoe magnet placed below the solar surface. When we look at the total light of the sun, the contributions of spots of different polarity tend to cancel.

Thus, we might not anticipate magnetic fields in other stars. Nevertheless, in the 1950s, H.W. Babcock detected the Zeeman effect in several stellar spectra. Certain stars of spectral class A, particularly those showing spectroscopic variations, possess magnetic fields of the order of several thousand gauss. The field intensity changes with the period of the star's spectrum and light variation. One star, HD 215441, has a field of 30 000 gauss.

Such fields can be produced in terrestrial laboratories. Indeed, fields as strong as a million gauss can be created briefly in tiny volumes. But the stars can outperform any earthly physicist! In 1970, James C. Kemp, J.B. Swedlund, J.D. Landstreet, and R. Angel found circular polarization in white light from a white dwarf star, Greenwich + 70° 8247, a result which implied a surface magnetic field of at least one hundred million gauss! Heretofore unidentifiable spectral lines could be attributed to hydrogen subject to a Zeeman effect so huge that the normal appearance of the spectrum was completely obliterated. Individual Zeeman components could be shifted thousand of ångströms. The surface field of this white dwarf is probably of the order of five hundred million gauss! Extremely strong magnetic fields are attributed to pulsars; they are involved in the acceleration of high-energy charged particles such as cosmic rays.

Where do such monstrous magnetic fields come from? A good clue is that fields of millions of gauss can be found only in compact objects such as white dwarfs and pulsars. These fields are not generated on the spot by some incredible dynamo, but are 'fossil' fields left over from the collapse of a previously magnetized star that had a field of thousands of gauss. We can think of a magnetic field as being made up of numerous 'lines of force' all running along the direction of the field. The stronger the field the greater the number of lines of force per unit area. In a highly ionized gas the lines of force are anchored into the material so that if the star contracts and its density increases, the lines of force are drawn ever closer together and the magnetic field strength builds up.

What of electric fields? Laboratory experiments show that they can split atomic energy levels and can cause a line to be broken up into a number of components. Large-scale electrostatic fields cannot be built up in a star as there is no way of keeping charges of opposite sign apart. Momentary large-scale electric fields can be produced by rapidly changing or moving magnetic fields. These can accelerate protons and other nuclei to high energies. On the other hand, sub-microscopic electric fields do occur in stellar atmospheres but they occur on the scale of individual atoms and ions since they are produced by fluctuations in the number and positions of charged particles near a given atom. Magnetic fields, in contrast, are likely to be on the scale of many kilometers or even of stellar dimensions.

Molecules and their spectra

The world we live in is a world of molecules. The book you are reading, the hand that holds it, the chair you are sitting in, are all constructed of molecules. The hot stars, on the other hand, are worlds of atoms, ions, and electrons, where the complexity of coolness is replaced by the simplicity that accompanies high temperature. The link between our world and that of the stars is to be found in the cooler stars, where the pace is leisurely enough to allow atoms to unite in the fellowship of molecules. Even there, however, the atomic organizations are relatively simple. Carbon and nitrogen atoms join to form a fragmentary molecule, CN, which chemists call the cyanogen radical. Similarly, oxygen and hydrogen unite in the hydroxyl radical, OH, carbon and hydrogen in the radical CH, and so on. The most abundant conventional molecules include H_2 (hydrogen), CO (carbon monoxide), and H_2O (water). It is only under the comparatively frigid conditions on earth that atoms are permitted to give full reign to their organizing talents. The carbon atom on the earth, for example, is a master in the art of forming complex molecules. Some atoms of carbon, hydrogen, and oxygen cluster together in hexagons, others in long chains, like popcorn on a string. Carbon forms the base of all compounds found in living creatures; it is the atom that is mainly responsible for the complexities of the living world. It is in the cooler stars that simple molecules involving abundant elements may form.

Even the simplest molecule radiates a wonderfully intricate spectrum, whose appearance depends not only on the details of the molecule's structure but also on the local temperature. Like atoms, molecules can exist only in certain special energy states, and, like atoms, they emit light as they revert from a state of higher energy to one of lower energy. The energy states that are permitted a molecule, however, are vastly more numerous, and the relations between them more complicated, than those of any atom.

The two atoms of a diatomic molecule are bound together by strong attractive electrical forces to form a system resembling a tiny dumbbell with a slightly elastic connecting rod (Fig. 3.9). The molecule may rotate bodily in space about the axis DC, and the two atoms may vibrate toward and away from each other along the connecting line AB. In additon, the electrons within each atom may pursue a modification of any one of the many orbits normally permitted them. At any instant, the total energy of the molecule will depend not only upon the energies of the revolving and spinning atomic electrons, but also upon the distance between the two

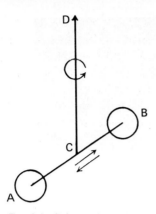

Fig. 3.9. Schematic model of a simple diatomic molecule. The arrows indicate that the molecule can rotate about the axis DC and vibrate back and forth along the line AB.

atoms and upon the speed of rotation of the molecule as a whole. Consequently, in place of each atomic line, corresponding to a particular electron jump, the spectrum of a molecule contains a system of bands, each consisting of a number of fine lines that converge slowly to a point known as the band head. The part of the spectrum in which the whole set or system of bands falls depends on the change of electronic energy in the molecule; the separations between the individual bands of a band system arise from changes of the vibrational states of the molecules, while the separation of the individual fine lines within each band is due to differences in rotational velocities.

In Fig. 3.10, the spectrum of cyanogen (CN) affords a beautiful example of the behavior of a typical diatomic molecule. All the bands in the figure and several others that are not seen in the photograph are analogous to one atomic line. We shall see later that molecules like cyanogen play an important role in the spectra of the cool stars.

Although the atomic spectra produced by two different isotopes of the same element differ so slightly that they rarely can be separated in stellar spectra, molecules involving different isotopes produce easily distinguishable spectra. The vibrational frequency of a molecule AB depends not only on the strength of the interatomic force but also on the masses of A and B. (A mechanical analogy would be a mass suspended by a spring and set in vertical motion; the heavier the mass the slower will be its rate of vibration.) Compare, for example, a C_2 molecule composed of two ^{12}C atoms with one composed of a ^{12}C and a ^{13}C atom. The binding forces are almost exactly the same in the two molecules, since they depend only on the distribution of electrons in the outer cloud, but the masses involved are different. The $^{12}C^{13}C$ molecule will vibrate more slowly than the $^{12}C^{12}C$ molecule; hence the whole band system will be displaced.

In the very coolest stars, some of which can be studied only in the infrared, triatomic molecules such as H_2O are found, but the most exciting developments pertain to the interstellar medium. Transitions between rotational states or spin-

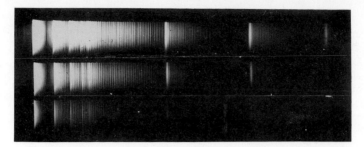

Fig. 3.10. The spectrum of cyanogen (CN). Note the individual lines that go to make up the bands. The overlapping of successive lines produces the phenomenon of band heads. There are three exposures, of different lengths, since the lines differ greatly in intensity. (Harvard Physics Laboratory.)

flops of individual electrons cause radiation in the radio-frequency range. We would expect to find molecular hydrogen, and indeed it is found, as is also the hydroxyl, OH, radical.

What is remarkable is the tendency of reactions in interstellar space to favor production of organic molecules such as HCN, CH_3OH, H_2CO, HCOOH, and even more complex structures such as fragments of amino acids, even though the amino acids themselves have not been detected. Such molecules (see Appendix F) are found in dusty clouds where they are well shielded from stray starlight.

In regions of star formation or demise, molecules of OH, H_2O and SiO sometimes show a curious tendency to radiate copious amounts of energy in certain spectral lines; the intensity is so high that it cannot be produced by any heated mass of gas. We call such sources masers (see Chapter 10). Their intensities vary with time and some are seen to be in rapid motion.

4

The climate in a stellar atmosphere

In the preceding two chapters we saw how matter hidden away in the far corners of the universe is forced to reveal its chemical identity by means of its spectrum, and that the spectroscope can even measure the speeds of the stars and reveal their duplicities. But the story of its almost magical gifts of detection has barely begun. Indelibly recorded on every photograph of a stellar spectrum is a detailed account of the atmospheric conditions at the surface of a star. Strictly speaking, the spectrum tells us only which radiations the atoms are absorbing or emitting and how intensely. The atom, however, is a creature of climate; its ability to swallow up light depends upon the atmospheric conditions to which it is exposed. With present knowledge of atomic structure, the astronomer may now predict just what influence the stellar climate exerts on a particular atom, and thereby infer the stellar atmospheric conditions from the spectrum.

How hot are the stars

The most important attribute of stars, and indeed the one that makes it possible for us to see them at all, is high temperature. The stars are so hot that their material cannot possibly exist in solid or liquid form but must be entirely gaseous. We shall see that the effects of high temperature on the deportment of matter are often spectacular.

In physics and astronomy we employ what is called the absolute or Kelvin, temperature scale, which is reckoned from the lowest temperature that it is theoretically possible to attain. The absolute zero falls 273 °C below the Celsius zero or freezing point of water. Thus temperatures are expressed on the Kelvin or absolute scale by adding 273 ° to the Celsius value; for example, the normal boiling point of water, 100 °C, is 373 K.

A body at any temperature above the absolute zero always radiates energy. Although such emission is insignificant at low temperatures, it becomes very important for hot bodies, in accordance with Stefan's law:

Rate of emission of energy = constant × (Absolute temperature)^4, that is

$$E = \sigma T^4,$$

where σ is called the Stefan-Boltzmann constant, see Appendix B. For example, the average temperature of the earth is about 300 K or $\frac{1}{20}$ that of the sun, which therefore radiates 20^4 or 160 000 times more energy per unit surface area than the earth. We can measure the amount of energy received on the earth from a given star and if we also know the star's distance and its size, as is the case in certain eclipsing systems, we may calculate how much energy is leaving each square centimeter of the surface. This quantity in turn is related to the surface temperature by Stefan's law, and hence we have a method of finding the temperatures of stars whose brightnesses and angular sizes are known; see Appendix E.

For some stars of known apparent magnitude, it is possible to measure the angular diameter by means of the Michelson stellar interferometer, by the 'intensity' or photon-correlation interferometer, or by the speckle interferometer. In this latter device, one compares about ten thousand individual 0.01 second exposures made on the target with a known point source. Thus the technique can be employed only with very large telescopes.

If the star's parallax is known, then in addition to the temperature, we may also find the star's diameter since

> Diameter in astronomical units = (Distance in parsecs) × (Angular diameter in arcsecs).

Fortunately for our purposes, not only the amount, but also the quality, or color, of radiation is governed by the temperature. Everyone is familiar with the way in which the coil of an electric stove changes color as the current is increased. At first the coil glows a dull red, then turns a bright cherry color, and, if the current is imprudently increased still more, the color changes successively to orange, yellow, and white. This does not imply that only a single color is being emitted in each case, for we have already seen that an incandescent solid radiates light of all colors. But the proportions of the different colors are altered as the temperature increases.

We obtain a better insight into what happens by studying, with the spectroscope, light sources of different temperatures. With the aid of a suitable energy-measuring device, we could determine how much energy is contributed by each wavelength interval or color over a range of temperatures from, say, 4000 to 20 000 K. Fig. 4.1 illustrates the types of curves that would be obtained; actually, these curves are calculated by Planck's radiation law (see Appendix B), since stable terrestrial sources of radiation at accurately known temperatures in the range 4000–20 000 K are not easy to provide. Notice, that the shape of the energy curve changes with the temperature; the wavelength at maximum energy output decreases with increasing temperature, which means that the light as a whole appears bluer. For this reason the overheated coil appears to run the gamut of the spectrum as the temperature rises. Good absorbers of radiation are good emitters, and vice versa – a statement known as Kirchoff's law. An ideal radiator, when cold, would appear as a perfectly black object. Hence, energy curves calculated by Planck's law are often called black-body curves. Experimentally, black-body or Planckian radiation may be obtained by uniformly heating an enclosure and allowing the radiation to escape through a small aperture.

Spectral-energy scans prove useful for evaluating stellar temperatures. Instead of

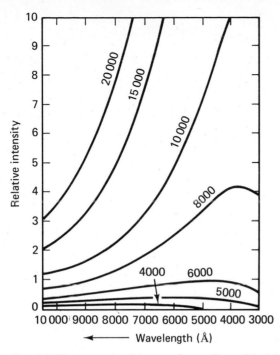

Fig. 4.1. Energy emitted by a perfect radiator (black body). The curves indicate the relative energies radiated in different wavelengths for different temperatures given in absolute or Kelvin degrees from 4000 to 20 000 K. The range in wavelength is that over which the energy of a star can be measured from the surface of the earth.

using a spectrograph with a narrow slit on whose edges much of the starlight might be lost, the astronomer uses a slot broad enough to admit all the starlight and scans the spectrum with a photocell. With the same apparatus he or she also scans the spectrum of a standard lamp whose accurate energy distribution has been established by comparing it with that of radiation from an enclosure maintained at some known, uniform temperature, for example the melting point of gold. Spectral scans of the star taken at different altitudes above the horizon permit a determination of the transparency of the earth's atmosphere, which is often troublesome to evaluate, especially in the ultraviolet. These data enable the astronomer to calculate the true energy distribution in a stellar spectrum from the flux of the radiation measured at the telescope.

Stellar spectral-energy distributions have been observed with satellites flown above the earth's atmosphere, such as the Orbiting Astronomical Observatory, OAO–2, and the International Ultraviolet Explorer, IUE. The data have proven extremely useful, particularly for the hotter stars. At the other end of the spectral range, infrared observations, particularly those obtained with the IRAS (Infrared Astronomical satellite), have opened up a window that has revealed some most amazing objects. These data are of crucial importance for the cooler stars, such as the OH-IR objects, highly evolved stars so cool that they radiate no measurable energy in the ordinary optical range.

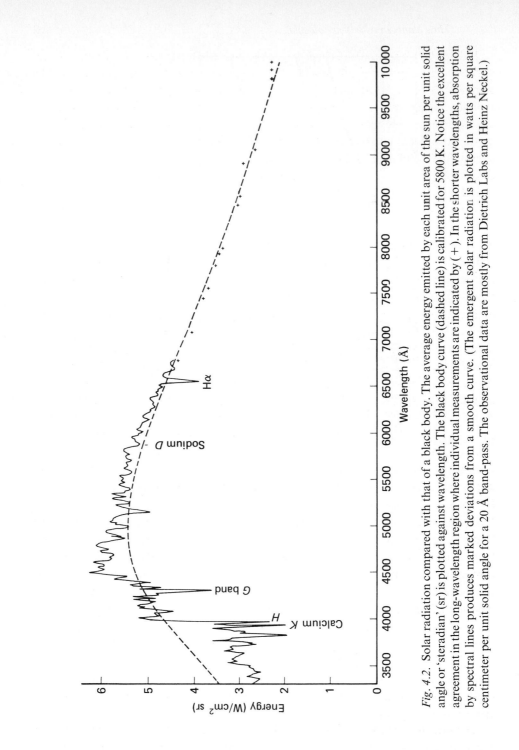

Fig. 4.2. Solar radiation compared with that of a black body. The average energy emitted by each unit area of the sun per unit solid angle or 'steradian' (sr) is plotted against wavelength. The black body curve (dashed line) is calibrated for 5800 K. Notice the excellent agreement in the long-wavelength region where individual measurements are indicated by (+). In the shorter wavelengths, absorption by spectral lines produces marked deviations from a smooth curve. (The emergent solar radiation is plotted in watts per square centimeter per unit solid angle for a 20 Å band-pass. The observational data are mostly from Dietrich Labs and Heinz Neckel.)

In Fig. 4.2 we compare the energy curve observed for the sun with a theoretical curve for a temperature of 5800 K. Although the shapes of the two curves are somewhat similar, the deviations are real and extremely significant in terms of the structure of the sun's outer layers, as will be made clear in the next section of this chapter. Quite generally, when proper allowance is made for distortions produced by the absorption lines which are important in cool stars, it is found that the stellar energy curves differ from those of an ideal radiator, that is, from black-body curves calculated by theory according to Planck's law (Fig. 4.1). The stars do not radiate as black or even gray bodies. There are two reasons for these deviations. One is that temperature increases with depth in the stars, so radiation from deeper layers corresponds to higher temperatures. The other is that the material in the star's atmosphere is not gray but may even be strongly colored; another way of saying this is that the emissivity depends strongly on wavelength. What then do we mean by the temperature of a star? By making the best fit of the energy curve to the theoretical black-body curve we could get some kind of color temperature but we would get different temperatures by making fits in different spectral regions. An alternative, but even less satisfactory method of temperature determination consists in ascertaining the color in which the star radiates the most energy. The sun (temperature 5800 K) pours out the greatest amount of energy in the green region near 4800 Å. Altair, whose temperature is about 8100 K, has a maximum in the violet. Energy maxima of yet hotter stars fall in the 'rocket and satellite' untraviolet beyond 2900 Å. It is rather remarkable that the sun's energy curve can be represented as well as it can by a theoretical curve for 5800 K, inasmuch as the outgoing radiation comes from layers that vary in temperature between about 4400 and 8000 K. The answer must be, of course, that the highest layers are too cool and rarefied to contribute very much, while the radiation from the very deep and hot regions is largely absorbed before it reaches the surface.

Another type of temperature, called an excitation temperature, can be inferred from the appearance of the line spectrum, as we shall describe shortly. However, from many points of view the most satisfactory definition is the effective temperature, which is the temperature of a perfectly black sphere of the same size as the star that would have exactly the same total energy output. Effective temperatures can be measured directly for very few stars. We would have to measure the total energy received from the star above the earth's atmosphere and the star's angular diameter, and we would have to correct for extinction by dust in the interstellar medium (see Chapter 10). This may prove to be impossible for many hot stars. The best method is to determine the energy received in the vicinity of the earth from stars whose angular diameter can be measured by devices such as the Michelson, photon-correlation, or speckle interferometers.

Consider first, stars hotter than the sun. Most of the energy may be emitted in the ultraviolet. With satellites, we can detect this radiation but only down to the hydrogen Lyman limit, beyond which absorption by interstellar hydrogen extinguishes all ultraviolet radiation. Most of the radiant emission from a star at 30 000 K falls in this eternally inaccessible ultraviolet. We try to calculate a model for the stellar atmosphere, predict its emergent radiation, and estimate what fraction we can actually measure.

Radiation from cool stars is severely blocked by heavy extinction in the earth's atmosphere (see Fig. 2.5). Nevertheless, important observations can be secured through a few somewhat 'dirty' windows at a dry site at high elevations. Increasing emphasis is placed on measurements made from aeroplanes flying at high altitudes (above nearly all terrestrial water vapor absorption) and from satellites such as IRAS.

For distant stars, extinction by interstellar smog can become important and its effects have to be evaluated. In very cool stars solid (essentially smoke) particles appear and often complicate interpretations of far-infrared measurements.

Thus, commonly used visual or *V*-magnitudes pertain to stellar brightnesses as perceived by the eye. They ignore ultraviolet and infrared radiation to which our eyes or simulating detectors are insensitive. For many purposes (see Chapters 6–8) we need a type of magnitude valid for the brightness the star would have if observed with a pefect detector that was equally sensitive to all wavelengths and if there was no atmospheric or interstellar extinction. Such magnitudes are called bolometric magnitudes. The bolometric correction is the quantity that has to be added to the visual magnitude to get the bolometric magnitude (see Appendix E). Improvements in both observation and theory make it possible now to secure seemingly accurate bolometric magnitudes from the hottest to the coolest stars. Note that for almost all stars, the bolometric magnitude is a quantity constructed from observational data. It is not a measured entity.

Now the absolute bolometric magnitude M_{bol}, gives the true stellar luminosity in terms of that of the sun by the equation.

$$\log L(*)/L(\text{sun}) = 0.4 \, [M_{bol}(\text{sun}) - M_{bol}(*)].$$

The total luminosity, $L(*)$, of a star is equal to its surface area $4\pi R(*)^2$ times the energy radiated per unit area, which equals $\sigma T_{eff}^4(*)$ by Stefan's law. Here T_{eff} (or T_e) is the effective temperature. This basic equation ($L = 4\pi R^2 \sigma T_{eff}^4$) relates stellar luminosity, L, radius, R, and effective temperature, T_{eff}.

In Chapters 7, 8, and 9 we shall see that calculations of the life history of a star (commonly called its evolution) give for each stage of its development the star's radius and luminosity and hence its effective surface temperature. For most stars the radius cannot be observed directly, but the effective temperature can be inferred from the color or spectrum. From the apparent magnitude and distance, the absolute magnitude can be found. In order to get the actual luminosity we must find the bolometric corrections, which depend on the temperature of the star (Appendix E).

A model solar atmosphere

Our discussion of stellar temperatures has proceeded on the implicit assumption that the intensity and spectral distribution of the radiation leaving the surface of a star can be represented by a single temperature. While this somewhat idealized view of stellar temperatures is most valuable for exploratory purposes, more detailed studies of stellar atmospheres must allow for the fact that the temperature increases with depth in the atmosphere.

Even an ordinary photograph of the sun in white light offers proof that the temperature of the sun increases inward. If the sun had a sharply defined radiating surface at a constant temperature, its brightness would be the same everywhere on the disk. Instead, the brightness decreases steeply toward the edge, or limb. As is explained in more detail in Chapter 5, the solar gases are highly opaque and therefore the radiation from the sun comes only from the outermost layers. These layers from which the white light radiation emerges are called the photosphere. Let us measure heights in the atmosphere from the point at which the gases are just starting to become appreciably opaque. The 'fogginess' of the photospheric layers increases so rapidly with depth that no radiation is received from below about 500 kilometers. It is within this 500-kilometer layer that both the continuous and dark-line spectrum of the sun is produced. At the center of the disk, the line of sight from the observer is radial and therefore extends deeply into photosphere, but as we approach the limb, the radiation emerges tangentially from higher and cooler photospheric levels, and hence the limb appears darker.

We call this phenomenon 'limb darkening', and it supplies the most elementary argument for a rise of temperature with depth in the sun. The degree of darkening diminishes markedly toward longer wavelengths, mostly because the contrast in brightness between parcels of radiating gas at two different temperatures becomes progressively smaller as the wavelength increases.

It is now clear why the observed solar energy curve cannot be fitted precisely to a theoretical curve with a single temperature value. Radiation emerging from the surface is a composite of contributions from all depths in the photosphere. The extent to which each layer supplies radiation to the total is determined by the absorptivity of the overlying layers. The problem is further complicated by the fact that the opacity of the solar gases varies with wavelength (see Chapter 5). If the temperature at each depth in the photosphere is known, together with the absorptivity of the gases at each wavelength, it is possible to predict from theory both the spectral energy curve at the center of the solar disk and the degree of limb darkening at each wavelength. Conversely, if observations of the energy and of the limb darkening at different wavelengths are available, the temperature gradient of the atmosphere and its absorbing properties may be found. Among the most useful of these solar atmospheric models are those calculated by H. Holweger and E.A. Müller and the Harvard–Smithsonian series. Table 4.1 is adapted from a model by J.E. Vernazza, E.A. Avrett, and R. Loeser. As our zero-point of the height scale, we choose that where the local temperature equals the effective temperature of the sun, 5780 K. Successive columns give the height in kilometers, the temperature in K, the gas pressure in atmospheres (1 013 246 dynes per square centimeter), the electron pressure (in units of a millionth of an atmosphere), and the density in grams per cubic meter. Notice that in the visible layers of the sun the total pressure is a small fraction of an atmosphere.

Table 4.1 *A model of the solar atmosphere. Adapted from Vernazza, Avrett, and Loeser (1973)*

Height (km)	Temperature (K)	Total gas pressure (atm)[a]	Electron pressure (10^{-6} atm)[b]	Density (g/m³)
400	4190	0.0026	0.21	0.011
360	4280	0.0057	0.32	0.016
320	4400	0.0055	0.48	0.023
280	4520	0.0079	0.69	0.033
240	4640	0.0126	1.0	0.046
200	4770	0.019	1.5	0.063
160	4890	0.029	2.2	0.088
120	5050	0.037	3.4	0.118
80	5270	0.049	5.2	0.16
40	5510	0.065	8.7	0.21
20	5640	0.074	12	0.23
0	5780	0.085	17	0.26
−20	5940	0.098	24	0.29
−40	6120	0.112	40	0.32
−60	6400	0.129	81	0.35
−80	7040	0.145	192	0.37
−100	7600	0.162	436	0.39

Notes:
[a] The total gas pressure is given in atmospheres. For our purposes we can take one atmosphere equal to a million dynes per square centimeter.
[b] The electron pressure is given in units of a millionth of an atmosphere, essentially in dynes per square centimeter (see Appendix B).

The relation between the temperature and spectrum of a star

Now that we can answer the question: 'how hot are the stars?' we should like to know whether there is any connection between stellar temperatures and the spectral classes that were described in Chapter 2. We recall that, purely on the basis of the appearance of the spectral lines, all stellar spectra could be arranged into one of the types O, B, A, F, G, K, M, R, N, and S. From the fact that O and B stars are blue in color, A stars are white, and G, K, and M stars are yellow, orange and red, respectively, we might suspect that the classes have been put in the order of decreasing temperature. Stellar temperature determinations by the methods described yield the results given in Table 4.2, which is based on the work of R. Hanbury Brown, A. Davis, L.R. Allen, A. Code, R. Shobbrock, T.G. Barnes, D.S. Evans, T. Moffett, D. Hayes, D. Morton, D. Popper, J. Oke, and others.

With the knowledge that each spectral class corresponds to a different temperature, we suspect that the weakness of the hydrogen lines in the O and M stars, the

Table 4.2 *Typical stellar temperatures*

Star	Spectral class	Temperature (K)	Name
Main sequence stars			
ζ Pup	O5	32510 ± 1930	
ζ Oph	O9.5V	31910 ± 2000	
τ Sco	B0V	29800	
α Vir	B1V	23930 ± 840	Spica
α Pav	B2.5V	17900 ± 700	
κ Hya	B5V	15000	
α Leo	B7V	12210 ± 310	Regulus
α Lyr	A0V	9660 ± 140	Vega
ε Sgr	A0V	9460 ± 220	
γ Gem	A0IV	9260 ± 310	
α CMa	A1IV	9970 ± 160	Sirius
β Car	A1IV	9240 ± 220	
β Leo	A3V	8850 ± 340	Denebola
α Aql	A7V	8010 ± 210	Altair
γ Vir	F0V	7000	
α Cmi	F5V	6510 ± 160	Procyon
β Vir	F8V	6120	
sun	G2V	5780	
κ Cct	G5V	5710	
τ Cet	G8V	5520	
δ Dra	K0V	5190	
61 Cyg	K5V	4320	
YY Gem	M0V	3520	
Lacaille 9352	M2V	3590	
Giants			
α Aur	G0III	5300	Capella
η Her	G5III	5100	
η Dra	G8III	4840	
β Gem	K0III	4670	Pollux
ξ² Sgr	K1III	4210	
α Boo	K2III	4070	Arcturus
87 Leo	K4III	3690	
α Tau	K5III	3590	Aldebaran
ν Cap	M2III	3440	
RZ Ari	M6III	3160	
Supergiants			
ζ Ori	O9.5Ib	29900 ± 2100	
ε Ori	B0Ia	24820 ± 920	
η CMa	B5Ia	13310 ± 560	
β Ori	B8Ia	11550 ± 170	Rigel
α Car	F0Ib	7450 ± 460	Canopus
δ CMa	F8Ia	6110	
α Sco	MIab	3560	Antares
α Ori	M2Ia	3450	Betelgeuse

former very hot, the latter cool, does not indicate a scarcity of that element; neither does the great number and intensity of iron lines in the spectrum of the sun necessarily point to an overabundance of iron. A more reasonable view is that the behavior of atoms, that is, their capacity for emitting and absorbing light, is regulated by the temperature. We shall see from what follows that temperature alone can produce the transformation from a rich M-type spectrum to a comparatively bleak spectrum of type B.

Temperature, radiation, and atoms

Suppose that in a large box, made of some hypothetical unmeltable substance, we placed an assortment of all kinds of elements – hydrogen, helium, oxygen, nitrogen, sodium, calcium, iron, chromium, lead. . . . – and that provision could be made for raising the temperature inside the box from absolute zero to perhaps 50 000 K. What would happen to the elements as the enclosure grew hotter?

At absolute zero all the matter is in the solid form; the individual atoms lie tightly packed, closely bound to one another in crystals or complicated molecular structures. The molecules are completely dormant, undisturbed by their neighbors or by any sort of radiant energy. As the temperature rises, the molecules begin to awaken from their lethargy and to stir about sluggishly, occasionally jostling one another. Soon more volatile elements, such as hydrogen, helium, oxygen, and nitrogen, become first liquid and then gaseous, driven by the ever-increasing speeds of their molecules. As the temperature becomes yet greater, the elements liquefy and vaporize one by one. The pace becomes faster. Molecules dash madly about, colliding with one another and loading each other's electrons with energy, which is later lost in the form of radiation. Each molecule is assailed by flying particles and rapidly oscillating radiation waves. The molecules cannot long survive such brutal treatment. Eventually, one after another is torn apart into its constituent atoms. Some molecules, like the hydroxyl radical, OH, or carbon monoxide, CO, are tied together more tightly than others, and may survive long after their contemporaries vanish from the scene. But they too are eventually disrupted, leaving only individual atoms with their electrons rapidly jumping back and forth between various excited levels, as each atom takes up energy from colliding electrons or ions and passes it on in the form of quanta of radiation. Some atoms, like hydrogen or helium, hold their electrons so tightly that only violent collisions or powerful pulses of energy are capable of raising the electron from its lowest orbit to a more distant one. Other atoms, like sodium, have only very loosely bound outer electrons, and much gentler encounters or weaker pulses of energy are sufficient to excite them.

As the gas grows still hotter, collisions become increasingly violent, and the supply of high-frequency radiation increases. The atomic electrons are now so heavily battered that one or more of them may be torn completely free of the parent nucleus, that is, the atom becomes ionized. In general, the metallic atoms, sodium, iron, and so on, are much more easily ionized than are the light gases, hydrogen, helium, oxygen, and nitrogen. The relative amounts of energy required to ionize several of the more abundant elements are listed in Appendix C. Notice that helium is nearly twice as difficult to ionize as is hydrogen, which in turn is bound about twice as tightly as calcium. This means that calcium, hydrogen, and helium tend to lose electrons at successively higher temperatures. It should also be noted that atoms that are easily ionized are also more readily excited than those that are difficult to ionize.

Thus far, in describing the influence of stellar climate on the behavior of atoms, we have made no mention of the pressure or density. Once an atom becomes ionized, it acquires a positive charge and does its best to retrieve electrons so that the charge will be neutralized. Whether or not the ionized atom has a good chance of succeeding in its quest depends upon the number of electrons in the vicinity, or, in other words, upon the electron density. An atom, therefore, is more likely to radiate in the ionized condition when the density is low, and in the neutral form when the density is high.

The picture that we have drawn was first established on a quantitative basis by the Indian physicist Megh Nad Saha, in 1920. Saha not only showed that ionization would be favored by high temperature and low density, but was able to calculate exactly what fraction of atoms of a given kind would be ionized under specified conditions of temperature and pressure. His findings may be summarized by the formula.

$$\frac{\text{Number of ionized atoms}}{\text{Number of neutral atoms}} = \frac{K}{\text{Number of electrons}}.$$

where K depends on the kind of atom and the temperature. The degree of ionization of any atom thus depends directly on the temperature and is inversely proportional to the number of free electrons. The Saha type of formula may also be employed to calculate the degree of disruption of molecules into atoms if the temperature is known. The number of neutral atoms is replaced by the number of molecules, and the numbers of ions and electrons by the numbers of the two constituent atoms into which the molecules are broken up (see Chapter 5). A detailed discussion of the ionization formula and its application will be found in Appendix C.

The meaning of the spectral sequence

We now ask: What are the consequences of this change in the structure of matter, from molecules to neutral atoms to ionized atoms, on the appearance of the spectrum at different temperatures? We have in a way already answered the question in Chapter 3. There we saw that the spectrum of a molecule, consisting of groups of closely spaced fine lines which blend together to form broad bands, is totally unlike that of an individual atom. Also the spectrum of an ionized atom is similar to that of a neutral atom with the same number of electrons, except that each ionized-atom line occurs much farther toward the ultraviolet than does the corresponding neutral-atom line. This fact was illustrated by the similarity between the spectrum of hydrogen and that of ionized helium.

With these facts in mind we may now venture an interpretation of the spectra

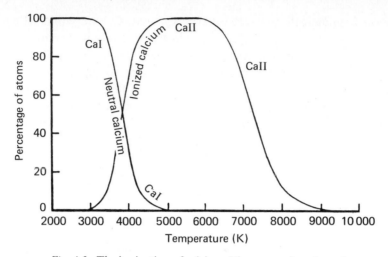

Fig. 4.3. The ionization of calcium. The curves show how the percentages of neutral and ionized calcium vary as the temperature rises. A constant electron pressure of 1/100 000 atmosphere has been assumed.

produced in stellar atmospheres. For the time being, we shall assume that all stellar atmospheres are of the same density, and consider only the effects of temperature. At a temperature of 2500 K, large numbers of atoms are still joined as molecules. Combinations such as titanium oxide (TiO), cyanogen (CN), and the hydrocarbon molecule (CH) imprint their intricate band patterns on the continuous spectrum. Of the elements that are present as individual atoms, those that are easily excited – the metals like calcium, sodium, and iron – are prominently featured. Surprisingly enough, in spite of the relatively large quantities of energy necessary to excite them, lines of atoms of hydrogen are also seen in the spectrum, and with considerable strength. The appearance of hydrogen must be due to its high abundance compared with other elements; in fact, as we shall see later, hydrogen accounts for about 90 percent of all the atoms in a star's outer envelope. Their great numbers compensate for the fact that at low temperatures only a small percentage are in a condition to absorb light.

As the temperature rises along the spectral sequence, more and more molecules become disrupted. At Class K0, the bands of titanium oxide have already vanished. Some of the more stubborn molecules, such as CN, CH, and OH, persist as far as G0; they are easily recognized in the sun. Meanwhile, as increasing amounts of energy become available, the lines of hydrogen steadily strengthen. Even at low temperatures, some of the more loosely held electrons are broken off from their atoms, as is shown by the appearance of the strong *H* and *K* lines of ionized calcium in even the low-temperature M stars. This pair of lines is strongest near K0, but from that point on, the calcium atoms begin to lose a second electron (Fig. 4.3); the *H* and *K* lines weaken and fade away entirely at temperatures greater than 10 000 K. They are still dominant, however, in Class G, as are the lines of neutral iron (Fig. 4.4), magnesium, and other metals, and the ever-growing hydrogen lines.

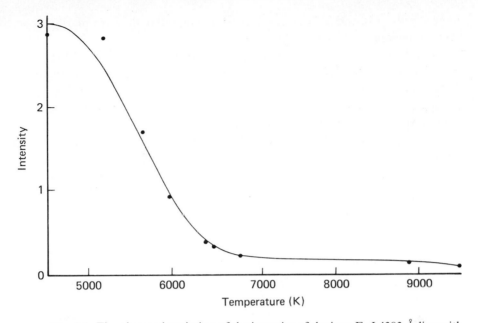

Fig. 4.4. The observed variation of the intensity of the iron Fe I 4383 Å line with temperature along the spectral sequence.

In Class F, at a temperature of about 6500 K, appreciable numbers of other metallic atoms part with their electrons; neutral iron and titanium weaken markedly; ionized iron and ionized titanium attain prominence until deprived of still a second electron, and then vanish as the temperature climbs beyond 10 000 K. At Class A0, hydrogen attains its greatest glory, completely overshadowing all other atoms. But the inexorable march of temperature soon strips great numbers of hydrogen atoms of their single electrons, without which they are impervious to radiation, and their lines begin to fade. Here again, however, at the upper end of the temperature scale, hydrogen remains visible through sheer weight of numbers of atoms.

The very hot stars, in Classes B and O, range in surface temperature from about 15 000 K to perhaps more than 50 000 K. The advent of high temperature is signaled by the appearance, in Class B9, of neutral helium, the most difficult to excite of all the neutral atoms. The helium lines acquire their greatest intensity in Class B3, and then rapidly weaken as the atoms become more and more ionized. The spectra of the B stars also exhibit singly ionized oxygen and nitrogen.

In the very hottest stars of Class O, hydrogen is about as conspicuous as in Class M. Under the violent conditions prevailing, neutral helium disappears completely, giving way to its ionized form. Spectral lines of elements that are stripped of more than one electron mostly fall in the far-ultraviolet part of the spectrum. Since light of wavelengths shorter than 2900 Å is completely absorbed by ozone and other gases in the earth's atmosphere, these radiations can be observed only from rockets and satellites. However, some lines of O III (doubly ionized oxygen), N III (doubly ionized nitrogen), and Si IV (triply ionized silicon) fall in the customarily observed spectral regions.

The practical classification of stellar spectra begins with Class O5 rather than O0, in order to leave a place for still hotter stars that might be discovered later. Theoretically, at temperatures near 100 000 K, all lines in the observable region of the spectrum should disappear, although the short-wavelength region from 100 to 2000 Å would be rich in lines of multiply ionized atoms. A star so hot that it shows no spectral lines would be placed in Class O0. Such stars are found in so-called planetary nebulae. However, in Class O many stars show bright emission lines indicative of extended atmosphere and breakdown of conditions under which the simple Saha theory holds.

In our quick reconnaissance we supposed for simplicity that all stellar atmospheres have the same density and chemical composition. Let us examine these assumptions in turn. Consider two volumes of incandescent gas of identical temperature and chemical composition but suppose that one of these volumes has a tenth the density of the other. Since the temperatures are the same, the ionization will be higher in the gas of low density. Although the probability of loss of an electron by the absorption of radiation will be the same for atoms in each volume, the chance of recombination will be much lower in the less dense gas. Thus the balance shifts to favor the ionized state. We can surmise the effect of density (more specifically the electron density) on the ionization of stellar atmospheres and the appearance of their spectra. This phenomenon can be used to good advantage to find the size and intrinsic luminosity of a star from its spectrum.

What about the chemical composition factor? Suppose, for example, that we have determined the chemical composition of the solar atmosphere (see Chapter 5). If we now take a series of theoretical stellar atmospheres and, with the aid of Saha's formula, predict their spectra at different temperatures from 2500 to 30 000 K, we find that we can reproduce the observed features of the spectral sequence if, and only if, we adopt roughly the same mixture of elements as in the sun. This result was established about 1925 by the thorough studies of Cecilia Payne-Gaposchkin. A somewhat analogous development occurred in the interpretation of the seemingly bizarre spectra of gaseous nebulae (see Chapter 10). Their spectra would be produced by material of essentially solar composition at extremely low densities, glowing in a vast volume, exposed to extremely attenuated radiation from a very hot star.

Nevertheless, incontrovertible evidence for real composition differences did exist. The most striking datum was the split in the spectral sequence between the M stars on the one hand and the carbon and S stars on other. Material of solar composition cooled to 2500–3000 K would lead to an M-type spectrum, never to an R, N, or S-type spectrum. R.H. Curtiss pointed out that if oxygen was more abundant than carbon in a cool atmosphere, nearly all the carbon would be tied up in carbon monoxide, CO, and oxygen would be left over to form compounds such as titanium oxide, TiO.

If carbon was more abundant than oxygen, however, essentially all the oxygen

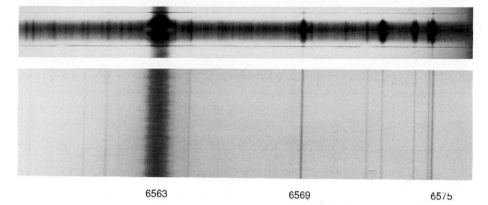

6563   6569   6575

Fig. 4.5. A comparison of photospheric and sunspot spectra. A sunspot spectrum (upper strip) is compared with that of the photosphere (lower strip) near the red hydrogen Hα line, 6563 Å. The whiskery appearance of the photospheric lines is due to vertical motions associated with solar oscillations (see Chapter 9). The lines at 6569, 6574, and 6575 Å are due to neutral iron (Fe I) and the line at 6572 Å arises from a combination of water vapor and neutral calcium (Ca I). Notice the strengthening of the lines in the sunspot, and the great increase in the number of weak lines due largely to fragmentary molecules. The bright streak in the center of Hα in the sunspot spectrum may arise from electromagnetic flare activity. (Courtesy Orren Mohler, University of Michigan.)

would disappear in CO. The spectrum would show bands of carbon compounds such as CN, C_2, etc, but no bands of oxygen compounds would appear. A somewhat analogous argument can be applied to show that in the S stars where ZrO bands are prominent, zirconium must be more abundant than in a star of solar composition. Other evidence appeared in the 1940s; J.L. Greenstein called attention to the helium-rich, hydrogen-deficient atmosphere of υ Sagittarii. Certain massive, very hot stars of the Wolf–Rayet type show bright lines of helium, carbon, oxgen and nitrogen with intensity patterns suggestive of two composition types, one rich in helium, carbon and oxygen, the other rich in helium and nitrogen with only a trace of carbon.

The Saha theory enables us to understand why the spectrum of a sunspot differs from the bright surface of photosphere of the sun. Whereas the solar photospheric temperature is about 5800 K, that of a typical sunspot is close to 4500 K. In the sunspot spectrum, lines of neutral atoms are strengthened, those of ionized gases are weakened, and the number and strength of molecular lines are enormously increased (Fig. 4.5). The electron pressure is lower in the spots than in the nearby regions of the solar photosphere, but its effect is overwhelmed by the marked temperature differences. The strong magnetic fields present in sunspots produce additional effects on spectral lines (Zeeman effect) but qualitative expectations from the Saha theory remain valid. We now turn to the question of how stellar chemical compositions may be ascertained. The relationship between the blackness, shape, or intensity of a line and the abundance of the relevant element is not a simple one.

5

Analyzing the stars

We have seen that, regardless of a star's chemical composition, the fraction of atoms capable of absorbing a spectral line is regulated by the temperature and density of its atmosphere. We have found in this way that, excluding the carbon-rich R and N, or 'C' stars, and the S stars, the main features of the spectral sequence are consistent with a series of stars of uniform chemical composition and varying temperatures and density. Having thus made the preliminary exploration, we may now fix our attention on detailed analyses of individual stellar atmospheres. Since the stars in each spectral class apparently share the same general physical characteristics, we have every reason to hope that studies of a small number of representative stars of each type will reveal the nature of the vast majority of galactic stars.

From the very beginning of stellar spectroscopy it has been recognized that matter everywhere in the universe is essentially alike, and that the same chemical elements that make up the earth are also found in other planets, in the sun, in the stars, and in distant galaxies. Although it was once believed that most stars have nearly the same composition as the sun, the divergences being confined mostly to certain very cool stars, we now know that there are marked differences which are important for theories of element building and stellar evolution. In particular:

(a) A knowledge of the original composition of the solar system can assist us in the development of geophysical and geochemical models of the earth and other planets. It may help us reconstruct the early history of the earth and solar system. Many volatile elements were lost from the earth; some elements sank to the interior. The solar chemical composition helps us reconstruct the original abundance pattern of the solar system.

(b) The abundance pattern of the elements gives clues to their origin and even to the early history of the universe.

(c) Chemical composition differences between stars and the relationship of abundance anomalies to evolutionary stage provide valuable checks on ideas of stellar evolution, including star formation and death.

Briefly, the observed composition differences are of two types: those that are caused by differences in the chemical composition of the medium out of which the stars were formed, and those that are produced by nuclear processes within the stars themselves, during the course of their lifetimes.

In making a detailed analysis of a stellar atmosphere, the problem facing us is to discover how the temperature, density, and chemical composition of the atmosphere may be deduced from the dark lines in the stellar spectrum. The problem is not an easy one. It is clear that the intensity, or blackness, of a spectral line is an index of the abundance of the element producing it. In order to absorb, let us say, the first line of the Balmer series, hydrogen atoms must be in the second energy level (Chapter 3) and the fraction of hydrogen atoms in this level will depend on the temperature and density of the atmosphere. The intensity of the hydrogen line will depend on local temperature and density in yet another way since these quantities affect the broadening of spectral lines. Finally, the intensity will depend on the actual process of spectral line formation.

The widths of spectral lines

There are several reasons why spectral lines appear broad. In the first place, the sharpness of the lines is limited by the fact that the slit of the spectograph is not infinitely narrow but has a very definite width. This 'instrumental effect' can be overcome in studies of the solar spectrum by supplementing a grating spectrograph with special devices that enable one to attain spectra on what amounts to a very large scale. Then it is found that the lines in the spectrum of the sun, or indeed of any incandescent source, have a finite, measurable, intrinsic width. But even if we could observe the radiation from a single atom through an infinitely narrow spectroscope slit, the line would still appear to have a finite width. In other words, the atom does not radiate solely at a single wavelength, but may also radiate (or absorb) energy at adjoining wavelengths. The line is said to possess a natural width, as shown in Fig. 5.1 in which intensity is plotted against distance in wavelengths ($\Delta\lambda$) from the line center. Notice that most of the radiation is at wavelengths close to the center of the line.

We may, if we like, visualize the atom as a tiny broadcasting station, and the spectroscope as a radio receiver. The station is usually assigned a specific broadcasting wavelength, but, owing to natural limitations on the broadcasting equipment, the wavelength of the signal is not perfectly sharp. There is one place on the dial where the reception is loudest, but the program may also be received, although less distinctly, at neighboring wavelengths on either side of the assigned wavelength.

Another important factor in line broadening is the Doppler effect. We recall from Chapter 2 that the wavelength of the light emitted or absorbed by a source that is in motion along the line of sight is displaced from the normal position by an amount proportional to the speed of approach or recession. The spectral lines of an approaching star are shifted toward the violet, those of a receding star toward the red end of the spectrum. The individual atoms in the atmosphere of a star are not at rest, but are flying about with different velocities (Fig. 5.2a). Some atoms are approaching the observer at the instant they radiate; others are receding. Radiation emitted by approaching atoms will have a higher frequency than if the atoms were not moving, and radiation from receding atoms will have a lower frequency. The

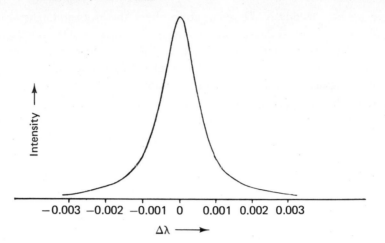

Intensity

−0.003 −0.002 −0.001 0 0.001 0.002 0.003

Δλ ⟶

Fig. 5.1. The natural breadth of a spectral line. The intensity of the emitted radiation of a wavelength(λ) 4383 Å iron line is plotted against distance in ångströms from the center of the line. The curve shows the shape that the emission line might have if we could observe the radiation from the atom at a temperature of absolute zero, a condition we might closely approach experimentally by cooling the discharge tube with liquid helium. Since capacity to emit is proportional to absorptivity, the curve also shows how the absorptivity varies in different parts of a spectral line.

velocities should be entirely random as far as directions are concerned and, since an observed spectral line is the sum of the contributions from a great number of individual radiating atoms, the spectral line will appear widened (Fig. 5.2b). The degree of blurring of a line depends on the velocities of the particles; thus hydrogen atoms move faster on average than other atoms, and hydrogen lines are widened more than those from heavier elements. At higher temperatures, also, the blurring is exaggerated because the atoms are moving more rapidly and the Doppler displacements are therefore larger. Even at laboratory temperatures, the physicist sometimes finds it necessary to cool the electric discharge tube with liquid air in order to narrow and thus separate spectral lines that are close together.

Electric and magnetic fields also widen the spectral lines emitted by atoms. Broadening and splitting of lines by magnetic fields (Zeeman effect) is observed in sunspots and in different types of magnetic stars (see Chapter 3). The magnetic fields are on a spatial scale very large compared with the size of any laboratory or may even exceed the size of the earth. We refer to such fields as 'macroscopic' fields.

The Stark effect (the name given to the splitting of lines in an electric field) is most pronounced in the lines of hydrogen and helium. Whenever the jumping electron is in a large orbit, and hence not firmly held by the attraction of the nucleus, it can be more easily dislocated by a passing charge, just as the outer moons of Jupiter are more seriously disturbed by the attraction of the sun than are the inner (Galilean) satellites. Thus the higher members of the Balmer series, Hδ (4101 Å), Hε (3970 Å), Hζ (3889 Å), Hη (3835 Å), Hθ (3797 Å), . . . , are more affected by Stark broadening than are the earlier members such as Hα (6563 Å) or Hβ (4861 Å).

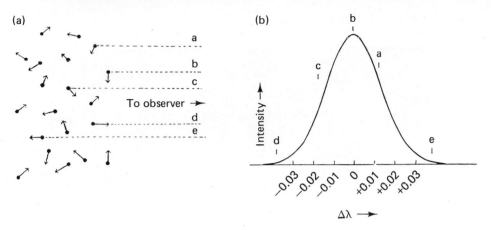

Fig. 5.2. Random atomic motions and Doppler broadening.
(a) Random motions of radiating atoms. Since atoms a and e are moving away from the observer, the wavelengths of their radiations are shifted toward the red end of the spectrum, whereas wavelengths from atoms c and d are shifted toward the violet end; the radiation from b is unchanged since it is moving perpendicular to the line of sight.
(b) The profile of the iron 4383 Å line for pure Doppler broadening for a temperature of 5700 K (appropriate for the solar atmosphere) is depicted. The letters a to e indicate wavelengths that are emitted by corresponding atoms in part (a) of the figure.

The Stark effect observed in the hydrogen and helium lines in stellar spectra differs from that produced in the laboratory in one important respect. Electric fields produced by laboratory apparatus are constant over volumes thousands of millions of times larger than those occupied by the individual atoms. In a star's atmosphere, each atom is subjected to an individual field of its own, produced by the electrons and ions that happen to be dashing about nearby. At higher temperatures, the space surrounding each atom is filled with rapidly moving positively charged ions and negatively charged electrons whose velocities and positions are quite random. Each charged particle produces a field of different intensity at the radiating atom. At one instant the separate fields due to the ions and electrons may nearly cancel at the radiating atom; at the next instant a charged particle may make a close approach and the field may become very large. Consequently, the simple Stark splitting of a line such as is produced in the laboratory (Fig. 5.3) is not observed in the stars, since the fields acting on the radiating atoms there are not uniform but fluctuate rapidly in character and differ from atom to atom. Hence the superposition of the radiations from the different atoms of the same element will not coincide, but will overlap to produce a broad, fuzzy spectral line.

Fig. 5.4, due to R.M. Petrie, shows how the appearance of the hydrogen lines depends on the gravity at the surface of a star. Very large (supergiant) stars have low surface gravities, relatively rarified atmospheres, and rather narrow, weak hydrogen lines. Dwarf stars, which do not differ greatly in size and mass from the sun, have

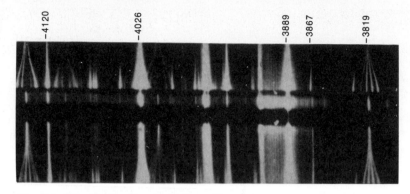

Fig. 5.3. The Stark effect in helium. Not only are the individual spectral lines separated into components, but the light from the components is polarized. The upper strip shows the lines polarized parallel to the electric field, and the lower strip those polarized perpendicular to the field. The field-free spectrum is shown at the center. (From a plate by J.S. Foster, courtesy of H.E. White.)

relatively broad, fuzzy hydrogen lines. The reason for this behavior is easy to guess. In the relatively dense atmospheres of a dwarf, radiating atoms and their disturbing charges are close together; consequently the momentary electric fields are larger and the lines become broadened. In the rarefied supergiant atmospheres, the density is usually so low as to render Stark broadening of minor importance. Hence the lines of hydrogen and helium, although broad and fuzzy in hot dwarf stars, are comparatively sharp and narrow in the supergiants.

From the observed shapes, or profiles, of the stellar hydrogen and helium lines it should be possible to obtain information about the temperatures and densities in the strata where these lines are formed. In order to make progress in this direction it is necessary to simulate (under laboratory conditions) the temperatures and densities obtaining in stellar atmospheres. Fortunately, it is possible to study the broadening of these lines under controlled conditions. Furthermore, considerable advances have been made in our understanding of the theory of hydrogen and helium line broadening. Helium shows particularly complex effects, some lines being much more sensitive to electric fields than others. A small, second-order, or 'quadratic', Stark effect occurs for lines of helium and heavier elements.

Various laboratory techniques have been devised for studying properties of ionized gases (commonly called 'plasmas') at high temperatures. A furnace cannot be used so ingenious procedures have to be devised so as to simulate conditions in a stellar atmosphere. One device is the luminous shock tube. In one version the experimenter uses a long, sturdy, iron tube in which a gas such as hydrogen or helium at high pressure is separated from a mixture of a 'noble' gas (such as argon) and hydrogen at low pressure by a thin membrane. When the membrane is pierced, a shock wave, traveling with a speed several times that of sound, rushes down the tube, strikes the far end, and is reflected back. The gas immediately behind the reflected shock wave is heated to incandescence and its spectrum may be observed. Pressure and temperature may not only be predicted accurately from the pressures, temperatures and gaseous mixtures chosen for the experiment, but may be measured

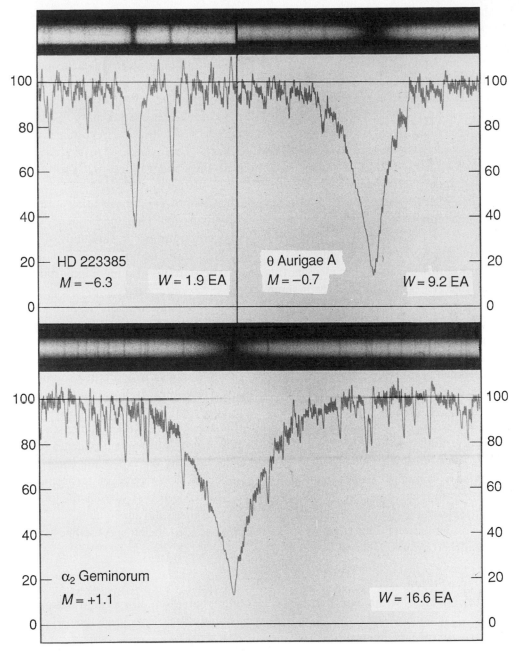

Fig. 5.4. Hydrogen lines in stars of differing surface gravities. Surface gravity and atmospheric density are very low in the luminous star, HD 223385, which is about 25 000 times as bright as the sun. The disturbing effect of charged particles on the radiating hydrogen atom is small and the line extremely narrow. In θ Aurigae A, a star about 150 times as bright as the sun, the density is higher, so the broadening effect produced by charged particles is greater. In Castor C (α₂ Geminorum), which is 30 times as bright as the sun, the atmospheric density is high, so the line is weakened. M = magnitude, W = equivalent width, EA = equivalent ångströms. (Courtesy Dominion Astrophysical Observatory.)

directly. In this way it is possible to ascertain the shapes of hydrogen emission lines under different conditions of temperature and density.

These and other experiments show that electric fields on the sub-microscopic scale are important in many stellar atmospheres. Can large-scale (macroscopic) electric fields also occur in stars? Indeed they can! In the neighborhood of sunspots, sudden changes in magnetic fields can induce momentary electric fields over volumes of thousands of cubic kilometers. Under these circumstances, charged particles can be accelerated to very high energies; in fact, sometimes even low-energy cosmic rays are produced. In the sun, such events are called flares. The accelerated particles and attendant X-ray and far-ultraviolet radiation can cause magnetic storms, aurorae, and magnetic disturbances on the earth.

In a relatively dense stellar atmosphere, such as that of the sun, radiating atoms may be bumped by passing neutral atoms, mostly hydrogen. Then the frequency of the emitted radiation is changed. Since these collisions occur randomly, the observed spectral line is broadened. In the sun, this collisional broadening is more important than natural broadening. In a given spectrum it is also more important for lines corresponding to electron jumps between the larger orbits.

In the early 1930s, O. Struve and C.T. Elvey found that the spectral lines in many giant and supergiant stars were broadened by the Doppler effect in such a way as to indicate that the radiating gases were moving with velocities sometimes as high as 60 or 70 kilometers per second (35 or 40 miles per second). Such velocities could not be attributed to the temperature of the gas, since these stars had temperatures of only 5000 or 10 000 K, and the temperatures necessary to reproduce the line shapes would be in the millions of degrees. Struve and Elvey suggested that the atmospheres of these stars are not orderly quiescent envelopes but are subject to violent, large-scale, chaotic motions, which they characterized as turbulence. Independent evidence for large-scale mass motions of radiating gases is provided by the supergiant components of eclipsing binaries such as 31 Cygni. A modest degree of turbulence (amounting to at most only a few kilometers per second) appears to exist in the solar atmosphere, but the phenomenon is most developed in certain supergiants.

To summarize: exclusive of instrumental imperfections, stellar spectral lines are broadened by two classes of causes.

Intrinsic causes

(a) Natural width, which is due to the fact that an atom, like a radio station, cannot radiate at one sharp frequency, because the energy levels themselves have a certain width;

(b) Doppler effect, which is due to random motions of atoms in any heated vapor (see also g, h, and i below);

(c) Zeeman effect, which is the splitting of spectral lines by magnetic fields, as in a sunspot;

(d) Stark effect, which is the splitting of a spectral line by an electric field; in stellar atmospheres the lines are broadened because the fields acting on any radiating atom are momentary and random;

(*e*) Collisional broadening, which originates because radiating atoms may collide with neutral atoms and suffer a change in their radiated frequencies;

(*f*) Hyperfine structure; certain lines of various elements are observed to be split into a number of very close components as a consequence of a magnetic interaction between the spin of the nucleus and the total angular momentum of the electron. The phenomenon is analogous to the interaction of the magnetic field of the spinning electron with the field produced by its orbital motion (see Chapter 3) except that it is on a scale that is roughly a thousand times smaller.

Extrinsic causes

(*g*) Turbulence, or large-scale vertical motions of large masses of radiating and absorbing gases in a stellar atmosphere;

(*h*) Rotation of the star itself, which broadens all of the spectral lines; rotational speeds as high as 200–300 kilometers per second have been observed in A and B stars, whereas G and K dwarf stars, like the sun, appear to rotate slowly;

(*i*) Loss of material to interstellar space. The atmospheres of many stars, particularly those of early spectral class, are subject to winds in which much material is lost to interstellar space. Sometimes, as in P Cygni or in Wolf–Rayet stars, material is blown out steadily; in exploding stars or novae, there occurs a violent outburst. These fast-expanding stellar atmospheres or winds produce broadened unsymmetrical lines.

Before we turn to the interpretation of spectral lines in stellar atmospheres we must emphasize that the intensity of a dark line is a relative rather than an absolute quantity. A spectral line appears dark by contrast because the intensity at a given point in the spectrum is less than that at adjacent wavelengths. Thus, the intensity of the line is always measured relative to that of the bordering continuous spectrum, and hence the interpretation of dark-line intensities must be based on a prior understanding of the process by which the continuous spectrum is formed.

The continuous spectrum

As already explained in Chapter 4, there is no sharp dividing line between the main body of a star and its atmosphere. Looking down through successively deeper layers of the atmosphere, we arrive at a point where the gaseous material is completely opaque. This level is what we commonly refer to as the surface of a star. The thickness of the atmosphere thus depends upon the absorptivity of its material. In a dwarf star, where the gases are greatly compressed, we can penetrate through only a relatively shallow layer of material, and the depth of the atmosphere is thus small. In a giant star, however, the density is so low that we can see through a great depth of the atmosphere, which is said to be extended.

In our discussion of the chemical composition of the solar atmosphere (later in this chapter), we shall find that a surprisingly small amount of solar material, about 2 grams per square centimeter of the sun's surface, suffices to conceal the radiation from below the surface. The total amount of matter in the solar atmosphere, which is 10^{17} (a hundred thousand million million) tonnes, is huge only because of the sun's great size, for it represents only one part in twenty thousand million of the whole solar mass. The inference is that the gases in stellar atmospheres are exceedingly hazy. If the earth's atmosphere, with its relatively great density, were as opaque, one could hardly see as far as 15 meters.

The chief reason for the fogginess of the atmospheres of the hotter stars is that gases in the process of being ionized are highly opaque. We know, of course, that the atoms in a stellar atmosphere strongly screen radiations in the neighborhood of absorption lines, because an atom raised to higher energy levels absorbs energy corresponding to discrete wavelengths. But when an atom becomes ionized it may absorb energy of *any* frequency greater than the minimum amount necessary for ionization. Thus the ionization of hydrogen atoms whose electrons are in the second orbit produces a continuous absorption spectrum stretching to the violet of the Balmer series limit at 3650 Å, while the ejection of electrons from the third orbit screens off energy at wavelengths shorter than the limit of the Paschen series at 8210 Å in the infrared. It is clear that hydrogen atoms cannot contribute very much to the opacity of stellar gases unless a fair fraction of them become excited to the second and higher levels. Only those atoms excited at least to the second level can absorb radiation at wavelengths smaller than 3650 Å, and only those excited to at least the third level can absorb radiation in the part of the spectrum from 8210 to 3650 Å. At a temperature of 5700 K, the theory of the excitation of atomic levels (see Appendix C) suggests that only four or five hydrogen atoms out of every thousand million are in the second level. Thus, despite its great abundance, atomic hydrogen contributes but slightly to the opacity of the middle and high layers of the solar atmosphere. In the deeper layers, where the temperature is higher, absorption by neutral hydrogen is undoubtedly important. In the much hotter Class A stars, many hydrogen atoms are excited to the second and higher levels, and the hydrogen becomes highly opaque. Fig. 5.5 shows the strong absorption at the limit of the Balmer series in a Class B star.

After it was established that the opacity of the atmospheres of the sun and of the cooler stars could not be accounted for by neutral hydrogen, it seemed logical to suppose that in these stars the ionization of metallic atoms was responsible. Alas, the abundances of the metals are too low to account for even a small fraction of the opacity of the solar atmosphere. The nature of the unknown source of opacity in the solar atmosphere was clarified in 1938 by R. Wildt, who pointed out that in stars as cool or cooler than the sun a neutral hydrogen atom may acquire a second electron and thus become a negatively charged ion. Negative hydrogen ions are voracious absorbers of energy in the visible and infrared regions of the spectrum. The attachment of the second electron to hydrogen is exceedingly weak and only 0.75 electron volt of energy is required to remove it. Hence, in any given volume of the solar atmosphere, the proportion of negative hydrogen to neutral hydrogen is only about one part in a hundred million. However, the hydrogen is so abundant that

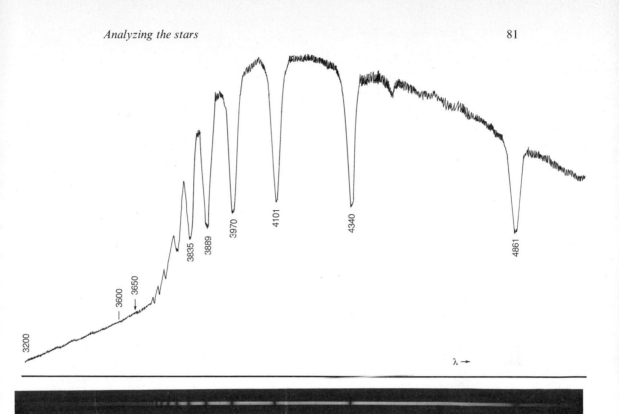

Fig. 5.5. The spectrum of ξ₂ *Ceti.* A scan obtained at Mount Wilson Observatory with a photoelectric spectrum scanner is compared with the spectrum of the same star photographed at Lick Observatory. Notice the sharp drop in intensity at the limit of the Balmer series.

enough negative ions are present to render opaque the atmospheres of the sun and the cool stars. Precise calculations by S. Chandrasekhar and others have revealed that the absorptivity of the negative hydrogen ion varies with wavelength in a unique and interesting fashion, as shown in Fig. 5.6. Since the energy required to detach an electron from the negative hydrogen ion is so small, these ions can absorb radiation of wavelength shorter than about 16 000 Å in the near infrared. Hence, the detachment of electrons from negative hydrogen will produce an absorption continuum covering the near infrared and visible region. The curve labeled 'Bound-free' in Fig. 5.6 shows that the absorption by this process of electron detachment increases toward shorter wavelengths, with a maximum at about 8600 Å, and then declines toward still shorter wavelengths. Once the electron has been removed, it may be thought of as moving in a hyperbolic rather than a circular or elliptic orbit. Such free electrons may also absorb radiation and pass to a hyperbolic orbit of higher energy. This process gives rise to another continuous absorption spectrum, the intensity of which increases progressively toward longer wavelengths, as shown by the curve labeled 'free-free' in Fig. 5.6 (Free-free absorptions and emission processes are discussed in Chapter 10.). Finally, the heavy curve represents the combined absorption of negative hydrogen ions as a result of both processes. The

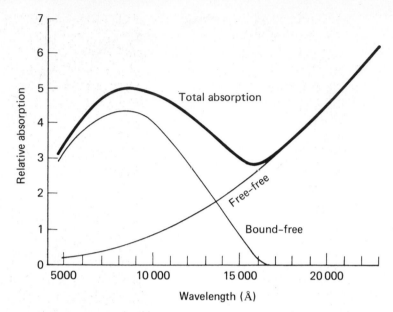

Fig. 5.6. The absorptivity of the negative hydrogen ion. Relative absorption is plotted against wavelength for the negative hydrogen ion for a temperature of 6300 K. The bound–free contribution corresponds to the photo-detachment of the electron from the negative hydrogen ion (this is analogous to ionization of an atom). The free–free contribution arises from the interaction of a free electron with a neutral hydrogen atom; when the electron comes close to the atom, the proton and the bound electron electrostatic interactions do not quite cancel so there is a net effect. The solid curve gives the total absorption produced by both of the contributions. (After S. Chandrasekhar.)

observed continuous absorption in the solar atmosphere may also be derived from observations of the energy distribution in the spectrum at the center of the disk plus observations of limb darkening at different wavelengths. The agreement of the observed curve with those based on theoretical curves of the type shown in Fig. 5.6, although not exact, leaves no room for doubt that the negative hydrogen ion is the principal source of opacity in the solar atmosphere. In the ultraviolet, continuous absorption by silicon and metal atoms assumes some importance, but here spectral lines do most of the blocking of the outgoing radiation.

The negative hydrogen ion is the principal absorbing agent in the stars of spectral classes G and K. In the hotter stars, the ionization of neutral hydrogen atoms is undoubtedly the major source of opacity, whereas in the very cool stars the overlapping of closely spaced atomic lines and molecular bands may contribute significantly to the opacity, in addition to that due to negative hydrogen. Theory agrees with observation in predicting that the atmospheres of the hotter stars should be more opaque than those of the cooler stars. Thus the atmosphere of an A star is about 20 times as opaque as that of the sun.

In the early days of solar spectroscopy, it was supposed that the sun had a sharply defined radiating surface from which the continuous spectrum was emitted, whereas the absorption lines were produced as the radiation from the surface worked its way through a cooler atmosphere, or reversing layer. It is now clear that this picture was grossly oversimplified. In reality both continuous spectrum and absorption lines are formed in essentially the same regions of the atmosphere. The photosphere, in which (by definition) the continuous spectrum originates, is not a sharply bounded surface but a layer with a thickness of about 400 kilometers. In the giant stars the layer is much more distended, whereas in the dwarfs it is highly compressed. We have already seen that in the sun and in the cooler stars the continuous spectrum is formed as a result of absorption by negative hydrogen ions. The radiation from the lowest layers of the photosphere, which is at a temperature of about 8000 K, is much more intense than the radiation from the higher levels, where the temperature is lower. The radiation from the deep levels is much more heavily absorbed than the high-level radiation. Hence it has to traverse a much greater thickness of absorbing negative hydrogen ions before it can escape into space. The result is that the radiation that finally does escape corresponds in color and quantity to an average temperature of about 5800 K.

How an absorption line is formed

The negative hydrogen ions are not the only hazards that must be faced by radiation working its way up to the surface. Stellar atmospheres contain, in addition to hydrogen, all of the chemical elements familiar to us on the earth. Each type of atom can absorb radiation in its own discrete set of wavelengths, and hence those quanta of radiation that are of the proper wavelengths are absorbed. It is true that every absorbed quantum is re-radiated, but, whereas the beam of energy from the photosphere flows outward along the line of sight, the radiation emitted by atoms may be thrown off in any direction whatever, backward and sideways as well as forward. Hence the original outward beam is depleted in the absorbed wavelengths. Eventually the radiation leaves the star but only after being mutilated by its encounters with absorbing atoms. When properly interpreted, the marks that the spectrum bears yield a history of the passage of the radiation through the atmosphere – the types and numbers of atoms encountered, as well as the temperature of the gas. In the sun, most of the absorption in the so-called 'wings' of a line occurs in the middle and upper regions of the photosphere, but the absorption at the centers of strong lines originates in the lower levels of the solar chromosphere.

The number of absorbing atoms

Intensities of the dark lines in a stellar spectrum are usually measured on photographic plates, with a charge-coupled device or with a photoelectric detector. Modern techniques permit very accurate measurements to be made. By the intensity of a dark line we mean the amount of energy that has been subtracted from the continuous spectrum at the position of the line. The unit of intensity is the equivalent width, expressed in ångström units. Thus, an equivalent width of 1 Å signifies the

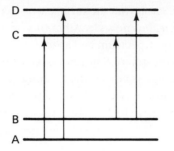

Fig. 5.7. Energy levels of a schematic atom (see text for explanation)

removal of an amount of radiation equivalent to that contained in 1 Å of the neighboring continuous spectrum. The line intensities are governed chiefly by three properties of stellar atmospheres, namely, the chemical composition, the temperature, and the density. The problem that confronts the astrophysicist is to derive these properties from the observed intensities. The intensity of a dark line must depend upon the number of atoms per unit cross-sectional area along the line of sight that are absorbing at the wavelength under consideration. This quantity is called the number of absorbing atoms and Fig. 5.7 attempts to clarify its physical meaning. Suppose that energy is being absorbed by a hypothetical atom with four energy levels, A, B, C, and D. Let us assume that N_a atoms are in level A and N_b in level B, and that the atoms are being struck by radiation of the correct wavelengths to excite them to levels C and D. Assuming that the impinging radiation is equally intense in all four wavelengths, A–C, A–D, B–C, and B–D, how many atoms per second will absorb each of the four radiations? In the first place, the number of absorbing atoms will be proportional to the number of atoms, N_a or N_b, in the lower energy level. We must allow for the fact that certain transitions have a greater probability of occurring than others. An atom in level A, for example, will not usually have an equal preference for levels C and D. One or the other will be more inviting. This preference may be expressed by assigning each line a number, usually less than one, which is known as the *f*-value for the line. The *f*-numbers depend only on the structure of the atom, and may be computed from theory or measured in the laboratory. They are defined in such a way that the number of absorbing atoms is proportional to the product of the number of atoms and the *f*-value, Nf. Thus the number of atoms absorbing the line A–C, is $N_a f_{ac}$. In hydrogen, for example, the *f*-values of successive members of the Lyman series, beginning with the first line, are 0.42, 0.079, 0029, 0.014, 0.0078, and so on. Since all the lines originate from the same level, the first, we see that the number of absorbing atoms, Nf, diminishes rapidly along the series.

The curve of growth

The first step in the analysis of a stellar atmosphere is to evaluate the quantity Nf for each line from the observed intensities in the spectrum. To do so we must have a numerical relation between the intensity of the line and the number of absorbing atoms responsible for its production. At first sight it might appear that these quantities should be directly proportional to each other. Actually the relation is much more complicated, and depends on the mechanism responsible for broadening the lines. We have already mentioned that spectral lines are never perfectly sharp. Each line has associated with it an intrinsic natural width due to the fact that the energy levels themselves are so broad (they are zones rather than simple lines), and also a so-called Doppler width arising from the random motions of the absorbing atoms.

To illustrate the argument we shall revert momentarily to the simplified model of a stellar atmosphere and suppose that above the star's surface, which radiates a continuous spectrum, there exists a layer of perfectly motionless absorbing atoms. We suppose that the density of the gas is so low that collisional broadening is not important and only the natural width of the spectral lines is significant. Consider what happens to the radiation from the surface as it passes through a vertical column of the absorbing atmosphere. At the wavelengths corresponding to absorption lines, the radiation will be depleted by atoms of the atmosphere. If we now increase the length of the absorbing column, how will the blackness of the line increase?

The curves shown in Fig 5.8a represent the shapes of absorption lines produced by successively greater relative numbers of absorbing atoms from 1 to 100 000, as calculated from theory. Since each profile is symmetric, only half of it has been plotted. A large number of atoms act to produce a given spectral line; most of them absorb near the center, and progressively fewer at greater distances from the center.

Referring back to Fig. 5.1 we see that, while the absorptivity of an atom is very high at the center of a line, it falls to about 2 percent of its maximum value at a distance of only 0.003 Å from the line center. Thus, even when relatively few atoms are present in the absorbing column, as illustrated by the curve labeled 100 in Fig. 5.8a, the centers of the lines are completely black, but away from the center the intensity falls rapidly to zero. However, as the number of absorbing atoms increases, more and more radiation is removed away from the line center, where the absorptivity *per atom* may be $1/100$, $1/1000$, or $1/10000$ of its maximum value. Sheer weight of numbers of atoms overcomes the disadvantage of small absorptivity per atom and results in the removal of much radiation away from the center of the line, in the wings. When the number of absorbing atoms is so small that the line center is not yet completely black, the intensity, which is measured by the total area under each line profile, increases in direct proportion to Nf. As the number of absorbing atoms increases past the point at which the line center becomes black, the intensity increases at a slower rate, as the square root of Nf. To double the amount of energy absorbed, four times as many absorbing atoms are required. The resulting relation between the intensity and the number of absorbing atoms is shown in Fig. 5.8b.

In most stellar atmospheres, the broadening of atomic energy levels by collisions between atoms is much greater than the so-called natural broadening. However, the collisional broadening of the lines of all elements other than hydrogen and helium results in profiles of very nearly the same shape as those due to natural broadening,

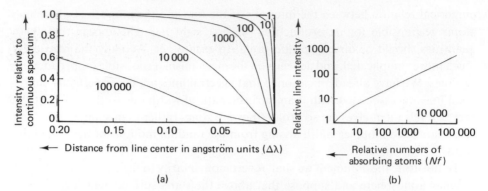

(a)   (b)

Fig. 5.8. Absorption lines with pure natural broadening.
(a) Profiles of absorption lines showing natural width only. The curves show the
change in profile as the relative numbers of absorbing atoms increases from 1 to
100 000. Since the profile is perfectly symmetric, only half of it is plotted. Notice that
as the number of atoms increases, very strong 'damping wings' come into
prominence.
(b) The relation between relative line intensity and the relative numbers of
absorbing atoms for pure natural broadening. Except when the atoms are very few
in number, the intensity is proportional to the square root of the relative number of
absorbing atoms, $\sqrt{Nf}$.

and hence the relation between intensity and numbers of absorbing atoms is also
similar. The curve lies higher, however; that is, for a given number of atoms, the line
is stronger than it would be for pure natural broadening.

The H and K lines of ionized calcium, which are the strongest lines recorded on
ground-based solar spectrum observations, are very good examples of lines with
pronounced wings due to natural and collisional broadening, although in dwarf
stars, like the sun, collisional broadening is very much the more important. Fig. 5.9
reproduces a small section of the solar spectrum in the vicinity of the K line. Note the
enormous extent of the line 'wings' and the multitude of weaker lines of other
elements.

We now consider a column of atoms in rapid motion, so that broadening by the
Doppler effect predominates (Fig. 5.10a). We may also assume for the present
illustration that collisions are absent and that the line has no natural width, each
atom absorbing only at a wavelength determined by the speed of its motion along
the line of sight. The shape of the resulting absorption line will therefore depend
upon the relative numbers of atoms absorbing at each part of the line. The
absorptivity is defined by a curve such as that in Fig. 5.2b. The shape of the
corresponding absorption line resembles that in Fig 5.10a. As in Fig. 5.8a the curves
have been drawn for successively larger numbers of absorbing atoms. Notice that
for large numbers of atoms the curves are bell-shaped, with flat tops and very steep
sides. The physical meaning of the shapes of the curves is that, in a random
distribution of speeds, numerous atoms possess velocities near the zero value but
only relatively few have excessivly large velocities. Fig. 5.10a shows that when the
number of absorbing atoms is small the line is not very black, but broad. The energy

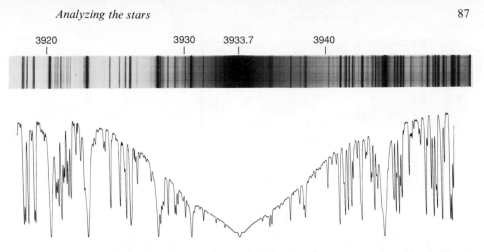

Fig. 5.9. The profile of the 'K' line of Ca II in the solar spectrum. The *H* and *K* lines of ionized calcium are the strongest lines recorded in the solar spectrum as observed from the surface of the earth. The wings are due to natural broadening and collisional broadening. The wavelength scale is given at the top of the figure.

absorbed is spread out over a wide range of wavelengths. This is because nearly as many atoms are absorbing slightly away from the center as at the exact center. Accordingly, when more atoms are added, a great deal of energy near the line center is still available for absorption, and the intensity of the line increases directly as the number of absorbing atoms. But the process does not continue indefinitely; the 'growth' slows down. Eventually, as more atoms are added, the line becomes black at the center, and, since few atoms have high enough velocities to absorb very far away from the zero position, the line becomes 'saturated'. In other words, no matter how many additional atoms are added to the absorbing column, very little more energy can be extracted from the continuous background. The corresponding relation between intensity and number of absorbing atoms, *Nf*, is shown in Fig. 5.10b. The shape of the curve evidently depends upon the temperature, for at high temperatures large numbers of atoms possess high speeds, more energy is available for absorption, and the line does not become saturated until a relatively high intensity is attained.

In reality, neither Doppler nor natural plus collisional damping operates independently. The two are combined, but in such a way that Doppler broadening prevails for small numbers of absorbing atoms and natural plus collisional broadening for large numbers of atoms (Fig. 5.10c). The resulting relation between intensity and number of absorbing atoms, which is known as the curve of growth, has the form shown in Fig. 5.11.

It will be noted that the curve of growth has three distinctive parts: (1) for very small values of *Nf*, the line center has not yet become completely black, and the intensity is directly proportional to *Nf*; (2) for intermediate values of *Nf*, the line center is black but absorption by the line wings has not yet become large and the intensity increases very slowly with *Nf*; (3) for very large values of *Nf*, the intensity is proportional to the square root of *Nf*. The relation between the various domains of

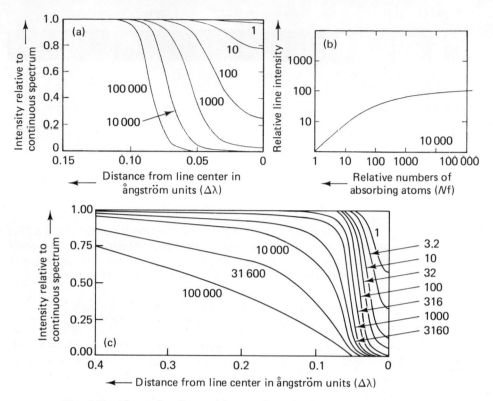

Fig. 5.10. Absorption lines with pure Doppler broadening and with combined effects of Doppler and natural radiation (plus collisional) widening.

(a) Profiles of absorption lines with Doppler broadening and no natural width. As in Fig. 5.8a only half of each profile is shown. Note that as the number of atoms becomes very large, the total absorption increases very slowly.

(b) The relation between relative line intensity and relative numbers of absorbing atoms for pure Doppler broadening. When the number of absorbing atoms, Nf, is small, the intensity is very nearly proportional to Nf, but when Nf is very large, the intensity increases very slowly as more atoms are added.

(c) This shows the combined effects of Doppler and natural (or natural plus collisional) broadening. Notice that initially the profiles resemble those of the Doppler effect acting alone. As the numbers of atoms continue to increase, the line develops the extended wings as displayed by the 'K' line in Fig. 5.9.

the curve of growth is determined by the relative importance of Doppler versus natural and collisional broadening. The position of the left-hand part of the curve is unaffected by the kind and magnitude of the line broadening. But the value of the intensity for which the curve begins to flatten out and the position of the right-hand part of the curve are determined by the ratio of the combined natural and collisional broadening to the Doppler broadening. Since the Doppler broadening, which is governed by the random speeds of the atoms, depends on the temperature, and the frequency of collisions on the density or pressure, the ratio will be larger for an atmosphere of high density and low temperature than for one of low density and

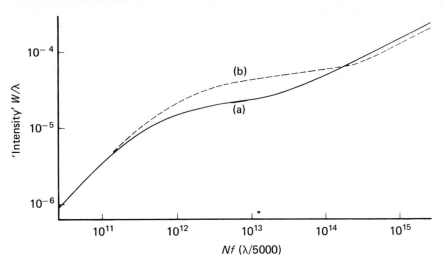

Fig.5.11. The theoretical curve of growth. In (a) we have an atmosphere, such as that of the sun, with a high density and relatively low temperature. In (b) we consider an atmosphere with a low density and a high temperature. Here, W/λ (called the 'intensity') is the ratio of the equivalent width of the line to its wavelength. It is ploted against $Nf(\lambda/5000 \text{ Å})$, where N can be interpreted as the 'number of atoms above the photosphere'. Notice the pronounced effect of collisional broadening in (a), whereas in (b) it is negligible. Note also the long interval in Nf over which the intensity of the line rises very slowly as the number of absorbing atoms increases. See also Appendix C.

high temperature. These two extremes are illustrated by the curves marked (a) and (b) in Fig 5.11. When the collisional broadening is very large, as in (a), the transition between Doppler and collisional broadening is relatively short. When the collisional broadening is very small, as in (b), a very large number of absorbing atoms is needed to build up the line wings. Hence, over a considerable range in Nf, after the center of the line has become saturated and before the wings have begun to grow, a large increase in the number of absorbing atoms has very little effect on the line intensity. Actually, the lines are not totally black at the center as simplified theory predicts, but have a small but measurable central intensity that is properly given by more realistic, physically correct theories.

The theory of the curve of growth has been verified by comparison with empirical curves derived both from laboratory experiments and from observations of stellar spectra. Fig. 5.12 shows a curve of growth for Si I lines in the sun.

Various refinements of the curve of growth theory must be introduced to allow for effects of other types of line broadening: Stark effect, Zeeman effect, hyperfine structure, and turbulence. The Stark effect is important for hydrogen and helium, whose lines cannot be handled by curve of growth theory, although the Stark effect found with metallic lines can be treated as a special case of pressure broadening. The Zeeman effect occurs in large-scale magnetic fields such as are found in sunspots. Similar magnetic areas probably occur in other dwarf stars, but in the so-called

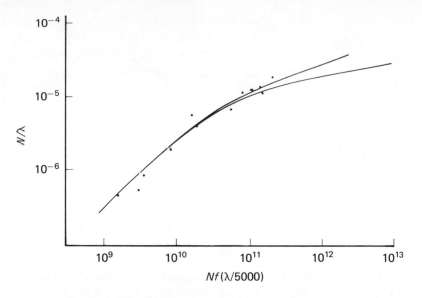

N/λ

$Nf(\lambda/5000)$

Fig. 5.12. Theoretical curve of growth for Si I in the sun. The ratio W/λ is plotted against $Nf(\lambda/5000\ \text{Å})$ for lines of neutral silicon. Two possible branches of the curve of growth are shown, corresponding to two assumptions about the relative importance of collisional and natural line broadening.

magnetic A stars, such as HD 125248 or α_2 Canum Venaticorum, H.W. Babcock found intense magnetic fields that cover large areas of the surface. Analyses of the spectra of magnetic stars need a major modification of curve of growth theory. Hyperfine structure is important for some metals such as manganese; it can be handled properly only by the spectrum synthesis method. Turbulence affects the shape of an absorption line in the same fashion as does Doppler broadening. In the calculation of the curve of growth it is necessary to use an appropriate average of the vertical velocities of the large-scale mass motions of the gas in addition to the contribution from the velocities of individual atoms. Stellar rotation tends to broaden stellar spectral lines; it makes the line profiles characteristically dish-shaped but has little effect on the shape of the curve of growth.

    In the simplified form of the curve of growth used for preliminary analyses, it is assumed that one may employ an average temperature and pressure for the atmosphere. The temperature of any given star may be estimated from its color and the pressure from the broadening of the hydrogen lines, so that the theoretical curve may be computed. An illustrative example is given in Appendix C.

    Hence, with the aid of the curve of growth one may read off the value of Nf corresponding to the observed value of W/λ. Since the f-value is known from laboratory measurements or theoretical calculations, one may then obtain N, the number of atoms in the lower of two energy levels corresponding to the observed line. Occasionally, a given element, potassium for example, is represented in the ordinarily observable spectrum by lines arising from a single lower energy level. Many atoms, such as iron and titanium, display lines arising from a large number of different levels. Then one can calculate the temperature of the stellar atmosphere, for the relative numbers of atoms that are excited to the various energy levels of an atom are governed by the local temperature (see Appendix C). At low temperatures the higher energy levels are sparsely populated, while at high temperatures there may be an appreciable number of atoms in these levels. In either event, if we know the population of even a single level, we may calculate the population of all levels in that stage of ionization.

In other words, we do not determine the total abundance of the element, but only the number of atoms that are neutral or ionized, depending on which lines are actually observed. For example, the overwhelming majority of sodium atoms in the solar atmosphere are ionized, but only the lines of the neutral element are observable. In Chapter 4 we saw that if the temperature and number density of free electrons were known we could solve for the ratio of neutral to ionized atoms by Saha's equation.

In the atmosphere of a star like the sun, whose temperature is known, we can estimate the electron density from the broadening of the hydrogen lines and from the relative number of ions and neutral atoms for metals such as calcium, iron, titanium, and barium, which show lines of both neutral and ionized atoms. Given the temperature and the ratio, say, N(ionized Fe atoms)/N(neutral Fe atoms) we compute the electron density, $N\epsilon$, from the ionization equation. Then, given the number of neutral sodium atoms we can compute N(ionized Na atoms)/N(neutral Na atoms) and thus get the total amount of sodium.

The important thing to realise is that the intensity of a spectral line in a given star is fixed by a number of factors:

(a) *The fraction of the total number of atoms of that element capable of absorbing the line in question.* This quantity is determined by the local temperature and density (especially electron density) at each point in the atmosphere.

(b) *Sources of line broadening*; see p. 78. The relative importance of these various factors varies from star to star and with depth in the atmosphere, and may change from line to line and element to element.

(c) *The continuous absorptivity.* For normal stars this depends primarily on processes involving the hydrogen atom or negative hydrogen ion. As a consequence, whether we use the curve of growth method or a more rigorous procedure, what we finally determine is the ratio of the abundance of the element in question to that of hydrogen.

(d) *The f-value*, a 'bias' factor that expresses the preference of an atom to undergo one transition rather than another.

(e) *The abundance of the element.* The determination of this quantity is usually the object of the game.

Application of the curve of growth is not confined to stellar atmospheres. It can also be used to study the concentrations of low-density gases in the interstellar medium, whose absorption lines are superposed on the spectra of more distant stars (see Chapter 10).

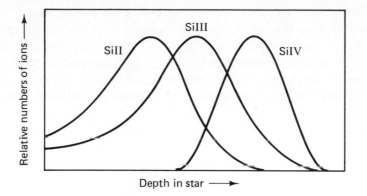

Fig. 5.13. Variation of ionization with depth in the atmosphere of a B star of effective temperature, 19 000 K. Notice that three stages of ionization can co-exist at one point. The ionization rises with depth as the effect of an increasing temperature overwhelms the counter-effect of rising pressure.

Refined methods of spectrum analysis

The method just described presupposes that absorption lines are formed in an atmosphere of constant pressure and density, whereas in reality both the temperature and density of a stellar atmosphere increase inward. In order to allow for these effects, we must construct a model atmosphere, as described in Chapter 4, wherein pressure and temperature are specified at each depth. For the sun it is possible to use measurements of the energy distribution at the center of the disk and the limb darkening at different wavelengths as a basis for calculating an empirical model of the solar atmosphere. For other stars, we must use theoretical models, requiring the atmosphere to be in strict mechanical equilibrium; the pressure at each point suffices to bear the weight of the overlying layers. Furthermore, there must be a constant outward flux of energy through each stratum. Such computations have been undertaken by many investigators. Then, one may obtain essentially a separate curve of growth for each spectrum line.

As an example, let us consider the formation of silicon lines in a B star with a temperature of about 19 000 K. In this star we observe silicon lines in three ionization stages, Si II, Si III, and Si IV. Fig. 5.13 shows how the relative numbers of silicon ions vary with depth in the atmosphere. We can also calculate the relative contribution of each layer to the total intensity or equivalent width of each line. See Fig. 5.14.

For very weak lines we find that the contribution from each layer agrees qualitatively with our expectations from Fig. 5.13. Because of the opacity or 'fogginess', each layer supplies less and less as we go deeper, until finally we receive no contribution at all. The Si II and Si IV lines are weak; their contribution curves behave just as we might expect them to do, but the shape of the Si III curve is unanticipated. Its maximum falls in shallow layers where we would expect Si II to dominate. Why? The reason is that the Si III 4552 Å line is very strong; hence the

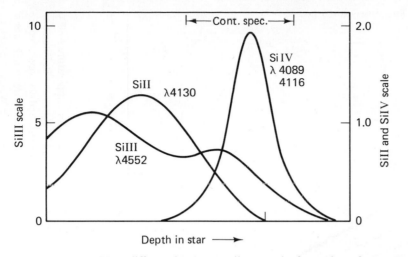

Fig. 5.14. How different layers contribute to the formation of a spectral line. The arrows indicate the zone in which the continuous spectrum is mostly formed. Notice that the Si II and Si III lines originate primarily above the layers making the main contribution to the continuous spectrum while the Si IV lines, 4089 and 4116 Å, are formed mostly in the same layers. The predicted line intensity is proportional to the area under the appropriate curve. The relative scale for Si III is given on the left, that for Si II and Si IV is given on the right. (Adapted from L.H. Aller and J. Jugaku, 1959, *Astrophysical Journal* **38** p. 109, fig. 1, courtesy of University of Chicago Press.)

absorptivity in most of the line profile is high. Consequently, 4552 Å light photons cannot travel far before they get scattered; most of them are thrown sideways or back into the star. The only photons near the line center that can escape to the outside directly come from the shallowest layers; only in the far wings of the line where the absorptivity is low can they escape from large depths. Photons of the weak Si II and Si IV lines escape relatively easily from all parts of the line profile.

Thus, in a refined treatment, we must consider contributions from all atmospheric layers. A fundamental factor is the ratio of the absorptivity in the line to the opacity in the continuum. The line absorptivity depends on the elemental abundance and the fraction of all atoms in a level capable of absorbing the line in question. In most stars the continuum opacity is produced primarily by hydrogen. Although, nowadays, using high-speed computers, we do not actually calculate the contribution curves, we do predict the equivalent width of each line for a sequence of values of $[N(\text{element})/N(\text{H})] \times f$-value. Thus we get what amounts to a separate curve of growth for each line and, by fitting the observed to the predicted value, we obtain the ratio $[N(\text{element})/N(\text{H})]$ for each line of known f-value.

The steps in the procedure may be illustrated as follows:

(a) We obtain energy scans of the spectrum, large-scale, high-dispersion spectra, and profiles of strong lines, particularly those of hydrogen.

(b) We then examine a grid of model atmosphere calculations such as the one prepared by R.L. Kurucz that gives theoretical fluxes and hydrogen line

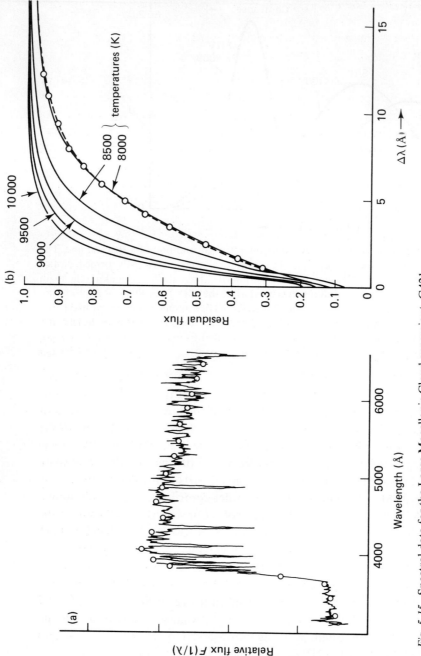

Fig. 5.15. Spectral data for the Large Magellanic Cloud supergiant, G421.
(a) Energy distribution in the stellar spectrum. The circles indicate theoretical fluxes from Kurucz's models for temperature = 8000 K and surface gravity = 0.0012 that of the sun. The observational data were obtained by B.J. O'Mara, J.E. Ross, and B. Peterson. (b) Profile of Hδ, a comparison of theory and observation for models with g = 0.0012 that of the sun and various effective temperatures. The observed data points, due to O'Mara and Ross, are marked by circles. The theoretical curves are from Kurucz. The observations were secured with the Anglo-Australian Telescope at Siding Spring Mountain, New South Wales, Australia.

shapes for different values of surface gravities and effective temperatures. We select the one that gives the best fit. See Fig. 5.15.

(c)   Then, with the model atmosphere chosen, we calculate theoretical curves of growth and determine elemental abundancies.

The procedure may be illustrated by the analysis of a supergiant A star in the Large Magellanic Cloud, designated as G 421 (see Fig. 5.15). The stellar energy distribution and the shape of Hδ seem to be well represented by the Kurucz atmosphere for an effective temperature of 8000 K and a surface gravity 0.0012 that of the sun. These values seem appropriate for a supergiant star. Theoretical curves of growth were calculated via a computer program written by J.E. Ross. A necessary condition for the model is that the same value for the iron abundance must be found from Fe I and Fe II. In this star, the metals seem to be depleted by a factor of the order of two to four, with respect to the solar metal/H ratio.

Most analyses of stellar atmospheres have used total intensities of spectral lines (equivalent widths) rather than line profiles. Until recently, limitations in spectroscopic resolving power have restricted line profile measurements to very broad strong lines. With modern techniques the profiles of even very weak lines can be ascertained. The exact shape of a line profile, however, tells us much more about the real structure of a stellar atmosphere and elemental abundances than does the total intensity alone. Different parts of a line profile give information on different levels of the radiating gases. In the cores of strong lines we are looking only into the very highest strata of a star's atmosphere (its chromosphere). In the wings we can look down into photospheric regions.

The procedure is to calculate the line shape, point by point across the relevant region of the spectrum, taking detailed account of line broadening effects, ionization, and excitation at each stratum through the atmosphere. The advantage is that one can allow for the effects of overlapping lines (blends) which cannot be handled by the curve of growth procedure. This advantage is particularly important for lines of rare elements. Fig. 5.16 illustrates an analysis of osmium lines in the sun. A limited section of the solar spectrum is reproduced by computing the overlapping profiles of nearby lines: in this instance, Fe I, Sm II, Sc II, Sr I, and Ti II, to get the solar abundance of osmium and also platinum. One has to adjust the elemental abundances multiplied by the *f*-values for each of the lines and calculate the theoretical collisional broadening until this small section of the solar spectrum is reproduced. Possible unknown blends and uncertainties in *f*-values and collisional broadening effects limit the accuracy. Since what we obtain is the product of the element abundance (in terms of hydrogen) and the *f*-value, the final accuracy of any abundance determination can be no greater than that of the *f*-value.

Some outstanding problems in stellar spectral analysis

Before we give a detailed account of the composition of the sun (the star which has been analyzed in most detail), let us look at some problems astrophysicists still face

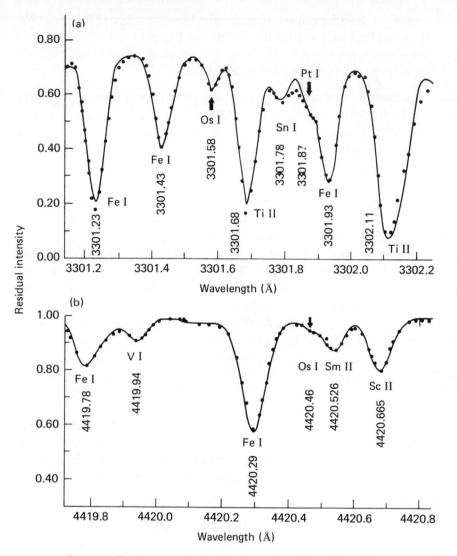

Fig. 5.16. Spectrum synthesis of solar osmium lines. The solid curves give the scans of the solar spectrum as taken from the Kitt Peak solar atlas of J. Brault and L. Testerman. The computed points (dots) were obtained by George Jacoby, who used a model atmosphere and spectral synthesis program developed by J.E. Ross. The platinum (Pt) 3301.87 Å line is also shown; Harry Burger studied the abundance of this element.

in trying to find stellar chemical compositions. The data on abundances of the various chemicals play crucial roles in assessment of ideas of stellar evolution and element building in stars.

Let us first consider stars that are much cooler than the sun. These objects fall into two broad categories (see Chapter 6) – small, faint dwarfs denser but less massive than the sun, on the one hand, and giant, subgiant, and supergiant stars on the other. The latter are characterized by extended atmospheres, low densities, and low surface gravities. These are highly evolved objects, whose atmospheres sometimes contain products of nuclear reactions and element building in their cores. Hence, chemical composition studies are of considerable interest. The spectra of cool stars are dominated by complex-appearing molecular bands, each of which consists of many fine components. Theoretically, each single line could be treated as one produced by a certain number of absorbing molecules in much the same fashion as atomic lines. The problem is complicated by the fact that the lines seriously overlap as the number of absorbing molecules increases. Furthermore, except in spectrograms of very large scale, the lines are too close together to be separately resolved. Under these circumstances, the curve of growth method is inadequate, and we must resort to some method of spectrum synthesis.

In analogy with the situation for atomic lines, the total blackness of a molecular band depends upon the abundance of the compound involved, and upon the temperature and density of the stellar atmosphere. In turn, the molecular abundance depends on the abundance of the constituent atoms. The problem is a classical chemical one, involving reactions between atoms and molecules. One assumes various mixtures of atoms and then predicts the numbers of molecules to be anticipated for each mixture.

Suppose that two atoms A and B, say titanium and oxygen, react to form the molecule AB, and that the molecule may also break down into its separate atoms. This reversible reaction may be expressed symbolically as

$$A + B \rightleftarrows AB.$$

These two inverse processes go on until the rates at which they occur just balance. We then say that the two reactions are in equilibrium. The resulting relative numbers of atoms and molecules depend on the concentration of atoms, the temperature, and the amount of energy required to dissociate the molecule. The process is similar to the ionization of atoms and may be handled by a formula similar to that derived by Saha for atoms, ions, and electrons (see Chapter 4):

$$\frac{(\text{Number of } A \text{ atoms}) \times (\text{Number of } B \text{ atoms})}{\text{Number of } AB \text{ molecules}} = K,$$

where K depends on the temperature and the kind of molecule. This equation is called the dissociation formula.

The first comprehensive study of molecules in cool stars was carried out by Henry Norris Russell. He first investigated an atmosphere somewhat similar to that of the sun, in which oxygen is more abundant than carbon. At temperatures well below that of the sun, molecular carbon, C_2, the hydrocarbon, CH, and cyanogen, CN, are very abundant. The most plentiful molecule of all is hydrogen, H_2, but its bands do not fall in the observable spectral region. At yet lower temperatures (about 3000 K), formation of the very stable carbon monoxide molecule, CO, the familiar and lethal constituent of car exhaust fumes, steals the carbon away from other molecules. These theoretical studies seemed to indicate that at least as far as the oxygen/carbon ratio is concerned, the K and M stars have roughly the same chemical composition as the sun.

Furthermore, Russell's theoretical calculations elucidated the greater strength of the CN bands in giant stars of low surface gravity than in dwarf stars of high surface gravity. Consider giant and dwarf stars of the same atmospheric chemical composition. Among cool stars of the same temperature, the CN bands should be stronger in giants than in dwarfs. The tendency of the molecule to become dissociated at the lower density of the giant star's atmosphere is more than offset by the excessive haziness of the dwarf star's denser photosphere. We see to much greater depths, that is, look through much more material in the giant atmosphere than in the dwarf atmosphere. When we compare stars of the same spectral class, we find that the CN bands are much stronger in the giants, because dwarfs are hotter than giants of the same spectral class, and the increased temperature favors dissociation of the molecules.

The compounds formed and the resultant spectrum observed are extremely sensitive to whether carbon is more or is less abundant than oxygen. If carbon is the more abundant in a cool star, practically all the oxygen will be tied up in carbon monoxide. There will be negligible amounts of titanium oxide, TiO, and the excess carbon will permit the formation of large amounts of CH, CN, and C_2. Thus Russell's calculations in the mid-1930s confirmed beautifully R.H. Curtiss' earlier suggestion that the R and N-type stars had an excess of carbon over oxygen. At higher temperatures, molecules disappear, so the spectrum of a carbon star may superficially resemble that of a 'normal' star of Class K, unless the composition is markedly abnormal.

Another remarkable group of stars are those of S spectral class, which show prominent bands of zirconium oxide rather than titanium oxide. Furthermore, atomic lines of zircon, and of nearby elements in the periodic table, niobium, molybdenum, and ruthenium, are strengthened. Even more remarkable is the presence of lines of technetium (atomic number = 43), an element that does not occur naturally on the earth but can be produced by bombarding neighboring elements with neutrons.

Exact quantitative theoretical treatments of atmospheres of cool stars are difficult for a number of reasons. First, such extended atmospheres are often unstable. There may be stellar pulsations or even winds carrying material away from the star (see Chapter 9). The chemistry becomes involved. Not only are more complex molecules formed, but solid grains (essentially smoke particles) may condense and block outflowing stellar radiation.

At the other end of the scale, stars of high temperature are also of interest for chemical composition studies. The strongest lines are those of light elements; the opacity of the atmosphere is due to atomic hydrogen and helium and to scattering of light by electrons. These stars are very young; their compositions relate to that of the interstellar medium from which they were recently formed. In principle, calculations are straightforward, but in practice verification is difficult because most of the stellar radiation is emitted in the far ultraviolet. Model atmospheres can differ substantially in structure and yet predict nearly the same spectral features for observable regions. Furthermore, in stars of spectral classes B0 and O, the Saha equation and the Boltzmann equation (which relate the population of different energy levels to one another; see Appendix C) cease to be applicable. Populations of atomic levels no longer depend simply on the local temperature and density, but are strongly influenced by details of the local radiation spectrum. Predictions of line shapes and intensities, although still possible, become very difficult.

Given the detailed energy-level diagram for each pertinent atom or ion, accurate *f*-values, collisional and recombination rates for excitation of each level, and the condition that we have a steady state, it is possible to devise a complex computer program that will enable us to calculate the population in each level of interest at various depths in the stellar atmosphere. Then we can predict the continuous spectrum, the intensities, and the profiles of spectral lines. These methods, due largely to Dimitri Mihalas and his associates, have been applied to the interpretation of the spectra of hot stars, where R. Kudritzki has made important contributions. They may also be applied to certain lines in the spectra of cooler stars.

Expanding stellar atmospheres and stellar winds impose another problem, particularly in massive and highly evolved stars. The hot, extremely tenuous outermost layer of the sun, the corona, is not in static equilibrium like a stellar photosphere or the earth's atmosphere, but is expanding with a speed that picks up with increasing distance from the sun. This solar wind has been measured by spacecraft, but it causes a negligible mass loss and has no influence on the observed spectra of stars like the sun. The situation is otherwise in many hot, massive, stars whose ultraviolet spectra show peculiar line shapes of the P-Cygni type (see Chapter 9), indicative of substantial mass flow in the photospheric regions where ordinary, dark, spectral lines are produced. Clearly, the calculation of a model atmosphere for such a star constitutes a challenge.

Even if there is no significant mass loss, a given stellar atmosphere may teeter on the brink of instability. It is remarkable that model atmosphere methods seem to work as well as they do for the Magellanic Cloud supergiant G421, whose surface gravity is scarcely more than a thousandth that of the sun.

Chemical composition analyses have been extended to the tenuous, evanescent envelopes of exploding stars or novae, and to gaseous nebulae, both of which emit bright-line spectra. The techniques are quite different to those used for stars, as we shall see in Chapter 10.

The chemical composition of the sun and the solar system

The solar chemical composition has been more thoroughly studied than that of any other star. The brilliant pioneering investigations by Henry Norris Russell in 1929, before the curve of growth was invented, clearly revealed the broad features of the abundances. The picture has been greatly improved over the years as more powerful analytical tools have been developed. Some relatively rare elements such as arsenic are missing. Their weak lines are masked by strong lines of abundant elements. It is necessary to infer the abundances of some elements from the solar chromosphere, the corona, or even the solar wind. The abundances of the noble gases, particularly helium, are poorly determined. Other missing elements, judging from their scarcity on earth and the nature of their spectra, are probably present in minute quantities in the sun, but in such small amounts that their weak lines cannot be found. An example is uranium which is rare and all of whose numerous lines have small f-values.

The solar abundance of isotopes, particularly heavy hydrogen or deuterium, is of great interest. Although there is some evidence that deuterium may be produced in solar regions of great electromagnetic activity in the neighborhood of sunspots, there is no secure determination of its solar abundance. The $^{12}C/^{13}C$ ratio may be the same in the sun as on the earth, but this subject merits further investigation.

A particularly engaging question is the chemical makeup of the material from which the solar system was formed. It has an important bearing on the story of the origin of stars and solar systems. The composition of the solar atmosphere ought to give us some good clues to the primordial solar system, assuming that the solar atmosphere has suffered no substantial modification in the last 4.5 thousand million years. Such modifications could arise in principle from:

(a)   accretion of interstellar material if the sun passed through one or more dense clouds;

(b)   the mixture of material into the surface layers from the deep interior (where hydrogen is being converted into helium); or

(c)   diffusion causing a settling out of heavier metals, such as lead, as compared with light metals, such as magnesium.

The presence of the solar wind makes process (a), accretion, unlikely unless the sun had encountered a very dense interstellar cloud. Upwelling of the products of nuclear reactions from the deep core of the sun is improbable for a relatively unevolved star. What, then, about diffusion?

Fortunately, some evidence from a different field of science is available just when we need it. Analyses of the earth's crust give some data that may be of more general interest, notably certain isotope ratios, such as $^{12}C/^{13}C$, and the abundance ratio of lanthanide elements, such as cerium, samarium, etc. Such metals are very difficult to separate chemically and would go along with one another in any kind of a geochemical process we can expect. Otherwise, analyses of rocks of the earth's crust are of limited usefulness. The chemical composition of the crust is certainly no more representative of the earth's average composition than is that of the slag in a smelter crucible representative of the composition of the original ore. In each instance, iron or nickel and the so-called noble metals like gold and platinum tend to sink to the bottom, while metals such as magnesium, aluminum or sodium tend to rise to the top. Hence it is very difficult to deduce from the earth's crust what the original composition of our planet must have been. During the involved process of its formation, the earth must have lost most of its hydrogen, helium, nitrogen, and other gases.

The moon rocks reveal a complicated history of crushing, melting, and exposure to energetic solar particles. Some information on isotope ratios, such as $^{12}C/^{13}C$, is given by atmospheres of the giant planets; the carbon isotope ratio seems to be the

Table 5.1 *Abundances of elements in the sun*[a]

At. no.	El.[b]	N(el)	At. no.	El.[b]	N(el)	At. no.	El.[b]	N(el)
1	H	1.0 (12)	24	Cr	4.8 (5)	56	Ba	135
2	He	9.8 (10)	25	Mn	2.5 (5)	57	La	15
3	Li	14	26	Fe	4.7 (7)	58	Ce	35
4	Be	14	27	Co	8.3 (4)	59	Pr	5.1
5	B	40	28	Ni	1.8 (6)	60	Nd	32
6	C	3.6 (8)	29	Cu	16 000	62	Sm	6.3
7	N	1.1 (8)	30	Zn	40 000	63	Eu	3.2
8	O	8.5 (8)	31	Ga	760	64	Gd	13
9	F	3.6 (4)	32	Ge	256	65	Tb	0.8
10	Ne	1.2 (8)	37	Rb	400	66	Dy	12.5
11	Na	2.1 (6)	38	Sr	800	67	Ho	1.8
12	Mg	4.0 (7)	39	Y	173	68	Er	8.5
13	Al	3.0 (6)	40	Zr	406	69	Tm	1.0
14	Si	3.7 (7)	41	Nb	26	70	Yb	12
15	P	2.9 (5)	42	Mo	83	71	Lu	6
16	S	1.6 (7)	44	Ru	69	72	Hf	8
17	Cl	3.2 (5)	45	Rh	13	74	W	13
18	Ar	3.6 (6)	46	Pd	49	76	Os	28
19	K	1.3 (5)	47	Ag	9	77	Ir	24
20	Ca	2.3 (6)	48	Cd	72	79	Au	10
21	Sc	1250	49	In	46	81	Tl	8
22	Ti	1.0 (5)	50	Sn	100	90	Th	1.3
23	V	1.0 (4)	51	Sb	10	92	U	<3

Notes:

[a] The abundances (N(el)) are given by number of atoms. The numbers in parentheses indicate the power of ten by which the preceding number is to be multiplied to give the abundance. For example, the entry for P is 2.9 (5), which means the abundance of this element is 290 000. The abundances are normalized to hydrogen as 1.0 (12) or 1×10^{12}.

The data are primarily from the compilation by N. Grevesse and E. Anders although other sources have also been consulted.

[b] For full names of elements (El.) see Table 3.1, p. 38.

same as on the earth. Possibly the best sample would be a chunk of a 'new' comet or of the satellite of an outer planet of the solar system, both of which would require sophisticated space probes to collect them.

Although meteorites impinging on the earth are of varying types and chemical compositions – irons, stony-irons, and achondrites (which resemble earth-like rocks), the most frequent falls are the stony chondrites. Carbonaceous chondrites, a type of stony chondrite, appear to have been subject to very little 'processing' since they were formed. That is, they have not suffered massive physical and chemical changes because of pressure or heating. They can be analyzed chemically with high precision. Going up the table of elements from magnesium to lead, they match closely the solar abundances. In the sun there appears to be no significant diffusion effect.

Table 5.1 gives the solar abundances by numbers of atoms on the scale of one

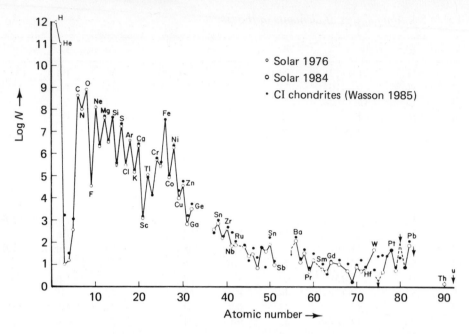

Fig. 5.17. The chemical composition of the sun. The number of atoms is plotted on a 'ten power' or logarithmic scale against atomic number and the solar values (denoted by open circles) are compared with values for the CI carbonaceous chondrites (stony meteorites that have suffered a minimum of physical and chemical processing; solid circles). The chondritic and solar data are fitted at silicon. The meteoritic data are from *Meteorites, their Record of Early Solar System History*, John Wasson, 1985, Freeman, San Francisco. For references to the solar data see *Spectroscopy of Astrophysical Plasmas* A. Dalgarno and D. Layzer (eds.) 1987, Cambridge University Press, p. 89. An almost identical diagram is obtained from the more recent data of Grevesse and Anders.

million million (10^{12}) hydrogen atoms. The total mass involved in the solar atmosphere is surprisingly small. Reverting to our simple model of an absorbing gas overlying a hot continuum-producing photosphere, we note that the solar spectrum is produced by a layer of gas amounting to less than two grams per square centimeter! So we base our analysis of the sun on an aliquot of about a twenty-thousand-millionth of its total mass.

Fig. 5.17 compares the composition of the sun with that of a CI-type (least processed) carbonaceous chondrite. For all non-volatile elements with reliable determinations, the agreement between the solar and chondritic abundances is impressive. Among the less plentiful elements there is indeed scatter, but here we must remember that the solar determinations are based on only a few lines, which may be seriously blended, or for which *f*-values may be poorly determined. Thus, in compiling a table of solar system abundances, as has been done recently by N. Grevesse and E. Anders, one would be guided by the solar data for gaseous and volatile elements and by carbonaceous chondritic data for the others.

Properly interpreted, the abundance picture in Fig. 5.17 constitutes a cosmic historical record of great significance. The following characteristics are to be noted:

(a) Hydrogen is more abundant than all other elements added together; helium is runner-up with a ratio of about one helium atom for approximately every ten hydrogen atoms.

(b) There is a deep minimum corresponding to lithium, beryllium, and boron.

(c) This deep minimum is then followed by a group of elements of much greater abundance, with peaks at carbon, oxygen, and neon.

(d) Following the oxygen and neon peaks there is an irregular decline until scandium is reached.

(e) A prominent abundance peak occurs at iron, following which the abundances drop off irregularly until we reach an atomic weight of about 100 (atomic number = 45).

(f) Beyond this point the decline is very gentle; there are minor peaks at barium and lead. After lead, the curve drops off rapidly. Bismuth is the last stable element.

(g) There is a pronounced difference between the abundances of elements with even atomic numbers and those with odd atomic numbers; the even atomic number elements tend to be more abundant. Thus, carbon, oxygen, and neon are more abundant than nitrogen, fluorine, or sodium, while silicon and sulfur are more abundant than phosphorus. This is the rule of Oddo and Harkins.

How is such a distribution to be explained? There is general consensus that hydrogen and most of the helium were made in the 'big bang' at the time of creation. Except possibly for a small amount of lithium all the other elements were produced later, mostly in stars as they approached the ends of their lives (see Chapters 9 and 11). Considerations of this problem lead us into questions of the origins, evolution, decay, and demise of stars. Before we address these broad topics we should look more carefully at properties and statistics of different types of stars and certain fundamental relationships such as those between mass, temperature, and luminosity.

6

Dwarfs, giants, and supergiants

'One star differeth from another in glory.'

The words of the Scripture apply not only to the apparent brightnesses of the stars as seen from the earth but to their true luminosities, sizes, and masses as well.

The total amount of radiation flowing from the surface of a star, that is, its intrinsic luminosity, is measured by its absolute magnitude, which is the apparent magnitude the star would have if it were at a distance of 10 parsecs, or 32.6 light-years (see Chapter 1). Since the energy radiated per unit area depends only on the stellar surface temperature (which can be deduced from color and spectrum observations), two stars of the same size and temperature should have the same absolute magnitude. Or, if we compare two stars of exactly the same temperature, their surface areas will be proportional to their luminosities.

'The whales and the fishes'

All stars, however, are not uniform in diameter; they vary from diminutive objects one-fiftieth the diameter of the sun to mammoth stars a thousand times larger. In 1912 Einar Hertzsprung noted that among the hotter stars there did seem to exist a correlation between temperature and true brightness, in the sense that the hottest stars were also the brightest. Among cooler stars the situation was otherwise.

For example, the brighter component of Capella is visually about 132 times as luminous as the sun. It is also cooler, so that the power emitted per unit area is only 60 percent as great; hence the ratio of areas is 220 and Capella must have a diameter about 15 times that of the sun, or about 21 million kilometers. Hence it may be called a giant star. Arcturus is about 520 times as bright as the southern dwarf star ε Indi; since these two Class K stars have the same surface temperatures, the diameter of Arcturus must be about 23 times as great as that of ε Indi. The fainter component of the famous binary 61 Cygni B is 3700 times fainter than β Ursae Minoris (Kochab); hence the radii of these two K stars of the same temperature differ by a factor exceeding 60. The M giant β Pegasi is 450 000 times as bright as the faint M dwarf Lalande 21185 of the same temperature. The radii of these two stars differ by a factor of 670!

If one also includes extremely luminous stars such as Antares, Betelgeuse, μ

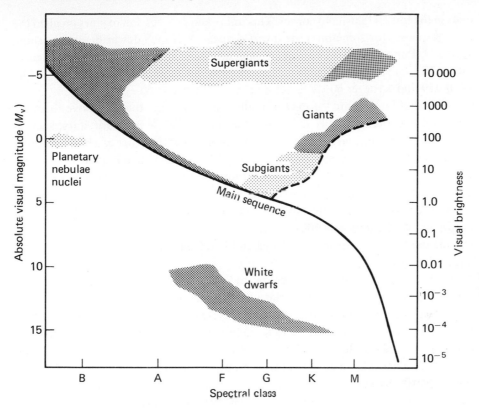

Fig. 6.1. The Hertzsprung-Russell diagram for stars in the solar neighborhood. The absolute visual magnitude (left-hand scale) is plotted against spectral class. The right-hand scale gives the visual brightness in terms of the sun.

Cephei, and VV Cephei, even greater disparities in dimensions are involved. Such stars are called supergiants. Even with the rather poor data available in 1912, Hertzsprung concluded that, as regards sizes, stars could differ from one another as the whales differ from the fishes.

The Hertzsprung–Russell diagram

In 1913, Henry Norris Russell, at Princeton, USA, made the first plot of a fundamental diagram which illustrates the great disparity of stellar sizes, temperatures, and brightnesses. In this diagram, known as the Hertzsprung–Russell or HR diagram, we plot spectral classes (or colors), which depend on temperature, against absolute visual magnitudes, which measure intrinsic brightness. Fig. 6.1 shows the results for a mixture of stars in our local region of the Galaxy. Notice that the stars are not scattered at random on the diagram but tend to be concentrated in certain well-defined zones or strips. The great majority of observed stars fall in a narrow continuous belt running diagonally downward across the diagram from very blue, hot, luminous objects at the upper left-hand corner to red, cool, intrinsically faint ones in the lower right-hand corner. This unbroken progression of stars is called the main sequence (or sometimes the dwarf sequence). It includes brilliant, hot stars, such as ζ Puppis and 10 Lacertae, as well as nearer, more familiar stars, such as Sirius, Vega, Procyon, α Centauri, the sun, 61 Cygni, and faint red, dwarf stars like Wolf 359 and Krueger 60.

Of special interest are the exceptionally luminous stars that lie above the main sequence and the unusually faint ones that lie far below it. Giant stars, such as Capella, Aldebaran, and Arcturus, have luminosities of the order of a hundred times that of the sun. They extend from Class G to M, and also include some representatives of spectral classes R, N, and S, i.e. carbon and heavy-metal stars. There is a gap, called the Hertzsprung gap, between the main-sequence and giant stars. Scattered sparsely across the top of the diagram, and including stars from about 300 times as bright as the sun to others 100 000 times as bright, are the supergiants. Examples include the brightest Orion stars, Rigel and Betelgeuse, and also Deneb, Canopus, and β Centauri.

At the opposite extreme, far below the main sequence, are dim stars whose surface temperatures are appropriate to the spectral classes at which they are plotted. These are the so-called white dwarfs. Other groups have been identified. Between the giants and the F, G, and K main sequence lie numerous stragglers which are called subgiants. Objects of this type often occur in eclipsing binary systems (see Chapter 8). On the left side of the diagram, below the main sequence but above the white dwarfs, are compact, hot objects found as central stars in planetary nebulae such as the Ring Nebula (Chapter 10). Certain exploding stars, called novae, are also found in this region.

Of great importance is the existence of certain regions of avoidance. For example, there is a large gap between the main sequence and the white dwarfs, although a few stragglers are found just below the main sequence. These objects are called subdwarfs. The dashed line below the giant region and to the right of the subgiant region defines the edge of an excluded zone in which no stars are found.

One additional fact must be emphasized. As usually plotted, the HR diagram presents a biased picture. Most of the stars visible to the unaided eye on a clear, dark night are brighter than the sun, yet the overwhelming majority of stars found in a nearby volume of, say, a million cubic light-years are fainter than the sun. Supergiants like Rigel are prominent at distances of a thousand light-years, but not a single Class M dwarf star is visible without optical aid. Thus, any catalogue of stars that goes down to some limiting apparent magnitude (and most star catalogues do) will be biased in favor of intrinsically bright stars. Except at the most extreme ranges of brightness, we can correct the HR diagram for this effect and produce a diagram appropriate to a unit volume in the solar neighborhood. Thus the main sequence is thinly populated at its brighter end but the number of stars increases steadily until the Class M red dwarfs are reached. The vast majority of all observed stars belong to the main sequence. The next largest group is the white dwarfs; then follow subgiants and giants. The rarest are the lonely, luminous supergiants that adorn the vast domain of our Galaxy.

One may construct HR diagrams for different regions of the Galaxy, for star clusters, or associations (e.g. galactic star clusters such as the Pleiades), for globular clusters such as M92, and for the central bulge of the Galaxy, and even for clusters in the Magellanic Clouds. The thus obtained HR diagrams often differ in important ways from one another and provide clues to the histories of stars and stellar systems. The establishment of such HR diagrams occupies a central role in modern astronomy, especially when they are interpreted with the aid of data on stellar masses, dimensions, temperatures, and chemical compositions.

Stellar masses and dimensions

The scientist is constantly on the alert to find correlations between independently measured quantities, such as the masses, luminosities, and diameters of stars. To illustrate the meaning of a correlation, suppose we measure the heights of a sample of boys of all ages, ranging from a few months to 20 years. We would find that on average the tallest boys would also be the oldest, although an occasional lad would be shorter, even though older than another. Clearly, the two observed quantities, age and height, are correlated. If we also measured their weights, we would find weight correlated with age and also with height. The advantage of such correlations is that we need only to measure one quantity in order to get an idea of the other. Studies of double stars have yielded the important result that, for main-sequence stars, mass and luminosity are correlated in the sense that the heaviest stars are also the brightest. Along the main sequence, luminosity is also correlated with surface temperature.

Table 6.1 gives the masses (as compiled by Daniel Popper) for a sample of nearby visual binaries with good data. The columns give for each star of a binary or multiple system (denoted as A, B, or C in order of brightness) the spectral class, the mass with its associated error, the luminosity, the radius, the density, the surface gravity (in terms of that of the earth), the distance in parsecs (pc), and the separation of the two stars in astronomical units (AU). Mass, luminosity, radius, and density are all given in terms of the solar values. Note that the luminosities are bolometric, they give the star's total power output as compared with that of the sun.

Since the parallaxes of these stars have been measured, and therefore their distances are known, the mean separation of the stars can be found. Orbits determined in units of arcsecs can be converted into astronomical units. Then, from the periods, we can find the masses. As noted in Chapter 1, an x-percentage of uncertainty in the parallax will produce a $3x$-percentage error in the mass.

For each star listed in Table 6.1, the spectral class permits an estimate of the temperature, from which we can compute how much energy is radiated by each unit of surface area. We may then work out how large each star must be to provide its observed luminosity. Next, from the mass and volume, we get the average density. Mass and radius yield the surface gravity. Inspection of the table shows several engaging facts. First, the white dwarf companions of σ_2 Eridani B and Sirius B are much too faint for their masses as compared with other objects. Notice further the

Table 6.1 *Mass and luminosity, with derived radius, density, and surface gravity for some dwarf stars*

Star	Spectral Class	Distance (pc)	Separation (AU)	Radius	Mass	Density	Luminosity	Surface gravity
η Cas A	G0V	5.81	70	0.98	0.91 ± 0.15	0.97	1.15	26.5
B	M0V			0.59	0.56 ± 0.15	2.73	0.074	45
σ₂ Eri B	WD A	4.8	33	(0.022)	0.43 ± 0.07	(40 000)	(0.0033)	25 000
C	M4.5V			0.22	0.16 + 0.03	15.3	0.0056	93
Sirius A	A1V	2.65	20	1.68	2.20 ± 0.2	0.464	23.4	21.8
B	WD Aᵃ			(0.020)	0.94 ± 0.1	(117 000)	(0.0027)	66 000
α Cen A	G2V	1.35	23.6	1.274	1.14 ± 0.05	0.55	1.318	30
B	K0V			0.944	0.93 ± 0.05	1.17	0.525	29
ξ Boo A	G8V	6.8	33	0.77	0.90 ± 0.20	2.0	0.537	42
B	K4V			0.55	0.72 ± 0.15	4.33	0.11	66
ζ Her A	G0IV	9.6	13	2.24	1.25 ± 0.3	0.50	5.50	7.0
B	K0V			0.79	0.70 ± 0.2	1.40	0.525	31
70 Oph A	K0V	4.9	22	0.79	0.84 ± 0.15	1.66	0.447	37
B	K5V			0.68	0.61 ± 0.10	1.97	0.145	37
Kru 60A	M3V	4.0	9.5	0.35	0.28 ± 0.03	6.71	0.0138	65
B	M4.5V			0.23	0.16 ± 0.02	13.3	0.0050	85
L726-8 A	M5.5V	2.6	5.35	0.16	0.11 ± 0.02	25.8	0.0015	117
UV Ceti B	M5.5V			0.151	0.11 ± 0.02	31.8	0.0010	134

Note:
[a] WD denotes a white dwarf. Surface temperatures are not as well determined for white dwarfs as for other objects; hence uncertainties in luminosity, radii and density are larger and these figures are given in parentheses.

high densities and surface gravities of these stars. Conversely, the subgiant, ζ Herculis, is too bright for its mass.

Mass determinations may be made also for main-sequence eclipsing binaries for which adequate spectroscopic measurements are available. Table 6.2 gives data from a compilation by Popper for a number of eclipsing stars for which the radii, separations, and masses may be determined. The masses and radii are given in solar units, and the separations of the stars in units of solar radii (696 000 kilometers) and also in units of 0.01 astronomical units. Fortunately, in eclipsing binary systems the derived dimensions are independent of the distance of the star; they depend only on the accuracy of the orbit. One may also use spectroscopic binaries where interferometric measurements give orbital data; size, eccentricity, and inclination to line of sight, data which are similar to those found for visual binaries.

Fig. 6.2 shows the mass–luminosity correlation for main-sequence stars based on the careful work of Daniel Popper. Note the strong dependence of luminosity on mass. A star twice as massive as the sun is about 16 times as bright; a tenfold increase in mass corresponds roughly to a rise in luminosity of about 10 000. Main-sequence eclipsing binaries tend to define a tight correlation. Some of the scatter arises from observational error, but some of it is real; stars truly deviate from a narrow main

Table 6.2 *Data for a number of eclipsing binaries (courtesy of D.M. Popper)*

Star	Spectral class	Period (days)	Separation Solar radii	Separation (0.01 AU)	Radius	Mass
YY Gem (Castor C)	M1V	0.81	3.87	1.79	0.62 ± 0.01	0.59
					0.62 ± 0.01	0.59
UV Leonis	G2	0.60	3.88	1.81	1.08	0.99
	G2				1.08	0.92
VZ Hydrae	F5	2.90	11.5	5.4	1.35	1.23
	F6				1.12	1.12
WW Aurigae	A	2.52	12.1	5.6	1.89	1.98
	A				1.89	1.82
RX Herculis	B9	1.78	10.1	4.7	2.44	2.75
	A0				1.96	2.33
U Oph	B5	1.68	11.0	5.17	3.43	5.16
	B5				3.11	4.60
Y Cygni	O9.8	3.00	23.2	13.1	6.0	16.7
	O9.8				6.0	16.7

sequence. Fig. 6.3 shows the dimensions of main-sequence eclipsing binary stars in comparison with the sun. The stellar radii listed in Tables 6.1 and 6.2 range from about a sixth that of the sun for UV Ceti to about six solar diameters for Y Cygni.

Another path to obtaining stellar dimensions involves making measurements of their angular sizes. Angular diameters (in arcsecs) were first measured by F.G. Pease with an instrument called the stellar interferometer. This device, invented by A.A. Michelson, is usable only for bright red giant and supergiant stars (Figs. 6.4 and 6.5). Fortunately, the 'photon correlation' interferometer, invented by R. Hanbury Brown and R.W. Twiss, provides complementary data on various bright, hot stars, such as Sirius or Vega. A method called speckle interferometry permits measurements of angular sizes of many stars; it does, however, require a large-aperture telescope.

If the distance of the star is known, its real diameter can be found at once from the angular diameter. This method is used to get the solar diameter. The sun's angular diameter, 1919.26 arcsecs or 0.009305 radian, is accurately measurable. The distance of the sun is also accurately known: 149 600 000 kilometers. Hence the diameter of the sun is $0.009305 \times 149\,600\,000 = 1\,392\,000$ kilometers ($=866\,000$ miles). Stellar angular diameters, however, range downwards from 0.056 arcsecs and can be measured directly with only relatively low accuracy. Alternatively, if we know the surface brightness of a star and its apparent brightness, we can calculate its

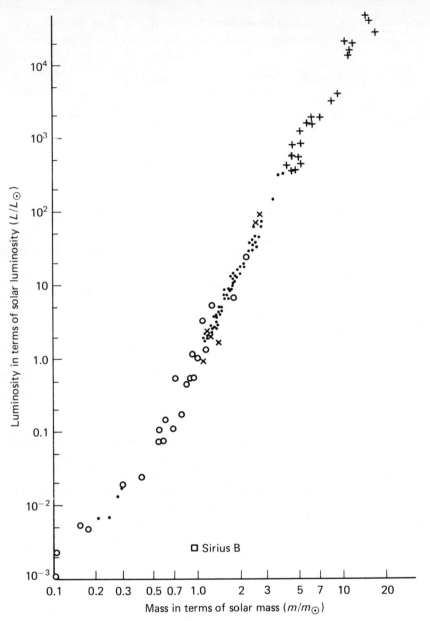

Fig. 6.2. The mass–luminosity relationship. From data assembled by D.M. Popper stellar luminosity is plotted against stellar mass for main-sequence stars. A small dot (.) indicates main-sequence eclipsing binary stars; + shows detached (not contact) main-sequence O, B stellar systems; × shows spectroscopic binaries which are resolved by interferometric methods; and ○ shows visual binaries. Note that small errors in parallax determinations can produce much larger errors in the derived parallaxes; hence the visual binaries tend to show a larger scatter than do many of the eclipsing binaries. The white dwarf, Sirius B, is included on this diagram to emphasize the severe deviations of these dense objects from the mass–luminosity correlation defined by the main sequence. Procyon B, whose mass is 0.65 that of the sun, is fainter than Sirius B.

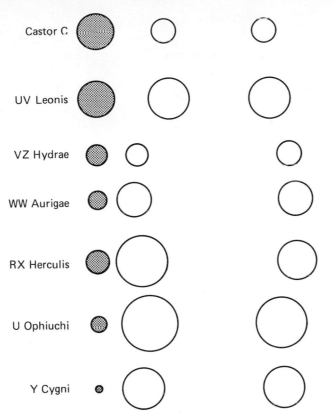

Fig. 6.3. Dimensions of main-sequence stars. The sizes and separations of typical eclipsing binaries are plotted to scale. In each instance the sun is represented by the shaded circle at the left. The data are from D.M. Popper.

angular size. The angular sizes so derived are in harmony with those found by interferometric methods.

As reference to Fig. 6.4 shows, giant stars live up to their names. The supergiant, Antares, however is more than 500 times larger than the sun (Fig. 6.5). On the other hand, some of the white dwarfs are no larger than the earth. Large stars tend to be massive, but the range in stellar masses does not begin to approach the range in sizes. Consequently, the stars show a startling variety of densities. The average density of the sun, for example, is somewhat greater than that of water, about equal to that of soft, lignite coal. Most main-sequence stars have densities ranging from about 0.1 to 3 times that of water. The radius of the supergiant Antares is about 560 times and its volume is 560^3 or about 175 million times that of the sun. The mass of Antares probably does not exceed 50 solar masses so its average density must be less than one-millionth that of the sun. On the other hand, although the white dwarf σ_2 Eridani B has only about nine-millionth the solar volume, its mass is about 43 per cent that of the sun, which gives the star the amazing density of about 40 000 times that of water!

A number of dwarf binaries are of particular interest. Looking at the column of Table 6.1 which shows separation in astronomical units, we note that many of these

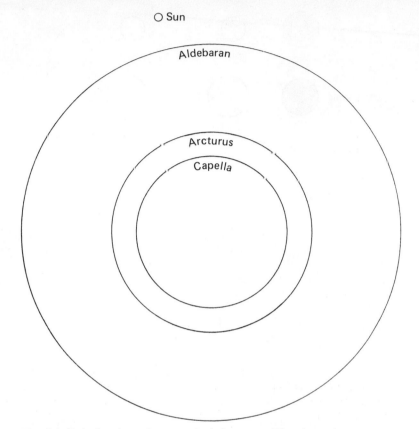

Fig. 6.4. Relative sizes of some typical giant stars. The pioneering measurements of the angular diameters of giant stars such as Aldebaran and Arcturus (and also of supergiants) were made many years ago at the Mount Wilson Observatory by F.G. Pease, who used the Michelson stellar interferometer. When the parallaxes, and therefore the distances of the stars are known, their linear diameters can be computed. This diagram shows the relative sizes of some giant stars and the sun.

systems are built on about the same scale as the solar system. α Centauri, our nearest stellar neighbor, consists of two stars with masses comparable to that of the sun, plus a small distant companion, Proxima, one-fifteen-thousandths as bright as the sun. Krueger 60B is one of the faintest stars whose masses are known, although several less massive, dark or 'brown' stars have been discovered. An hypothetical planet would have to be about 10.6 million kilometers from Krueger 60B to receive as much light and heat as the earth gets from the sun. It would be uninhabitable because tidal effects would quickly stop its rotation. To receive our present allotment of radiant energy from van Biesbroeck's star (which is one millionth as bright as the sun), our planet would have to be closer than half the distance of the moon from the earth. If Arcturus were to replace the sun, we would be comfortable just outside the orbit of Saturn, but with Betelgeuse as our central luminary, a distance about nine times that of Neptune from the sun would be appropriate. From this vantage point, cool Betelgeuse would appear about 30 times larger than the sun.

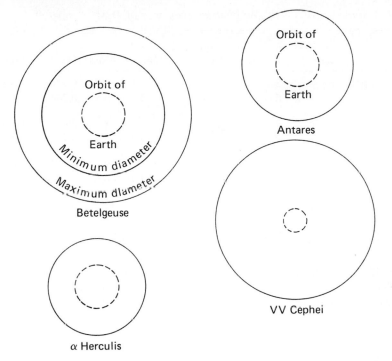

Fig. 6.5. The relative sizes of supergiant stars. In each example, the broken circle indicates the size of the earth's orbit (radius 149 600 000 kilometers). Betelgeuse varies irregularly in both light emission and radius. VV Cephei is a component of an eclipsing binary. All of these stars have strong stellar winds (see Chapter 9).

Before we leave, temporarily, the subject of binaries, some unusual objects are called to attention (see Table 6.3). Algol was once considered a textbook eclipsing binary, but notice a strange property of the fainter component. It is cooler than the sun, a bit larger in size than the brighter component, and less massive, but brighter, than the sun. Some other systems have a relatively normal B star paired with a supergiant. How can such seemingly strange systems come into being? The answer is provided in Chapter 8, where we discuss binary star evolution.

Relation between spectrum and luminosity

Fortunately, the spectra of stars supply important clues to their luminosities, so that if a spectrum of sufficient scale is obtained one can make a good estimate of the star's intrinsic brightness. If one compares carefully the spectra of giants or supergiants and dwarfs of the same spectral type, one finds that the two groups of spectra are not exactly alike, as membership in the same spectral class would imply. Although the general features of the spectra may match very well, there will be certain easily recognizable differences. For example, the lines of ionized strontium are not prominent in the sun, but they are extremely intense in the spectrum of ζ Capricorni, a supergiant 6000 times as bright as the sun. Similarly, a comparison of the spectrum

Table 6.3 *Some unusual eclipsing binaries (courtesy of D.M. Popper)*[a]

Star	Spectral class	Period	Separation	Radius	Mass	Density	Luminosity
β Persei (Algol)	B8	2.87	16.3	3.1	3.7	0.124	16.0
	G8	days		3.2	0.8	0.244	6.3
ζ Aurigae	K4Ib	972.16	1000	130	8	3.6^{-6}	1400
	B8	days		(3.5)	5.7	0.13	200
31 Cygni	K3Ib	3780	2630	135	10	4.1^{-6}	5500
	B3V	days		(4.7)	6.6	0.064	1700
VV Cephei	M1Ia	20.4	5440	1600	18	4.4^{-9}	
	B9	years		(50)	18		

Note:
[a] The radii, masses, densities, and luminosities of the components are all given in solar units. The separation of the two components are given in units of the solar radius (696 000 kilometers). Parentheses around numbers indicate they are less certain.

of the supergiant CE Tau with that of the dwarf HD 95735, both of Class M2, shows that the line of neutral calcium at 4226 Å is enormously stronger in the latter. These examples illustrate a general rule found by W.S. Adams and A. Kohlschütter, namely, that lines of certain ionized atoms tend to be strong in giant stars and weak in dwarfs, and that lines of certain neutral atoms behave in the opposite sense.

These luminosity or absolute-magnitude effects must be established empirically by a careful comparision of the spectra of stars that are known to differ in luminosity. The effects can be explained, qualitatively at least, in terms of the different physical conditions that prevail in the atmospheres of giant and dwarf stars. The chief difference is one of density. Table 6.1 shows that the stellar material of the intrinsically faint dwarfs is relatively closely packed, and that the bright stars are very much more tenuous. The densities that we have given are, of course, average values for the whole body of each star, but the dense dwarfs may also be expected to have shallow, compressed atmospheres, and the tenuous giants rarefied and extended ones. In other words, the density of a star's atmosphere appears to be correlated with its size and luminosity.

We recall from Chapter 4 that the density has an important influence on the appearance of the spectrum. When the density is low, and free electrons are few and far between, atoms are more easily maintained in the ionized condition than when the density is high. Consider for example, the behavior of an element such as calcium, which exists in both neutral and ionized forms in a great many stars. Neutral calcium atoms absorb a spectral line in the blue–violet region at 4226 Å; ionized calcium produces the well-known *H* and *K* lines in the near ultraviolet. Given two stars of the same temperature, one a tenuous giant and the other a dense dwarf, we would expect a greater percentage of calcium atoms to be ionized in the giant. Consequently, the line of neutral calcium should be weak in the giant and

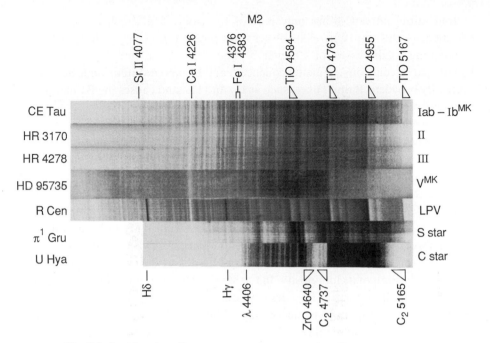

Fig. 6.6. Luminosity effects at spectral class M. This set of spectra compares cool stars of differing intrinsic luminosity; it includes a 'heavy-metal' S star, a long-period variable, LPV, and a carbon, C, star. The Roman numerals Iab to V on the right indicate decreasing luminosity. MK denotes the Morgan–Keenan spectral class. Establishing luminosity or 'absolute magnitude' effects require spectra of high quality. Note the great increase in the strength of the 4226 Å line of neutral calcium in going from stars of high luminosity (and low atmospheric densities) to the faint dwarf HD 95735 with its dense atmosphere. (From *Michigan Atlas of Objective Prism Spectra*, Nancy Houk and M.V. Newberry, 1984, University of Michigan.)

strong in the dwarf, whereas the converse should hold true for the lines of ionized calcium.

Now in practice we do not compare the spectra of two stars of the same temperature, but rather two stars of the same spectral appearance. The spectral type is judged from the intensities of the spectral lines and not from the intensity distribution in the continuous background. A given spectral type corresponds therefore to a certain degree of average ionization. High temperature as well as low density favors ionization; a dwarf star is hotter than a giant of the same spectral class, the higher temperature of the dwarf compensating for the lower density of the giant. If this compensation were identical for all elements, the spectroscopic discrimination between giants and dwarfs would be virtually impossible. Fortunately, this is not the case, and for some elements, such as strontium, the ionization is more sensitive to low density than to high temperature. Hence the ionized-strontium lines are strong in giants but weak in dwarfs of the same spectral class. Calcium is another example. The ionization of calcium is more sensitive to the low densities of the atmospheres of giants. Hence the line of neutral calcium at 4226 Å is stronger in dwarfs than in giants of the same spectral class (see Fig. 6.6).

Ionization differences are only part of the story. The much higher densities prevalent in the atmospheres of dwarf stars produce profound effects of density broadening as discussed in Chapter 5. The 4226 Å calcium line in dwarfs is broadened by collisions. In main-sequence A and B stars the interatomic Stark effect markedly broadens Balmer lines, such as Hγ and Hδ, and causes the Balmer lines to be more washed out than in supergiant stars (see Fig. 5.4, p. 77).

Atmospheres of giants and supergiants are often characterized by large-scale mass motions of the material, sometimes called 'turbulence'. These motions produce changes in the shapes of spectral lines that are easy to distinguish from the effects of density broadening but not always from the effects of stellar rotation.

Since the pioneer work of Adams and Kohlschütter, many investigators have studied effects of luminosity differences on spectral features. Atomic lines, molecular bands, and even features of the continuous spectrum have all been utilized. In modern work astronomers employ the Morgan–Keenan (M–K) system of spectral classification. The classical Henry Draper system assigned each star a spectral class with no indication of its luminosity. In the M–K system one adds a Roman numeral to indicate a luminosity. The designations extend from Ia for the brightest supergiants to V for main-sequence stars. Thus Rigel (B8Ia) is a very luminous supergiant, Betelgeuse (M2Ia) and Antares (M1Iab) are also supergiants, β Gruis (M3II) falls between supergiants and giants, and Arcturus (K2III) is a giant. A GIV or KIV star would be a subgiant, but a B5IV star (such as Achernar), an A0 IV star (γ Geminorum), or an F5V star (Procyon) are bright main-sequence stars. Vega (A0V), the sun (G2V), and 61 Cygni AB (K5V and K8V) are dwarf stars.

Luminosity classes can be interpreted in terms of absolute magnitudes only when they are calibrated. Thus we must know the true distances of stars whose spectra we compare. The nearby dwarf stars offer no problems since their distances are easy to determine. Giant and supergiant stars can be handled when they are members of star clusters whose distances can be found.

The colors of the stars

With the advent of the photoelectric photometer it became possible to measure the color of a star more accurately than one can establish a spectral class. Furthermore, with a given telescope one can measure magnitudes and colors of much fainter stars than one can observe spectroscopically.

The astronomer is immediately confronted with the need to decide on a type of photocell and appropriate filters with which to work. By using a carefully selected set of narrow band-pass filters, each one of which covers but a small wavelength range, an astronomer may be able to get more detailed information on individual stars than the observer who works with fewer, wide band-pass filters. The price paid is that he or she is not able to reach such faint stars. One of the most popular combinations of photocells and filters is the Johnson–Morgan system (see Appendix E). The observer measures three colors with the aid of three different filters. The system is often called the *U, B, V* (or ultraviolet, blue, visual) system. The *U* filter transmits a broad band in the near ultraviolet, the *B* filter in the blue region between about 3800 and 5000 Å, and the *V* filter, together with a conventional photocell, gives a color sensitivity very similar to that of the human eye; hence photoelectric *V*-magnitudes may be regarded as equivalent to visual magnitudes and we shall so regard them in subsequent discussions. With the brightness of a given star measured in this system, we may set up two different kinds of brightness differences or color indices: $U-B$ and $B-V$.

By combining the advantages of photoelectric photometry with those of the photographic plate, accurate measurements of colors and magnitudes of large numbers of stars are easily obtainable. With a conventional photoelectric phot-ometer it is possible to observe only one star at a time, while a photographic plate can record a whole cluster simultaneously. The astronomer measures photoelectri-cally the magnitudes and colors of a few selected stars over a range in color and brightness and uses these as benchmarks with which to compare perhaps hundreds of stars photographed on the plate. A relatively recent improvement on this technique is to employ what are called charge-coupled devices, commonly called CCDs. With them it is possible to take what amounts to 'electronic photographs' and bypass photographic methods entirely. A great disadvantage of a photographic plate is that the blackening of the emulsion is not proportional to the intensity of the light, that is, it is a non-linear detector, so photographic images cannot be added or subtracted. A CCD is a linear device; its response is directly proportional to the intensity. Thus, one can add images or subtract them. The sky background can be subtracted so that faint star images can be detected and measured more easily. A present limitation of a CCD is that only a small field can be observed compared with what can be recorded on a photographic plate.

Use of colors instead of spectral types has one severe drawback. The color, but not the spectral class, of a distant star may be reddened by particles in interstellar space, much as the setting sun is reddened by selective light-scattering by molecules in the earth's atmosphere (see Chapter 10). A hot O or B star can appear as yellow as a K star. By measuring three (or preferably more) colors one can often correct for this effect since the brightness of a star in a given spectral region will be affected in a different way by temperature than by a coloring produced by interstellar smog.

Below we describe how the three colors of the Johnson–Morgan system can be used to analyze data for a star cluster (see also Appendix E). From *U*, *B*, and *V* color measurements plots of $B-V$ against $U-B$ color indices are used to help assess effects of coloring and absorption by interstellar particles. Then one derives the corrected or $(B-V)_0$ colors and plots them against V_0 (the *V*-magnitude also corrected for space absorption) for members of the cluster or stellar association. Thus we have a plot $(B-V)_0$ against V_0. The next step is to determine the amount of the shift in V_0, keeping $(B-V)_0$ fixed, to superpose the cluster diagram onto the standard HR diagram, which gives $(B-V)_0$ against absolute visual magnitude M_v. In this way one gets the distance modulus of the cluster, V_0-M_v and hence its distance (see the following section).

Furthermore, for each value of the true color index $(B-V)_0$ the temperature may be established (see Appendix E). One may also assign for each value of $(B-V)_0$ and of M_v the corresponding spectrum and luminosity class. Hence a plot of $(B-V)_0$ against M_v can be converted to a spectrum–luminosity diagram. We can go further than that. For each luminosity class and temperature we can derive the bolometric correction needed to convert a visual absolute magnitude to a bolometric magnitude and hence obtain the true luminosity of the star (see Appendix E). Finally we can plot surface temperature against true luminosity, thus obtaining a relation that can be compared directly with the predictions of theory (see Figs. 6.7 and 6.8). As we shall see in Chapter 9, the theory of stellar evolution predicts for a star of given mass, rotation, and chemical composition the variation of its luminosity and radius with time. Since the luminosity depends on the surface area and temperature, a diagram giving luminosity and temperature can be converted to one giving luminosity and radius, or alternatively radius and luminosity can be converted to surface temperature and luminosity.

Let us summarize the steps needed to convert an observed color diagram of a star cluster to a meaningful array (since all members of the cluster are at the same distance from us, differences in apparent magnitude correspond to the same differences in absolute magnitude):

(1)   Given the U, B, V, color measurements for a cluster we compute the indices $U-B$ and $B-V$ for each star;
(2)   Using the known effects of space absorption on colors and magnitudes, we convert to $(U-B)_0$, $(B-V)_0$, and V_0;
(3)   We compare the cluster HR diagram, namely, $(B-V)_0$ plotted against V_0 with the standard HR diagram derived for similar kinds of stars to obtain the distance modulus, $V_0 - M_v$. The distance modulus, corrected for space absorption, is related to the distance by the equation $V_0 - M_v = 5 \log r - 5$, where r is the distance in parsecs.

The color of a star can be affected by factors other than temperature, luminosity, and space absorption. Not only the spectrum of a star but also its color can be affected by its chemical composition and to some extent by mass motions (turbulence) in its atmosphere. Now a color measurement, unlike a spectrogram, takes a bite of radiation over a big range in wavelength. If the star has numerous strong lines in this region, the energy received by the photocell will be diminished. If the lines are weak because of a low ratio of metal to hydrogen, the measured colors will be affected. Likewise turbulence effects, by modifying the amount of energy removed from the spectrum, can affect colors in the same way as would an increase in the metal-to-hydrogen ratio.

Although the U, B, V color system introduced by Harold Johnson has been of inestimable value in astronomy, many other photoelectric color combinations are also extremely useful. Examples include the six-color photometry of Stebbins and Whitford and, more recently, of Stebbins and Kron (which covers a wide range of wavelengths), and the system used by Strömgren (see Appendix E). In the latter system, a judicious selection of filters and photocells enables one to measure both narrow spectral intervals (a few tens of ångströms) and wide ones (100 Å or more), so chosen that by a suitable combination of the measurements one can obtain for each star its spectral class, luminosity class, and metal-to-hydrogen ratio, and the influence of space absorption.

It is even possible to estimate the star's evolutionary status (see Appendix E). Not surprisingly, for cool stars it is useful to add a broad band-pass red (R) filter, and use the index, $V - R$ rather than $B - V$.

Star clusters and associations

The power of the HR diagram as a tool in astronomical research is exhibited in studies of aggregates of stars and particularly of star clusters. In our Galaxy there are basically two types of star clusters – the 'open' or 'galactic' clusters and the globular clusters. They differ rather fundamentally from one another, not only with respect to size, distribution in space and numbers of stars, but also with respect to the kinds of stars involved.

Galactic clusters are found mostly near the plane of the Milky Way. They may contain only a handful of stars, as in the Ursa Major cluster, or hundreds or even a couple of thousand stars, as in h and χ Persei. A few galactic clusters – the Pleiades, the Hyades, Praesepe, and Coma Berenices – are visible to the unaided eye, and many more can be seen with a field glass or small telescope. Some, like the cluster NGC 2244 in Monoceros or the cluster in Messier 8, have associated with them clouds of dust and gas. In others, there is no longer any trace of dust remaining. Their diameters range up to a few light-years. Sometimes the stars are so thinly spread that the cluster can scarcely be recognized against the background. When we consider the kinds of stars involved, or more particularly the individual HR diagrams, we note a wide variety among open clusters.

From spectrum–magnitude plots constructed with apparent photographic magnitudes and spectra, R.J. Trumpler discovered a great diversity among galactic clusters. Some, like the Pleiades, had only a main sequence starting around B3 and continuing on to fainter stars. Others, like the Hyades or Praesepe, had no B stars, but possessed a few giants; the main sequence began near A or F. Class B stars and giants seemed mutually exclusive in galactic clusters.

A great step forward in the study of star clusters came with the development of accurate techniques for the measurement of stellar colors and magnitudes. Conventional color–magnitude arrays are plotted with $B - V$ (corrected for space absorption) against V. In Fig. 6.7, however, we have plotted absolute visual magnitude against spectral class rather than color, giving the results in schematic form for seven galactic clusters; Fig. 6.8 shows absolute bolometric magnitude plotted against effective temperature. In NGC 2362 and h and χ Persei the main sequence extends to very luminous stars. The latter has also a few red supergiants. Notice that in the Pleiades the bright end of the main sequence veers to the right, that is, redward of the main sequence defined by h and χ Persei and other galactic clusters containing highly luminous stars. The main sequences of M11, and the Hyades start with A or F stars;

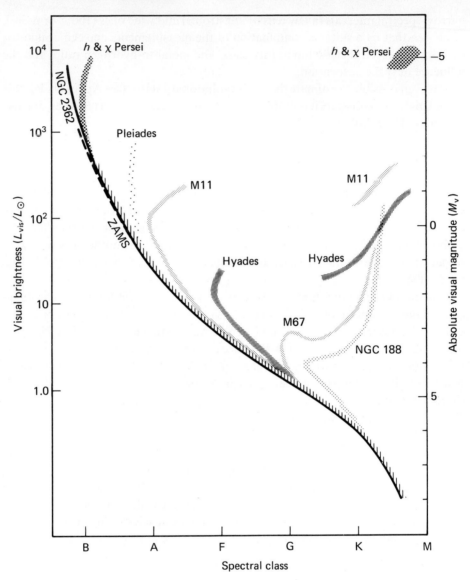

Fig. 6.7. Composite Hertzsprung–Russell Diagram for galactic clusters. Visual brightness L(vis)/L(sun), or absolute visual magnitude, is plotted against spectral class equivalents of measured $(B-V)_0$ colors. The line corresponding to the lower edge of the main sequence defines what is called the 'zero-age main sequence' (ZAMS).

there is a gap between them and a handful of red giants. In M67 or NGC 188 the main sequence does not extend to bluer or brighter stars than G stars like the sun, but the 'giant' branch is continuous with the main sequence.

We can define a sort of limiting main sequence that corresponds to that of NGC 2362. The younger the cluster, the more closely do the main-sequence stars adhere to this limiting main sequence, which is referred to as the 'zero-age main sequence' or ZAMS.

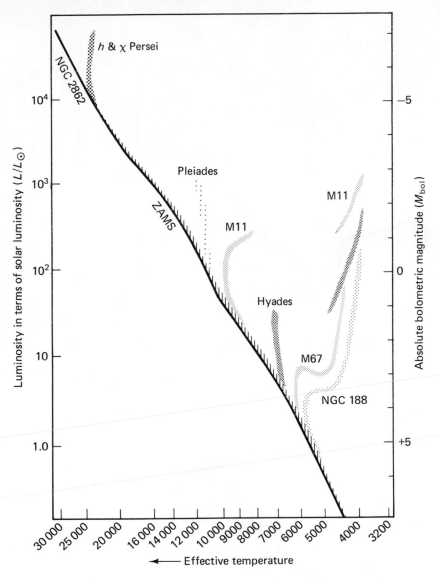

Fig. 6.8. A 'transcribed' Hertzsprung–Russell diagram for galactic clusters. Intrinsic luminosity or its equivalent absolute bolometric magnitude is plotted against the effective surface temperature. Although the diagram is qualitatively similar to Fig. 6.7, there are pronounced differences, especially conspicuous for very hot and very cool stars. The conversion from $(B-V)_0$ or spectral class to effective temperature and bolometric magnitude is made with the aid of tables in Appendix E.

Before we elaborate further on these observations, let us discuss briefly the globular clusters. Whereas there are probably thousands of star groups in the Galaxy that can qualify as open clusters, the number of known globular clusters is only about a hundred. Most of them are concentrated toward the central bulge of the Galaxy but they pay no attention to the galactic plane. A typical globular cluster will have a diameter of 40 or 50 light-years and may contain upward of 100 000 stars.

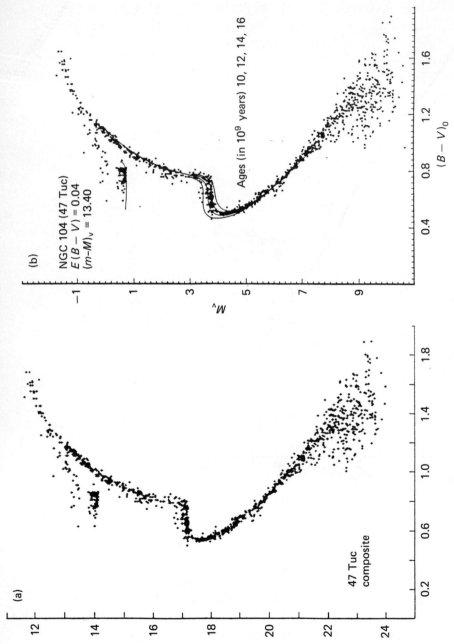

Fig. 6.9. A color-magnitude diagram for the globular cluster 47 Tucanae.
(a) The plot of observed *V*-magnitude against *B − V* color was constructed from data obtained with charge-coupled devices. Notice the narrowness of the upper main sequence. Faint stars are difficult to measure. They scarcely lie above the sky background.
(b) Correction for interstellar extinction gives a color excess, $E(B − V) = 0.04$, which implies very little reddening. A fit to the standard main sequence gives a distance modulus, $(m − M)_V$, or $(V_0 − M_V) = 13.40$, and hence a distance to the cluster of 4600 parsecs. Solid lines show theoretical evolutionary tracks for stars with a helium mass fraction of 0.24 by weight, an Fe/H ratio by numbers 0.22 that of the sun, and an O/H ratio by numbers of 0.45 that of the sun. Tracks are plotted for ages of 10, 12, 14, and 16 thousand million years. The best estimate of the age is 13.5×10^9 years. (Based on work by J.E. Hesser, W.E. Harris, Don A. Vandenberg, J. Allwright, P. Shott, and P. Stetson.)

It is in stellar content, however, that the globular clusters differ most strikingly from the galactic clusters.

In 1944 W. Baade called attention to the existence of two stellar population types. Population I is associated with spiral arms and particularly with clouds of gas and dust grains such as are found in the region of Orion. This population contains bright supergiants and main-sequence O and B stars. Population I stars are prominent also in the Large Magellanic Cloud, in the spiral arms of the Andromeda Spiral M31, and throughout most of the Triangulum Spiral, M33. Population II contains no blue main-sequence stars. The main sequence breaks off in spectral class F and is joined to a giant branch which reaches an absolute magnitude as bright as -2. Population II stars are characteristic of the thin halo of stars enveloping our galactic system, elliptical galaxies such as the companions to Andromeda, and the giant galaxy M87 (NGC 4486) in Virgo, which greatly exceeds our own stellar system in size. Population II stars are associated with relatively little dust and the few high-temperature stars that exist in this population tend to fall off the standard main sequence.

Data on stellar motions are sometimes useful for separating the two populations. Our Milky Way system is in rotation; the sun is traveling about the center in an orbit that is probably not far from circular with a velocity of about 220 kilometers per second, completing a revolution in a period exceeding 200 000 years. Population I objects moving in similar nearly circular orbits will appear to have relatively low velocities as observed from the earth. Consider now stars moving in greatly elongated orbits (thin ellipses) about the galactic center. As they cross the orbit of the sun their motion will be almost entirely inward or outward, so that with respect to the sun they will appear as high-velocity objects. These stars belong to Population II; they are associated with distant regions of our Galaxy. Often they travel to large distances above the galactic plane or to the central bulge. Nearly all high-velocity stars belong to Population II, but not all members of Population II necessarily have high velocities. More particularly, the globular-cluster stars are the prototypes of Baade's Population II, whereas the stars found in open clusters represent the prototypes of Population I.

Baade's original population type II is sometimes called the halo population since it is associated with the galactic halo and represents an extreme type of very old star. Stars in galactic clusters are much younger. The globular clusters, although much older than galactic clusters, do show a considerable age range. Fig. 6.9 shows a color-magnitude diagram for a globular cluster, 47 Tucanae, while Fig. 6.10 compares HR diagrams for several clusters.

There are virtually no main-sequence stars brighter than $M_v = 3.7$, that is, more than about one magnitude brighter than the sun. The giant sequence is joined to the main sequence by an almost vertical bridge and there is sometimes a narrow branch of white and blue stars about absolute magnitude 0. When color–magnitude arrays of globular clusters are compared, significant differences are noted. For example, in M92 the vertical and giant branches are shifted to the blue as compared with M13, M3, and 47 Tucanae. These systematic displacements of the giant and subgiant

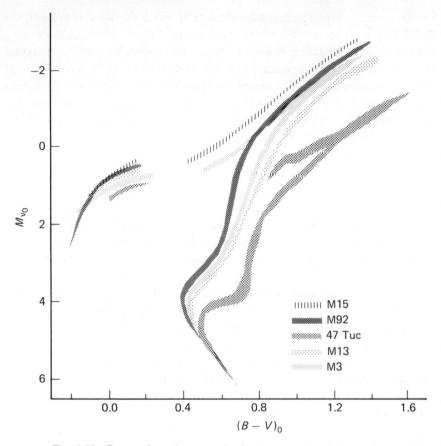

Fig. 6.10. Composite color–magnitude diagrams for globular clusters. The $B-V$ colors and V-magnitudes are corrected first for interstellar extinctions. Then the resultant V_0 magnitudes are converted to absolute visual magnitudes, M_v, by fitting the main sequences (where evolutionary effects have not yet taken place) to a standard main sequence. This step is uncertain, for if the main sequences of different clusters differ intrinsically because of chemical composition differences such a fitting is unjustified. The gap between $(B-V)_0$ values of 0.2 and 0.4 indicates the position of the RR Lyrae stars. The branch (on the left) near absolute magnitude about 0.5 is called the horizontal branch. The data are compiled from work by many investigators; much of the early work was done by H.C. Arp, H.L. Johnson, A. Sandage, W.G. Tifft, and R. Wildey.

sequences seem to be associated with the metal-to-hydrogen ratio, as we shall now explain.

The spectra of the brighter stars in many galactic clusters can be investigated in considerable detail, but individual stars in globular clusters can be studied only with difficulty. Following the pioneer work of D.M. Popper on ω Centauri, examination of various clusters with the 200-inch telescope showed that the spectra of stars of a given intrinsic brightness in different globular clusters differed not only from those of stars near the sun but also from one cluster to another. The differences were in the sense that the metal-to-hydrogen ratio often was smaller in globular-cluster stars than in the sun. This metal deficiency was more marked in some clusters than in others, ranging from a virtually 'normal' (i.e. solar) composition to a metal-to-hydrogen ratio a hundred times lower than the solar value.

Stars of similar metal-deficient composition, the so-called subdwarfs, have been recognized in the neighborhood of the sun. When the metal-to-hydrogen ratio is a hundred times lower than in the sun (as in the star HD 140283), a star of solar temperature (5800 K) may show such weak metallic and strong hydrogen lines as to invite classification as an A star (temperature = 8000–9000 K). Considered as an A star it falls well below the main sequence; properly considered as a G star it will fall near the main sequence. Hence the misnomer 'subdwarf'; such objects should be called 'metal-deficient dwarfs'.

Although detailed abundance studies require spectrograms secured with high dispersion, it is fortunate that metal-deficient stars can be recognized from their colors. The spectrum of a yellow star like the sun is terribly crowded with metallic lines below 4000 Å. Hence it is much fainter when it is observed with a U rather than with a B filter. A metal-deficient star has fewer and weaker lines in this region, with the result that the difference between magnitudes measured with U and with B filters is smaller. That is, metal-deficient stars of about the sun's temperature are brighter in the near ultraviolet than are stars of the same B (blue) magnitude that have normal metal abundances. Also the $B - V$ color indices are affected.

Thus, differences in the color–magnitude arrays for clusters such as 47 Tucanae and M92 arise largely from differences in the metal-to-hydrogen ratio. These differences enter in a subtle and complicated form. Not only does the relation between the $U - B$ and $B - V$ color indices and the effective temperature depend on the metal-to-hydrogen ratio but the actual internal structure of the star may be likewise affected (see Chapters 7 and 8).

Interpretation of cluster color–magnitude diagrams

As explained above, some of the disparities between the color–magnitude diagrams for different globular clusters can be understood in terms of differences in chemical composition. Can any interpretation be placed on the shapes of the color-magnitude arrays themselves? Indeed it can. These curves can all be explained in terms of stellar evolution (see Chapter 8).

The concepts can be grasped more easily by looking at the HR diagram for galactic clusters from the following point of view. Star clusters represent groups of stars of differing masses that were formed at about the same time. Suppose that the masses range from 10 solar masses to 0.1 solar mass, and that all the stars initially were shining on the main sequence. Then their luminosities would be correlated with their masses, but the more massive the star the more rapidly it would liberate energy (recall Fig. 6.2). Each gram of the most massive stars in our group would liberate energy about a hundred times as fast as a gram of the sun.

Now let us assume that the total amount of energy that can be squeezed out of each gram is the same for all matter everywhere. The more massive stars will exhaust their fuel more quickly, leave the main sequence, and eventually disappear. In this interpretation the cluster NGC 2362 is the youngest and in h and χ Persei the brightest stars are beginning to depart from the main sequence. As we proceed to the Pleiades, M11, the Hyades, M67, and finally NGC 188, the main sequence is 'rolled back'. We interpret these as increasingly older clusters in which the brightest, most profligate stars have used up their energy resources and disappeared, at least from the main sequence. We might speculate further and guess that giants and supergiants were stars that had evolved from the main sequence and that white dwarfs represented the final stage of evolution.

One important implication must be mentioned. Although we have not assigned ages to any of these clusters, it is quite clear that NGC 188, which contains no main-sequence stars bluer than the sun, must be much older than, say, the Pleiades, which in turn is older than h and χ Persei. Therefore, star formation must have been occurring in a more or less continuous fashion; presumably it is going on right now. A significant clue is that all these clusters belong to Population I – which is associated with dust and gas of the interstellar medium. So we might speculate further that stars are formed out of the smog and gas of the interstellar medium, shine for a while as main-sequence stars and then evolve away from the main sequence. To investigate these ideas we have to turn to a rather fundamental problem: how are stars constructed and what makes them shine?

7

What makes the stars shine?

In preceding chapters we have seen how the surface temperature, the atmospheric density, and the chemical composition of a star are learned from its spectrum. If we are lucky enough to find that this star is an eclipsing-binary system, we can also often obtain its dimensions, total mass, and average density. Since all sorts of stars are found in binary systems, we have a fair idea of the masses, luminosities, surface temperatures, radii, and compositions of our celestial neighbors.

From spectral studies we can learn much about the atmosphere of a star, but of the vast bulk of it, that is, its interior, we really have little direct information. We seek to know the origin of the energy of the radiation that the stars are so generously pouring out into space, and through what kinds of processes it is produced. In order to answer these questions properly we shall have to consider rather carefully the ways in which astronomers and physicists have sought to understand the processes occurring in stellar interiors. Then we shall try to reconstruct the life histories of typical stars and shall show how the concept of stellar evolution leads to a fitting together of data from many diverse sources into a coherent story. Finally, we shall show how the origin of the elements themselves is to be sought in processes in stellar interiors.

Celestial powerhouses

Our whole cosmogony, all our speculations on the history and future of the physical universe, depend upon the answer to the question: 'What makes the stars shine?' Only when this question is answered is it possible to reconstruct the life history of a star such as the sun, and to interpret many of our basic data such as the mass–luminosity correlation and the Hertzsprung–Russell diagram.

It is not difficult to get a good estimate of the power output of the sun. As C.S. Pouillet did in 1838, one may observe the rate of heating of a blackened flask containing water and exposed to sunlight. The loss of heat by convection and radiation may be gauged by observing the rate of cooling of the heated flask when it is shielded from the sun, and the effects of atmospheric absorption may be roughly measured by doing the experiment at different altitudes of the sun. In a refined version of the observation, the heating effects and losses are more accurately assessed and the fact that atmospheric extinction depends on wavelength is also taken into account. Thus it is found that the energy received by each square centimeter outside the earth's atmosphere is 1.96 calories per minute. Since the distance of the sun is known, one can calculate the total power output of the sun; it amounts to 3.84×10^{23} kilowatts or 5.06×10^{23} horsepower. Even if we assume energy to cost only 1 cent per kilowatt hour, this means that the sun, a fairly modest star, radiates a million million million dollars' worth of energy every second.

Such a figure does not convey much meaning, except perhaps in the context of armaments race budgets. Post-World-War-II military expenditures have squandered a million million dollars in the course of a few years. This is an enormous waste by human standards but on a cosmic scale a million million dollars' worth of energy would run the sun for but a millionth of a second. All of the stars of the Galaxy will emit of the order of a thousand million times as much energy as the sun.

By what means do stars radiate so much energy? How long have they been shining and how long may they continue to do so? We can answer the second question for the sun. By radioactive dating of the earth's crust and of meteorites we find that the solar system was formed about 4.6 thousand million years (4.6 'aeons') ago. More importantly, the radioactive clocks indicate that the oldest rocks in which appear traces of the earliest life forms are about three thousand million years old. During all this time the sun must have been shining very much as it is now for life is a fearfully fragile phenomenon. It functions over only a very small range of temperature between 0 and 100 °C. Survival is possible at temperatures lower than 0 °C, but living is not possible. If the solar energy output were to change by as much as 10 percent, either upward or downward, other factors remaining unchanged, life on the earth would probably be extinguished. So, what source of energy has enabled the sun to shine so dependably for more than three thousand million years?

An elementary calculation shows the hopeless inadequacy of any ordinary source of power, such as chemical combustion, that is, burning. Even if the sun were made of pure carbon, with just enough oxygen present to sustain combustion, it would have burned to ashes in a few thousand years. A more efficient, but still inadequate, source of energy is gravitational contraction. As a large, distended body contracts under the pull of its own gravity, the outer parts literally fall toward the center, and the energy of the falling material is converted into heat and light. H. Helmholtz and Lord Kelvin (William Thomson) suggested, nearly a century ago, that an annual contraction of the sun's radius by 140 feet (42.7 meters) would be sufficient to account for the observed rate of heat liberation. Further calculations show, however, that by shrinking from an almost infinite size to its present dimensions, the sun could shine at its present rate for less than fifty million years. Twenty million years ago the sun would have been at least as large as the earth's orbit, and at that time our planet presumably had essentially modern types of life on it.

One copious source of power, which looks very promising, is the conversion of matter into energy. Early in the present century, Albert Einstein showed that mass and energy were related by the simple equation

$$E = mc^2,$$

where E (in ergs) is the energy that is obtainable from the complete conversion of m grams of matter, c being the velocity of light, 3×10^{10} centimeters per second. In order to keep the sun shining at its present rate, 4 256 000 tonnes of material would have to be transformed into energy every second. Yet the sun is so massive that its mass would thereby be diminished by only 0.1 percent in fifteen thousand million years.

What are the operations whereby stars may convert mass into energy? Several possibilities present themselves. First, as with radioactive substances, a small fraction of the mass may be automatically converted into energy. Second, as in certain laboratory experiments, some atoms may be transmuted with the transformation of roughly 1 percent of the mass into radiant energy. The third possibility, that of the conversion of all the matter of some atoms into energy, seems unlikely. In fact, there is no experimental evidence that justifies a belief in the total annihilation of matter, so this possibility can be excluded.

The first and most obvious suggestion is that stars continue to radiate because they contain great quantities of radioactive material. Experiments in the laboratory show that the disintegration of uranium into radium and eventually into lead is accompanied by a release of considerable amounts of energy in the form of high-speed particles and radiation. The rate of conversion is slow; a piece of pure uranium will be converted into equal parts of lead and uranium in about four thousand million years. But the rate of disintegration is always the same, whatever the nature of the surroundings; we would therefore expect the luminosity of a star that is dependent on its radioactivity to be directly proportional to its mass. The mass–luminosity relation (Chapter 6) shows, however, that the luminosities increase more rapidly than the masses. A star twice as massive as the sun is about sixteen times as bright. It is highly improbable that the more massive stars would have been stocked with greater sources of radioactive materials. Furthermore, even a pure uranium sun would not provide enough energy to maintain its observed rate of radiation. We might imagine it to contain super-radioactive elements, but experimental nuclear physics shows that materials with the required properties do not exist. It would be extremely difficult to construct a star of uranium, thorium, and other radioactive elements that would supply anything near the required energy without having it explode like an atom bomb.

The second hypothesis, that of a transmutation of elements, with a bit of the masses of the interacting atoms consumed to provide energy, seems much more promising. Accordingly, we shall look into this possibility.

The anatomy of a star

As is frequently true in science, the answer to our question 'What makes the stars shine?' depends upon the answers to other questions, those relating to the structure of stars and atomic nuclei. The physicist may probe into the nuclei of atoms in the laboratory, but the astronomer can penetrate only the very outermost skin of the star, its atmosphere. The task of exploring the interior of a star is not as hopeless as it might seem, however, for the physicist has supplied us with the necessary tools.

Our problem is as follows: given the mass, luminosity, chemical composition, and radius of a star, and certain laws of nature, such as those of gravitation, radiation, and gases, what are the densities, pressures, and temperatures at various depths within the star? Let us suppose that the star is rotating so slowly that we can neglect the effects of rotation on its structure. The luminosity of a stable star will equal the total rate of energy output in its interior. With a specified radius and luminosity, the surface temperature will adjust itself so that the surface area times the amount of energy radiated per unit area will equal the total amount of energy generated. If a bright star is relatively small, it will have a high surface temperature; if it is large, it will have a low surface temperature. Let us, for the moment, suppose that the star is of uniform chemical composition throughout. This is probably true for all stars at the beginning of their lives and we shall see later how it is possible to allow for changes in the hydrogen-to-helium ratio brought about by evolutionary effects, that is, the aging of the star.

We can write down the conditions that a stable star must fulfill. Some of these conditions are obvious. Clearly, the model of a star must fit physically reasonable limitations. We cannot have a star with a hollow core or a core whose density increases without bound, a temperature decreasing with depth, or a stellar surface under a substantial pressure. Any physically possible model fulfills these conditions of course, but this alone does not guarantee that it is acceptable. The computed radius of the star should equal the observed radius, and the calculated luminosity should fit the observed luminosity. From the observed stellar luminosity we can easily calculate the rate at which it generates energy, for if a normal star is to remain stable, the rate at which energy is radiated at the surface must equal the rate at which it is being released in the interior. If the liberated energy does not escape but is stored up in some fashion, the mounting pressure of heated gas and radiation will eventually cause the star to explode.

A star can be stable only if at each point in its interior, gas and radiation pressure, which depend on the temperature and chemical composition of the interior according to known laws, exactly balance the weight of the overlying layers. At the surface the pressure is essentially zero, but as we go deeper into the star it rises as the weight of the upper layers increases. Fig. 7.1 shows how we calculate the pressure increase over a small increment h at a distance r from the center of the star. The pressure rise, $P - P'$, exactly equals the density (ρ) multiplied by h and the gravity (g) at the point in question, since the weight of the little box of unit cross-section and thickness h is $\rho h g$. If we can find some way to calculate the variation of the density, ρ, we can calculate the pressure P throughout the star. But even without this information we can still get some idea of the pressures attained in stellar interiors. The total pressure is the sum of gas and radiation pressure and it increases with depth in the star. Actually, the vast bulk of the sun has a density less than that of water and a temperature in excess of a million degrees.

Of great importance to the luminosity and structure of a star is its central temperature, since, as we shall see shortly, this temperature determines the rate of energy generation. The chemical composition of a stellar interior also plays a decisive role, for three reasons: first, because the chemical composition largely determines the transparency, and hence the ease with which energy flows to the

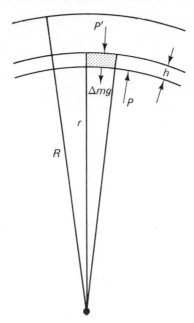

Fig. 7.1. Changes of pressure with depth in a star. Consider a tiny box of unit area, thickness h, and mass $\Delta m = \rho h$, where ρ is the density. The pressure exerted by the gas above the box is P', that at the bottom of the box is P, so that $P' - P = \Delta mg$, g being the acceleration of gravity at the point in question.

surface; second, and more important, because the central temperature depends on the composition; and third because the rate of generation of energy depends intrinsically on the chemical composition of the material involved.

The pressure, p, exerted by a gas is proportional to the temperature and the number, n, of individual particles per unit volume in accordance with the expression

$$p = nkT,$$

where k is the Boltzmann constant (Appendix B) and T is the absolute temperature. For a gas composed of neutral hydrogen atoms, the average molecular weight (the total mass divided by the total number of particles) is 1. We speak of the molecular weight of a gas even though we are dealing with atoms, for it measures the mass divided by the number of particles, the mass of the oxygen atom being taken as 16.00. (In ordinary usage, the molecular weight of a gas is the number of grams of the substance contained in 22 400 cubic centimeters at atmospheric pressure and 0 °C.) But if the hydrogen is ionized, as it is in a stellar atmosphere, there are twice as many free particles – nuclei plus electrons – with no change of mass, and the molecular weight becomes 0.5. A completely ionized carbon atom, with mass 12, yields seven particles, six electrons and a nucleus, so its molecular weight is 12/7, or 1.72. Each unit mass of hydrogen contributes two particles, each unit mass of carbon 7/12 of a particle. A carbon star and a hydrogen star that were alike with respect to size, mass, and density variation would differ in internal temperature, because the particles of the carbon star would have to work harder, that is, move faster, than the more numerous hydrogen particles to support the weight of the overlying layers.

Thus the star composed of heavy elements is hotter inside than the hydrogen star. Unless there is a deficiency of the necessary energy-producing fuel, energy will be generated more rapidly in the star with the hotter interior, and it will shine more brightly. As long as its supply of fuel lasted, a metal star would be about a hundred times as bright as a hydrogen star. If the sun were composed of pure hydrogen, its central temperature would be somewhere near ten million degrees. The central temperature would be about forty million degrees for heavy atoms, whereas a pure helium composition would require a temperature in the neighborhood of twenty-seven million degrees. Therefore, the chemical composition of its interior will profoundly affect the structure and total luminosity of the star. If the star is not homogeneous in composition, its central temperature will differ from these figures. Early workers who were unaware of the overwhelming preponderance of hydrogen and helium assigned central temperatures that were much too high.

It is a straightforward process to calculate the model of a star of uniform chemical composition, a 'zero-age main sequence' (ZAMS) star. One adopts the relative proportions of hydrogen, helium, and heavy elements from an analysis of the star's atmosphere. Then, with the aid of the known mass of the star, the gas laws, the radiation laws, and the law of gravitation, we proceed to calculate how the pressure, temperature, and density increase toward the center of the star. The march of temperature and density with depth will depend on the manner of energy transport through the layers in question.

In stars of normal density, energy may be transported in a stellar interior either by radiation, each quantum being passed on from one atom to the next, or by large-scale mass motions, commonly called convection. Let us consider these processes in turn.

Throughout most of the interior of a star, the temperature is well above a million degrees. Hence most of the radiation occurs in the form of X-rays, and most atoms are stripped of all electrons down to the inner shells. Absorption of radiation occurs as photoelectric ejection of electrons occurs; the phenomenon is similar to the photoionization of hydrogen from the second level which produces the absorption at the head of the Balmer series in the spectra of the hotter stars. At yet higher temperatures all electrons may be stripped off and the chief hindrance to the outward flow of radiation may be scattering by free electrons. Since the lighter atoms lose their inner electrons more easily, and heavy atoms such as iron may retain their innermost electrons at very high temperatures, the ability of the material to block outgoing radiation, that is, its opacity, will depend on its chemical composition. For any assumed mixture of elements, the opacity depends in a complicated way on temperature and density. Extensive calculations have been carried out, for example by Arthur Cox and his associates, for all temperatures and densities likely to be encountered in stars.

Convection is a familiar process for the redistribution of heat, in which the energy is carried by matter in mass motion. A stove heats a room primarily not by radiation but by warming a mass of air in its neighborhood, which then rises and moves across the top of the room. Meanwhile, its place is taken by cool air which in turn becomes heated by the stove. In a similar fashion, deep within a star, a blob of heated gas rises toward the surface, while cooler material from a higher layer sinks downward to replace it; the process is a continuous one. These large-scale convection currents may be orderly, completely disorderly (turbulent), or something in between.

Energy flows outward in a star because there is a steady decline of temperature from the center to the surface. The rate of radiation of energy by the star is given and this energy outflow must be provided by radiation or by convection currents. If the rate of fall of temperature (temperature gradient) required for radiative flow is too great, convection will set in. Convection is a more efficient mode for moving energy outward than is radiation, but it can occur only in certain regions of the stellar interior where the ionization or temperature gradient is just right. It is usually easy to decide whether a given layer in a star is characterized by one kind of energy transport or the other, but if convection currents are at work it is trickier to estimate the rate of energy flow and the change of density and temperature with depth. In the visible layers of the sun's atmosphere the outward flow of energy takes place almost entirely by radiation, but just below the photosphere a convective region starts and extends downward about a fifth of the solar radius. In yet deeper regions, the flow of energy is entirely by radiation. In cooler, less massive, main-sequence stars the convection zone becomes more important, and in faint red dwarfs it appears that throughout the whole star the energy is carried outward by convection currents.

Proceeding to more massive main-sequence stars, we find that somewhere near spectral class F5 the outer convection zone suddenly shrinks to a small size. An intrinsically very luminous main-sequence star, such as Spica, develops a convective core, although throughout the vast bulk of the star's energy is carried outward by radiation. These differences in the internal structures of stars have important consequences on their life histories.

To calculate a model for a star like the sun we would start with a chemically homogeneous body of the same mass and calculate its internal structure, including the central temperature and density, and predict its luminosity and radius. If such a program is carried out for the sun, using the best atmospheric estimates of the total hydrogen content and hydrogen-to-helium ratio, one finds the wrong radius and a luminosity that is too low. As we shall see in the next section, normal stars shine by converting hydrogen into helium. Our chemically homogeneous object corresponds to a star of zero age. Hence one must allow for the depletion of hydrogen and repeat the calculations until the computed radius and luminosity agree with the observed ones.

The condition that a normal star must be stable to small disturbances excludes many otherwise possible models. If it is perturbed, for example by the gravitational attraction of a nearby star, the star must rebound to restore the status quo. Oscillations may take place, as indeed occurs in Cepheid variable stars and even in the sun, but these oscillations must remain finite. There is an important exception. The requirement of a condition of stability may be relaxed or discarded in the last stages of a star's life (see Chapter 9).

The theory of stellar structure has advanced greatly in recent years. On the one hand, there have been striking advances in our knowledge of the underlying basic physics; on the other, improved computers have made it possible to carry out calculations that were wholly beyond our capabilities until quite recently.

An important outcome of the early work was that, even without knowing the specific mechanism of energy generation (assuming only that it depended on the density and temperature), it was possible to derive a mass–luminosity law, on the assumption that all stars considered were built on the same model. The argument goes somewhat as follows; the central temperature must be governed by the star's mass and its chemical composition, since the pressure must suffice to sustain the weight of the overlying layers. The greater the mass and the greater the effective molecular weight, the higher will be the temperature. The higher the central temperature, the greater the outward flow of energy either by convection or by radiation. Hence the luminosity of the star must depend on its mass and chemical composition. The mass–luminosity relationship essentially reflects the fact that a star is hot inside and therefore energy must leak out; the hotter the core of the star the faster the rate of energy flow and the brighter the star. An energy-generation process is required only to keep the furnace hot.

With plausible chemical compositions and the known masses of main-sequence stars, astronomers were able to predict stellar radii and luminosities to within the legitimate uncertainties of the basic physics employed (for example, the absorptivity of matter for the X-rays prevailing at temperatures of millions of degrees can be obtained only by theoretical calculations). Hence one could have confidence that the temperatures (10–35 million K) and densities (20–200 grams per cubic centimeter) predicted for the centers of main-sequence stars were physically meaningful, although accurate values could not be stated for any one star. It turns out that, at precisely these temperatrues and densities, things begin to happen to the nuclei of light atoms such as carbon and nitrogen when they are placed in a medium of hydrogen. What happens is that atoms are transmuted and energy is liberated in just the requisite amounts to explain the observed radiation of these stars.

The transmutation of elements

We have seen how, guided by the observed masses, luminosities, and diameters of the stars, and by well-known laws of nature, astronomers have deduced the physical conditions obtaining within a star. In searching for processes that will convert mass into energy, physicists have looked for one or more that will liberate enough energy at the predicted central pressures and temperatures of main-sequence stars to reproduce their observed luminosities.

Years ago the suggestion was made that the stars shine by converting hydrogen into helium. The atomic weight of hydrogen is 1.00813, and that of helium is

4.00386. Therefore, if four hydrogen atoms could be converted into one helium atom, 0.02866 unit of mass, or 1/141 of the original mass, would appear as energy. The stars of the main sequence and most giants, subgiants, and supergiants probably do shine by converting hydrogen into helium, but the process is not so simple as jamming four protons together to form a helium nucleus. In order to learn the conditions under which the transmutation of elements takes place, we shall turn to the experiments of the physicist.

From Chapter 3 we recall that the nuclei of atoms are composed of protons and neutrons. (We need not be concerned here with the even more fundamental entities such as quarks that make up protons and neutrons.) The number of protons determines the charge of the nucleus and therefore the kind of atom; the number of neutrons determines the isotope of the element. We saw, for example, that an ordinary carbon atom, of atomic weight 12, has a nucleus consisting of 6 protons and 6 neutrons, while a carbon isotope of atomic weight 13 contains 6 protons and 7 neutrons. Chemically, the two atoms are similar.

The heaviest naturally occurring elements, uranium and thorium, spontaneously break down into less heavy atoms, such as radium and mesothorium, and ultimately into lead. But it is possible, by bombardment with high-speed protons, helium nuclei (alpha particles), or neutrons, to disintegrate other atoms and thus achieve the transmutation of the elements. Devices such as the electrostatic generator or the cyclotron enable the physicist to speed up bombarding particles to enormous velocities and to fire them at atoms.

What kinds of particles are most effective? Protons and helium nuclei are useful only in the bombardment of light elements. Positively charged atomic nuclei repel the similarly charged protons and helium nuclei, and, since heavy nuclei have very large positive charges, they strongly repel the incoming hydrogen or helium nuclei and drive them away before they can penetrate the nucleus. But the neutron possesses no charge and consequently may easily penetrate the heart of any atom.

One may obtain neutrons from a radioactive 'pile' in an atomic-energy plant or from the heavy isotope of hydrogen, deuterium, whose nucleus consists of a single proton tightly bound to a neutron. When heavy water, that is, water made with heavy hydrogen, is bombarded with deuterium, each pair of colliding deuterium atoms is converted into one helium atom of atomic weight 3 and one neutron.

If we bombard nitrogen (atomic weight 14) with neutrons, we obtain boron (atomic weight 11) and helium (4). We may write the reaction in the form of an equation:

$$^{14}N_7 + {}^1n_0 \rightarrow {}^{11}B_5 + {}^4He_2,$$

where the superscript denotes the atomic weight and the subscript the charge. Similarly, iron bombarded by neutrons is transformed into a manganese isotope with the ejection of a proton:

$$^{56}Fe_{26} + {}^1n_0 \rightarrow {}^{56}Mn_{25} + {}^1H_1.$$

Reactions of this latter type appear to be of great importance in certain stars, not for the production of energy, since little energy is actually liberated, but rather for the building up of elements of higher atomic number. If a sufficient supply of neutrons can be produced, elements like titanium and iron can be built up by neutron capture into elements such as zirconium, barium, or even lead.

Several possibilities are open when a nucleus is bombarded by a proton. First, the proton may simply remain in the nucleus and produce a new nucleus with a mass and a charge each greater by one. Deuterium (heavy hydrogen) when bombarded by protons yields the helium isotope of atomic weight 3, plus radiant energy:

$$^2H_1 + {}^1H_1 \rightarrow {}^3He_2 + \text{radiation.} \tag{1}$$

Second, a proton colliding with a nucleus may be converted into a neutron, with the ejection of a positive electron, or positron, which has the same mass as a negative electron but a charge of the opposite sign. (P.A.M. Dirac predicted its existence from theory, and Carl Anderson at the California Institute of Technology later found it experimentally.) The resulting nucleus retains the same charge but has a greater mass. For example, in the case of hydrogen, a proton–proton collision may form a heavy-hydrogen nucleus or deuteron, consisting of a proton and a neutron:

$$^1H_1 + {}^1H_1 \rightarrow {}^2H_1 + \epsilon_1^+ + \nu, \tag{2}$$

where ϵ_1^+ stands for the positive electron (or positron) that is created. The ejected positive electron then encounters an ordinary negative electron, the two annihilate each other, and the excess energy appears as radiation. The symbol ν denotes an additional particle that is ejected, the neutrino, neutral particle of negligible mass that is emitted along with a positive electron. It carries away momentum and energy. Hence, not all annihilated mass appears as energy; some is carried away by neutrinos. There exists also an anti-particle called an anti-neutrino associated with the emission of negative electrons in naturally occurring beta decay.

Third, the bombarded nucleus may break up into two or more parts, one of which is a helium nucleus. Lithium (atomic weight 7) bombarded by a proton breaks down into two helium nuclei:

$$^7Li_3 + {}^1H_1 \rightarrow {}^4He_2 + {}^4He_2. \tag{3}$$

The light nuclei of beryllium and boron are particularly vulnerable to proton collisions, as shown by the reactions

$$^9Be_4 + {}^1H_1 \rightarrow {}^6Li_3 + {}^4He_2, \tag{4}$$
$$^{11}B_5 + {}^1H_1 \rightarrow 3{}^4He_2. \tag{5}$$

Note that the sums of the nuclear charges and of the atomic weights must be equal on both sides of each equation.

Fig. 7.2 shows the tracks of a negative and a positive electron in a bubble chamber. Charged particles produce bubbles as they are slowed down in liquid hydrogen. These bubbles, illuminated and photographed, show the paths of particles that are themselves much too small to be seen. A magnetic field has been applied, so that the path of the negative electron is bent in one direction and that of the positive electron in the other. Fig. 7.3 shows how a particle of high energy, say a proton of a thousand million electron volts energy, may collide with an atom and produce a host of secondary high-energy particles; these in turn collide with other

Fig. 7.2. Tracks of negative and positive electrons, formed in a bubble chamber. This photograph shows one of the most important results of modern physics – the direct conversion of energy into matter. A gamma ray is converted in the field of a hydrogen nucleus into an electron–positron pair, which forms the upper V. The triplet consists of a similar pair and an additional electron, which was knocked out of a hydrogen atom when the gamma ray was converted in the field of the orbital electron. A strong magnetic field is impressed perpendicular to the plane of the tracks. Consequently, at the point of decay the electron spirals off to the left and the positron spirals to the right. Note the tightening of the spiral tracks, which is caused by the loss of energy by both positron and electron. As the velocity decreases the curvature of the path steadily increases. (Courtesy Lawrence Radiation Laboratory, University of California.)

atoms and thus produce a shower of high-speed electrons and evanescent particles called mesons. In showers such as these, the nucleus that is hit is actually shattered and great quantities of energy (some of it in the form of short-lived particles) are liberated. Particles of such high energy are associated with the cosmic rays (see Chapter 12).

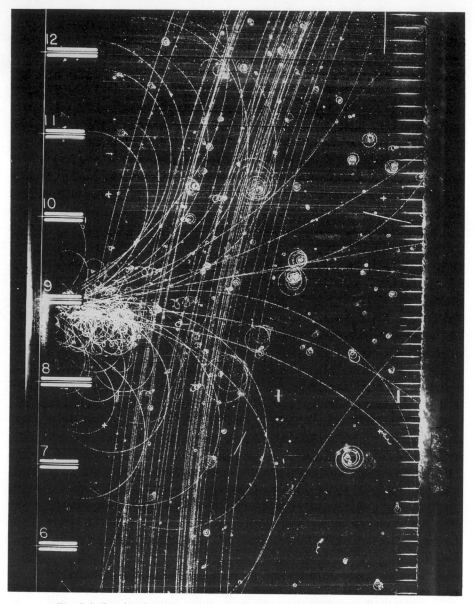

Fig. 7.3. **Production of secondary particles by a high-energy cosmic-ray particle.** This photograph, obtained with the 72-inch bubble chamber, shows what is called a cosmic-ray shower. The high-energy primary particle enters from the left, collides with an atom, and emits a shower of various particles. The impressed magnetic field (whose lines of force are perpendicular to the plane of the paper) causes the paths of the charged particles to be curved; the lower the speed of the particle the greater the curvature. (Courtesy of the Lawrence Radiation Laboratory, University of California.)

Reactions involving alpha particles (helium nuclei) are also of interest. Thus

$$^{12}C_6 + {}^4He_2 \rightarrow {}^{16}O_8,$$
$$^{16}O_8 + {}^4He_2 \rightarrow {}^{20}Ne_{10}, \tag{6}$$

and so forth. These reactions require very high energies (or temperatures). At sufficiently high densities and temperatures two alpha particles may momentarily coalesce to form a nucleus of $^8Be_4$; this nucleus is unstable and disintegrates spontaneously, within 10^{-14} second, into two alpha particles, but, if before it has time to fly apart another alpha particle is captured, a $^{12}C_6$ nucleus is formed. This is known as the triple-alpha process.

With yet higher temperatures (energies), reactions such as

$$^{12}C_6 + {}^{12}C_6 \rightarrow {}^{23}Na_{11} + {}^1H_1$$
or \hfill (7)
$$^{12}C_6 + {}^{12}C_6 \rightarrow {}^{20}Ne_{10} + {}^4He_2$$

may occur. In Chapter 9 we shall find that some of these reactions can become important in stars in the late stages of their lives.

How energy is produced in main-sequence stars

At the temperatures of millions of degrees encountered in stellar interiors, highly ionized atoms are moving about so rapidly that occasional encounters between protons and nuclei should be sufficiently violent to produce nuclear transformations. It is important to realize that the various nuclear transformations do not operate effectively at the same temperatures. Consequently, a nuclear process responsible for the generation of energy in one star may not work in another star with a different central temperature.

In stars such as the sun, whose central temperatures are in the neighborhood of thirteen to fifteen million degrees, the principal source of energy is the proton–proton reaction originally proposed by C.L. Critchfield and H. Bethe. In the first step, two protons collide to form a deuteron as in reaction (2); at the same time a positive electron and a neutrino are formed. Then the deuteron captures a proton to form a nucleus of $^3He_2$ by reaction (1) and two ^3He nuclei may collide to form a nucleus of ^4He and two protons:

$$^3He_2 + {}^3He_2 \rightarrow {}^4He_2 + {}^1H_1 + {}^1H_1. \tag{8}$$

Thus, in the course of events, four protons are combined to form an alpha particle; the two positrons that are also created are annihilated when they encounter ordinary electrons. The two neutrinos formed in reaction (2) steal parts of the energy and escape from the star. It is of interest that the first step in the reaction–the formation of the deuteron–has never been verified experimentally, but depends entirely on theory. Nevertheless, its validity seems well established. Alternative channels of the reaction are possible and have been discussed by a number of physicists. Fortunately, a direct experimental check on the predictions of theory is available.

Once a ^{3}He nucleus is formed, one of the possible reaction sequences is

$$^3\mathrm{He}_2 + {}^4\mathrm{He}_2 \rightarrow {}^7\mathrm{Be}_2 + \gamma,$$
$$^7\mathrm{Be}_4 + {}^1\mathrm{H}_1 \rightarrow {}^8\mathrm{B}_5 + \gamma,$$
$$^8\mathrm{Be}_5 \rightarrow {}^8\mathrm{Be}^* + \epsilon^+ + \nu,$$
$$^8\mathrm{Be}_4 \rightarrow {}^4\mathrm{He}_2^4 + {}^4\mathrm{He}_2.$$

That is, a ^{3}He nucleus collides with a ^{4}He nucleus to form ^{7}Be with emission of a gamma ray; ^{7}Be captures a proton to form ^{8}B and another gamma ray is emitted; ^{8}B decays first to an excited nuclear state of ^{8}Be with the emission of a positive electron and a neutrino (the * indicates the ^{8}Be is unstable), and the ^{8}Be promptly decays to two alpha particles.

The neutrinos escape forthwith from the sun since they interact only very slightly with matter. It is precisely this weak interaction that makes them so hard to detect. A possible reaction is

$$^{37}\mathrm{Cl}_{17} + \nu \rightarrow {}^{37}\mathrm{Ar}_{18} + \epsilon^-{}_{-1},$$

that is, a ^{37}Cl nucleus captures a neutrino to form an ^{37}Ar nucleus which ejects an electron. As a detector, Raymond Davis used a vast quantity of carbon tetrachloride (CCl_4) and he made his observations in a deep mine to avoid extraneous reactions produced by cosmic rays.

The observed rate is $2.1 \pm 0.3 \times 10^{-36}$ reactions per ^{37}Cl nucleus per second ($= 2.2$ solar neutrino units, SNUs), as compared with a predicted value of 7.6 SNUs. The discordance is distressingly large. Tinkering with models of solar structure, initial helium content, and other elemental abundances does not seem to help much. The fact that neutrinos were detected from supernova 1987A (see Chapter 11) shows that they cannot decay or transform themselves into some undetectable form in the 8-minute transit time from the sun to the earth. A recently proposed experiment will measure the neutrino flux at high energies, while one involving ^{71}Ga should yield neutrino flux at the low energies associated with the first step of the proton–proton reaction. Curiously, this experiment gave no counts, which suggests that solar neutinos lose energy by encounters with electrons and are transformed into what are called muon neutrinos (see Chapter 12). These are not detected by the gallium counters.

In stars somewhat more massive than the sun, the main source of energy is no longer the proton–proton reaction. When the central temperature approaches sixteen or seventeen million degrees, carbon is transformed and a remarkable process that leads to the production of helium from hydrogen takes place. The cycle is described in Table 7.1.

A carbon-12 nucleus captures a proton and becomes radioactive nitrogen-13 with the emission of radiation ($\gamma =$ gamma rays). Nitrogen-13 then disintegrates into carbon-13 with the ejection of a positive electron and a neutrino. The next proton collision transforms carbon-13 into ordinary nitrogen of atomic weight 14 with the emission of radiation. The nitrogen nucleus, struck by a proton, emits radiation and becomes oxygen of atomic weight 15. This nucleus is unstable, that is, radioactive,

Table 7.1 *The carbon–nitrogen–oxygen cycle*

Reaction	Energy in MeV	Reaction time (at 15 million K)
$^{12}C_6 + {}^1H_1 \rightarrow {}^{13}N_7 + \gamma$	1.94	1 000 000 years
$^{13}N_7 \rightarrow {}^{13}C_6 + \epsilon_1{}^+ + \nu$	2.22	15 minutes
$^{13}C_6 + {}^1H_1 \rightarrow {}^{14}N_7 + \gamma$	7.55	200 000 years
$^{14}N_7 + {}^1H_1 \rightarrow {}^{15}O_8 + \gamma$	7.29	200 000 000 years
$^{15}O_8 \rightarrow {}^{15}N_7 + \epsilon_1{}^+ + \nu$	2.76	3 minutes
$^{15}N_7 + {}^1H_1 \rightarrow {}^{12}C_6 + {}^4He_2$	4.97	10 000 years

and disintegrates into nitrogen of atomic weight 15 with the ejection of a positive electron and the inevitable neutrino. When the heavy nitrogen nucleus captures a proton, it nearly always splits into an alpha particle and the original carbon of atomic weight 12. By this process four hydrogen nuclei have been converted into one helium nucleus. Carbon, which plays a role analogous to a chemical catalyst, can be used over and over again until all the hydrogen has been converted into helium. The cycle is broken one time out of every 2500 when the nitrogen-15 nucleus captures a proton to form oxygen-16. The table lists the liberated energy expressed in MeV for each step of the carbon–nitrogen–oxygen (CNO) cycle for a temperature of 15 000 000 K (near the lower end of the range where this cycle is important). Note that nitrogen-14 is the longest-lived of any participant in the process. At higher temperatures the CNO cycle proceeds more rapidly, but nitrogen-14 remains the bottleneck. Thus most of the carbon originally present in the energy-producing strata will be converted to nitrogen, a process that can modify the composition of the star.

This CNO cycle, discovered independently by H. Bethe of Cornell University and C.F. von Weizsäcker in Germany, explains fairly well the luminosities of the stars along the brighter part of the main sequence. It may make important contributions to giant and supergiant stars also. The faint main-sequence stars appear to be explained adequately by the proton–proton reaction. The CNO cycle is probably responsible for less than 10 percent of the solar power output.

One very important point must be emphasized. As long as we require the stars to be chemically homogeneous, or nearly so, throughout their interiors, stellar model calculations yield objects that fall on or near the main sequence. This situation became apparent in the 1920s when it was pointed out that giant stars constructed on the same model as the sun would have relatively low central temperatures. As we shall see in the next chapters, the internal structures of giant and supergiant stars can become quite complicated. At certain stages it is possible for a star to have an inert carbon core surrounded by a thin shell where helium is being converted to carbon, an inert helium shell, and then a final energy-generation shell where hydrogen 'burns' into helium.

8

The youth and middle age of a common star

The problem an astronomer faces in trying to construct the life history of a star was described neatly more than a century ago in a classical essay by Sir John Herschel. Imagine the citizen of a giant city who had never even seen a tree to be turned loose in a virgin forest for an hour and required to bring back an account of the life history of a tree. An observant person would quickly perceive a dead snag or rotting log as the last chapter of a tree's history, and he or she might even deduce that small trees evolved into large ones with the same types of leaves or needles, but other stages, particularly the origin of trees, would be difficult to establish. The analogy to the astronomer's problem will be evident. Within a period comparable to the age of the earth, many stars must show appreciable aging, or as we often call it, 'evolution'. Evolution is a term generally applied to the history of a species, a society, a culture, or a civilization, but not to the normal life and aging of an individual. The universe or even the Galaxy evolves, but, strictly speaking, a star cannot be said to evolve; its life history is predestined by its mass, chemistry, rate of rotation, and whether or not it is a member of a binary system. The term is so ingrained, however, that it cannot be dethroned.

If the lifetime of a typical star like the sun is measured in thousands of millions of years, it may appear difficult to recognize evolutionary effects. Certain stages of a star's evolution are passed over quickly, but aside from the catastrophic supernova phase, which may affect only a few stars in the course of their lives, most stars evolve so slowly that any effects that may occur could scarcely be recognized.

Throughout most of its life a star shines by converting hydrogen into helium. To keep on shining at the present rate, the sun, an unassuming dwarf, must convert 564 000 000 tonnes of hydrogen into 560 000 000 tonnes of helium every second. The sun has been shining for at least 4.5 thousand million years and should continue to shine for several thousand million more. On the other hand, Y Cygni, which is burning up hydrogen a thousand or so times as rapidly as the sun, seemingly cannot last more than about 100 million years longer, no matter how we juggle the hydrogen content. If the hydrogen content of Y Cygni is 80 percent, a rather high estimate, the star should be only about 35 million years old. But the universe has probably been in very much its present state for at least ten thousand million years. Therefore, if nuclear reactions are responsible for the energy generation in stars, objects like Y Cygni must either be recent creations or have been kept from shining for most of their lifetimes. Hence we may conclude that not only do stars evolve, but the processes of star formation may be going on before our very eyes.

The formation of stars

In some fashion the cold dust and gas of the interstellar medium must be gathered into a blob or cloud of sufficient density to permit it to pull itself together under the attraction of gravity. The first step in the formation of a star is extremely difficult to understand fully. The material must be protected from dissipative forces: gravitational disruption by passing stars, tidal forces of the galactic bulge, and radiation pressure, until it can attain a sufficient density for gravitation to pull it together. Theoretical studies by C. Hayashi, R.B. Larsen, and many others all seem to agree that once gravity has a chance to take hold, the subsequent evolution of the star will be relatively rapid.

The speed of evolution will depend on the mass; the more massive objects will contract relatively rapidly, whereas the less massive ones will pull themselves together slowly. As the mass contracts, the interior becomes hotter and hotter until finally the object glows as a bona fide star. The star will continue to contract until the central temperature becomes high enough for nuclear reactions to occur. The central temperature stabilizes at a value that permits the star's entire energy output to be supplied by the conversion of hydrogen into helium. It continues to shine as a main-sequence star until hydrogen is exhausted in the core and it evolves away from the main sequence as a subgiant, giant, or supergiant. A body containing, say, only 1 percent of the solar mass would never develop a central temperature high enough to support nuclear reactions and would settle down as a cold, planet-like object without ever having tasted the glory of being a star.

It seems most rewarding to look for star formation among giant molecular clouds such as occur in the Orion nebular complex, in Sagittarius B2, in the dark clouds of Taurus, and in other regions of high obscuration. Nearly all young, bright, blue stars are found in or near regions of extensive interstellar matter. We would not expect to find luminous stars always in intimate association with dark clouds of absorbing matter. Once a bright star is 'turned on' its radiation profoundly affects neighboring gas and dust. The hydrogen and metals in the immediate neighborhood of the star become ionized. The expanding heated gas produces a shock wave that rushes through the cooler surrounding region and eventually causes the dissipation of much of the original interstellar cloud.

Sometimes very massive stars may end their lives as supernovae and produce substantial shock waves in any nearby interstellar material. These shock waves compress regions of the interstellar medium and may help to trigger star formation. 'Star chains', seen for example in a galaxy such as the Triangulum Spiral (Messier 33), can be accounted for by such events. Even luminous stars formed in dark clouds may spend their entire lives hidden from view. The very first stage in the formation of any star is likely to be obscured by dust. Only infrared and radio-frequency observations can help us in such circumstances.

Star formation in our own and other galaxies is believed to occur largely in 'bursts', producing what are often called associations, that is groups of stars that appear to have been formed relatively recently in a compact volume. Two types of association were recognized by V. Ambarzumian, who called attention to the significance of the phenomenon. The so-called 'O associations' consist of hot, luminous stars, and the 'T associations' of relatively faint objects which appear to evolve into main-sequence stars not unlike the sun. The T associations owe their designation to the great number of variables of the T Tauri type – dwarfish stars immersed in vast clouds of smog and gas. These objects have been studied in great detail by A.H. Joy, G. Herbig, and G. Haro. All present evidence indicates that they are stars in the process of formation, not yet settled down to the main sequence.

Stellar associations are not clusters, although it is possible that an open cluster may form occasionally from an association. The stars are simply formed in a region; thereafter they spread apart as fast as their velocities will carry them. Studies of the space motions of these stars, derived from their proper motions and radial velocities, give in many instances the rates of expansion of these associations; knowing their distances, we can then determine their ages. In this way A. Blaauw found the Lacerta association to have an age of about seven million years, while the Scorpio–Centaurus association appears to represent two or three associations with ages of about twenty million years.

The material of the proto-star is originally cool, perhaps as cold as 20 K, and consists largely of molecular hydrogen with some cold grains made of abundant elements. The condensing mass contracts, not in a spherically symmetrical mass, but as a blob surrounded by a disk of spinning material, a so-called accretion disk. The central blob (which contains most of the mass) contracts further, heats up until the center is hot enough for copious amounts of energy to be liberated so that it starts to shine as a star and moves towards the main sequence. Finally, nuclear fusion reactions start to occur.

The early stages of a star's life, while it is contracting from a blob of interstellar material, seem to be a time of considerable instability. Stars form as condensations in interstellar clouds and material is both collected from the cloud and ejected back into it. By the time the star has tapped the nuclear energy sources, it may have experienced all sorts of interesting events, such as the building of light elements in events like solar flares. Sometimes the slowly rotating, contracting material may break into two or more fragments as it spins faster and faster and a double star results. In other instances the accretion disk might evolve into a system of planets. Several well-established stars, such as Vega and β Pictoris are surrounded by disks of solid particles, grandiose versions of ring systems akin to those of Saturn. Stellar evolution may be complicated by the presence of magnetic fields which may do bizarre things to the involved stars and may even act as a brake on some of the spinning blobs.

The early stages of a star's pre-main-sequence life may be illustrated by the cluster NGC 2264 which has been studied by M.F. Walker (see Fig. 8.1). Here V (essentially visual) apparent magnitude is plotted against $B-V$ color. Note that although the

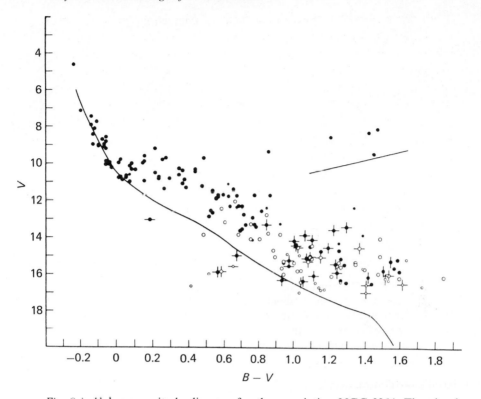

Fig. 8.1. Color–magnitude diagram for the association NGC 2264. The visual magnitude, or rather its more accurate photoelectric equivalent, V, is plotted against the blue minus the V magnitude, i.e. the $B - V$ color index. The data plotted are the actually observed magnitudes and colors; no correction has been made for possible effects of space reddening. The filled dots represent photoelectric observations; the open circles are photographic measurements, which are calibrated with the aid of photoelectric measurements. Vertical lines through the points indicate known stars with variable light, and the horizontal lines denote stars in which G. Herbig has found Hα emission. The solid lines give the standard main sequence and the position of the giant branch. (Courtesy M.F. Walker.)

brightest stars fit the main sequence defined by normal stars, fainter ones show pronounced abnormalities. Note also that beyond color index zero, corresponding to spectral class A, nearly all stars fall above the main sequence defined by nearby normal stars. The fainter, redder stars are mostly variables with bright hydrogen emission, indicating extended chromospheres or involvement in local nebulosity. Furthermore, the variable stars have an excess of radiation in the ultraviolet, as was pointed out by Herbig and Haro.

Evidently, we are witnessing the formation of stars from the grains and gas of the great interstellar Milky Way clouds. As the stars pull themselves together by gravitational contraction, they slowly move to the left in the V against $B - V$ diagram, but the evolution is not a smooth process. Instabilities, evidenced by the marked variability and colors of these objects, abound. Theoretical attempts to

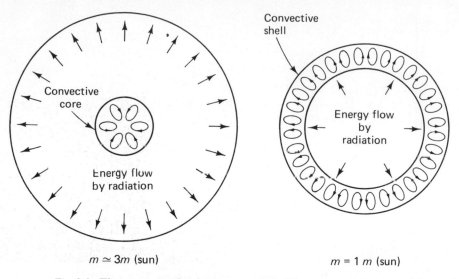

Fig. 8.2. The structure of main-sequence stars. The more massive star, of three solar masses, has a core of about half the sun's mass which is in vigorous convection, while energy is transported in the outer layers by radiation. In the star of solar mass, convection occurs in the outermost layers, while energy is transported throughout most of the interior by radiation.

handle the contraction of a star to the main sequence naturally assume an orderly development, so it is not surprising that theory cannot follow observations closely.

Massive proto-stars may reach the main sequence in a few tens or hundreds of thousands of years, but less massive objects may take a million or even tens of millions of years. 'Brown dwarfs', objects with masses less than 0.08 that of the sun never initiate thermonuclear reactions but slowly cool off to dense 'black' dwarfs or Jupiter-like objects.

The course of stellar evolution

As described above, we envisage stars as beginning their lives as huge, tenuous masses, slowly contracting under gravitational attraction from condensations formed in the interstellar medium. At this stage the temperature is so low that nuclear reactions cannot occur. As soon as the temperature and density rise to the point where nuclear transformations can take place, enough energy is produced to raise the internal temperature and gas pressure enough to stop the star from contracting further. The first energy-producing reaction to occur is the familiar proton–proton (PP) reaction, or alternatively the CNO cycle (Chapter 7). At this juncture the star has reached the main sequence and it remains there until the hydrogen in the interior has been seriously depleted. The star's internal structure (see Fig. 8.2) and its further evolution will depend primarily on its mass. In a star like the sun, hydrogen is first exhausted in the center, and the hydrogen-to-helium conversion (hydrogen 'burning') will gradually work outward in an energy-producing shell. At this stage the star will leave the main sequence and eventually evolve into a giant. On the other hand, the core of a massive star keeps itself stirred up by convection. Hence hydrogen is depleted at an almost uniform rate in a core having about a tenth of the stellar radius, and containing about a tenth of the stellar mass. In any event, the onset of an inert core has the following effect. Energy is produced by thermonuclear reactions only in a thin shell. The core continues to shrink in size even though it grows in mass, but the overall size of the star increases. Its total luminosity may increase but the surface temperature falls. At this stage the star becomes a giant or, if it is sufficiently massive, a supergiant.

In the course of these events the stellar interior undergoes complex changes. Eventually, helium is 'burned' into carbon until the helium core is exhausted. At this stage the star may have two shell sources, an outer hydrogen-burning shell overlying an inert helium zone, which in turn overlies an inner helium-burning shell that incloses a core of carbon 'ash'. If the star is massive enough, carbon can eventually be burned into oxygen, neon, etc. The evolution follows a complicated scenario but the last stage of a star's life is well identified. It becomes a white dwarf, an extremely dense, small faint object (see Chapter 9).

With this broad-brush picture as a guide, let us now examine the process in more detail. Extensive calculations have been undertaken by M. Schwarzschild, C. Hayashi, L.G. Henyey, B. Paczynski, R. Kippenhahn, A.V. Sweigart, I. Iben, A. Renzini, and many others. Let us consider the star from the moment it reaches the main sequence and starts 'burning' hydrogen into helium. The astronomer must not only calculate the initial model of the star (when it is chemically homogeneous throughout) but carry the calculation forward in time, allowing for the gradual change of hydrogen into helium.

The problem of calculating stellar evolution in anything but a schematic fashion was not tractable until the advent of modern computers. We have to consider not only the effects of changing hydrogen into helium and eventually burning helium to carbon upon the mechanical stability of the star, but also such effects as the contraction of the inner core and the expansion of the outer regions as the star bloats up to become a giant or supergiant. Energy is liberated by core contraction, but work has to be done against the gravitational field of the star to allow it to expand. This energy is extracted from the radiation passing through the expanding layers on its way from the core of the star to the surface. At certain phases, a large fraction of the energy produced in the interior is absorbed in the expanding outer layers.

The method for successfully resolving this problem was first developed by Henyey. In principle, the procedure is straightforward. Initially we start with a star which has just arrived at the main sequence and calculate a model by procedures outlined in Chapter 7. Starting with this zero-age main sequence (ZAMS) model one divides the stellar interior into about 30 or 40 concentric shells in each of which the temperature and density are known and the rate of conversion of hydrogen into helium can be computed. Initially, of course, each shell has the same chemical composition.

Consider first a star of about one solar mass at time t_0. Within all of the energy-producing regions the gases are stagnant, that is there is no mass motion, no mixing from one layer to another. The energy is carried outward by radiation. This model of the star is called $m_0(t_0)$.

We now take a time interval, $t_1 - t_0$, which may be as short as a few million years or as long as a thousand million years, depending on the mass of the star and thus on the hydrogen-to-helium burning rate. At the end of this period, the helium-to-hydrogen ratio in each of the interior shells will have increased, the change in composition being greater, the closer the shell is to the center. Then, with this new composition, we calculate a new model for the star. The computational procedure is so designed that if the changes between this and the preceding model are too large, the program will take a smaller time interval and repeat the calculation. When a satisfactory model, $m_1(t_1)$, is obtained we repeat the computation for a new time interval, $t_2 - t_1$, and get a new model $m_2(t_2)$, and so on. The number of shells may be varied, and in the computational procedure we put in such effects as energy liberation by core contraction or work done by expansion of the envelope or outer regions of the star.

If the mass of a star exceeds about two solar masses, its energy-generating core is a region where such vigorous convection occurs that out to about 10 percent of the radius of the star, mixing is complete. One still calculates a succession of models for different time steps, but mixing of the material in the central region causes the chemical composition of the core to be uniform at any instant of time. The reason massive stars have convective cores is that energy generation there takes place by the CNO cycle, which depends very steeply on temperature. Thus, a slight rise in temperature can cause a huge rise in energy output, so that the layers become mechanically unstable and vigorous convection sets in. Stars of the order of one solar mass derive their energy from the proton–proton reaction which depends much less steeply on temperature.

Initially, the star departs but little from its $m_0(t_0)$ model, but as time goes on, the hydrogen near the centre will be more and more severely depleted. A star such as the sun develops a central inert core that gradually grows outward as the 'helium rot' takes over one shell after another. The convective core of a luminous main-sequence star continues to burn hydrogen into helium until all of the hydrogen is gone. When hydrogen is exhausted in the core, the star continues to shine by burning hydrogen in a relatively thin shell around this inert region. The star is now built on a different model to that before. It finally develops a convective layer outside the region of the shell, whatever the previous mechanical state of the material there might have been.

Now the outer part of the star expands. The star begins to depart from the main sequence, becoming redder and more luminous as it evolves. At a certain stage it may move rapidly to the right on the Hertzsprung–Russell diagram and become a red giant or supergiant. The core continues to shrink in size but increases in mass as new material is added from the outer envelope.

In the advanced evolutionary stages of stars of low mass, the material in the core becomes what is called a 'degenerate gas'. The electrons no longer obey the perfect gas law, $p = nkT$. The temperature ceases to appear in the gas law; pressure depends only on the density of electrons. The only way to remove degeneracy is to supply

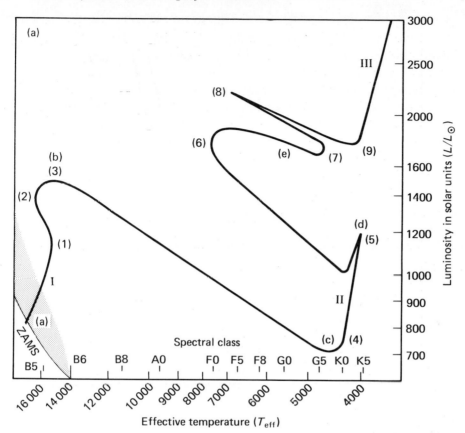

Fig. 8.3. Stellar evolutionary tracks.
(a) Evolution of a star of five solar masses. The luminosity (expressed in solar units) is plotted against the effective temperature. Corresponding spectral classes are also indicated. The cross-hatched area denotes the main sequence, and ZAMS = Zero-Age Main Sequence. The zero-age position of the star is denoted by (a). At (1) the core hydrogen is exhausted; the star contracts from positions (1) to (2) where it starts to burn hydrogen in a shell above the core of helium. From (2) to (3) the hydrogen-burning shell becomes compressed because the core temperature is not high enough to support the weight of the overlying layers. From (3) to (4) the radius of the star increases and the luminosity falls because the hydrogen-burning shell becomes thinner and the star has to do work against gravity to expand the outer envelope. The star moves very rapidly during this phase which corresponds to the Hertzsprung gap. At (4) or (c) the envelope becomes totally convective. Energy now flows more efficiently; the process is aided by the reduced opacity of the outer layers.

energy, for example by a sudden release of it in a shell source as may occur in what is called the 'helium flash' in certain stages of evolution. Breakdown of the perfect gas law and the onset of degeneracy is an important complication among the many complications that make stellar evolution calculations sporting.

Fig. 8.3a shows the evolution of a star of five solar masses (near spectral class B5), based on calculations by Iben, and by Kipperhahn, H. Thomas, and A. Weigert. The star evolves off the main sequence and establishes a hydrogen-burning shell.

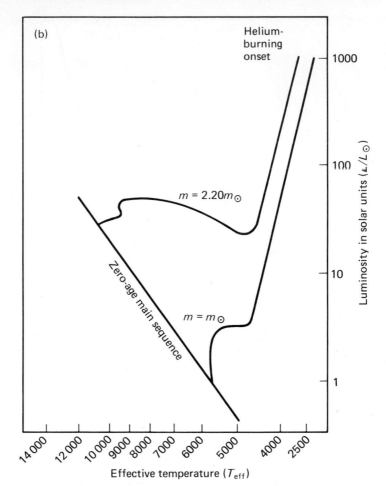

Caption for fig. 8.3 (*cont.*)

By the time the star has reached (d) or (5), the central temperature is a hundred million degrees and the density is 60 000 grams per cubic centimeter. In the core of the star, helium is now burned to carbon by the triple-alpha process. Part of this energy is absorbed in the core and the luminosity drops slightly. With a central core generating energy, convection ceases in the envelope. Helium burning, however, is far less efficient than hydrogen burning. From (6) to (7) the core exhausts its helium. Now there is a helium-burning shell source and the envelope again expands. From (7) to (8) the helium-burning shell increases in importance, the hydrogen-burning shell disappears, and at (9) the central temperature attains six hundred million degrees and carbon is now burned to heavier elements (cf. Fig. 9.1). The symbols I, II, and III are interpreted in Fig. 8.7; the schematic structure of the star at (a), (b), (c), (d), and (e) is shown in Fig. 8.4.

(b) Evolution of the sun and a star of 2.2 solar masses. As in part (a) luminosity is plotted against the effective temperature of the star. There are considerable quantitative differences between the two tracks. We may compute a family of such evolutionary tracks and compare the results with color–luminosity diagrams for globular clusters, noting that – strictly speaking – the latter do not represent stellar evolutionary tracks. See Fig. 8.3(c).

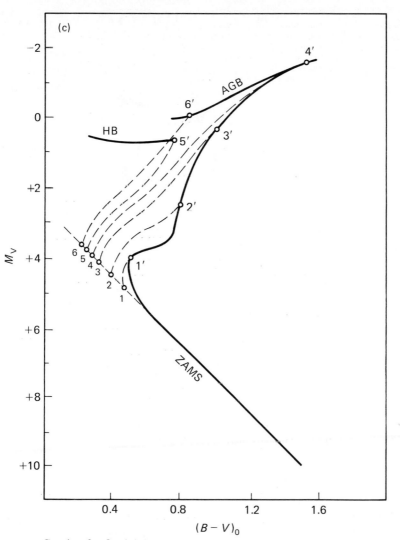

Caption for fig. 8.3 (*cont.*)

(c) A schematic interpretation of a color–luminosity diagram for a globular cluster. The solid lines define the distribution of stars in a color–magnitude diagram for a globular cluster such as 47 Tucanae (cf. Fig. 6.9). The giant, asymptotic giant (AGB) and horizontal (HB) branches are all populated by stars that have evolved from the main sequence. Thus a star at (1′) was once a zero-age main sequence (ZAMS) star at (1), a star at (2′) came from a ZAMS star at (2), etc. The dotted lines are drawn merely to connect the points; they **DO NOT** represent any evolutionary tracks. The originally intrinsically brighter and more massive stars on ZAMS have evolved furthest. Some of them have already perished as white dwarfs. Thus color–magnitude diagrams for these clusters represent what we call the loci of the stars that have evolved from the main sequence. This type of diagram was drawn many years ago by A. Sandage.

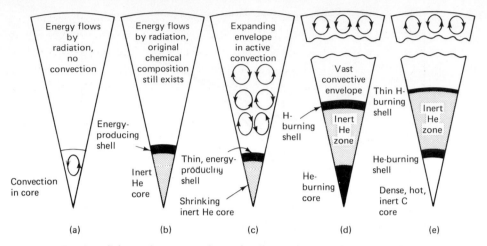

Fig. 8.4. Schematic cross-sections of stellar evolution. The letters refer to the stages of evolution depicted in Fig. 8.3a.

(a) This sector shows the zero-age main sequence model with a convective core. In the outer 'envelope' the energy flows by radiation.

(b) The star has now left the main sequence. A thick hydrogen-burning shell overlies the compressed helium core whose density and temperature are rising.

(c) The greatly expanded envelope has become convective, while the temperature and density of the core have risen.

(d) The core temperature has now reached 100 000 000 K; the central density is 60 000 times that of water. The triple-alpha process is now ignited in the core, but the star still has a hydrogen-burning shell which continues to produce most of the energy.

(e) The star now has a double shell source, an outer thin hydrogen-burning shell that will soon disappear, an inner, quiet, helium zone and then a helium-burning shell overlying an inert carbon core. As the core temperature approaches 600 000 000 K, carbon will be burned into heavier elements and the evolution will become complicated.

Then its core shrinks as the outer part of the star expands. It becomes a K-type supergiant and again rises in luminosity. Thereafter, it follows a zigzag evolutionary path for a time as it taps more refractory and ever less efficient modes of energy generation, burning helium into carbon, and carbon into heavier elements. Its life is short now, because the energy yield per unit mass is much lower than in hydrogen-burning reactions, and the star's luminosity is much higher than when the star was on the main sequence. Fig. 8.4 gives schematic sketches of the internal structure of the evolving star as it ages. More massive stars have even more dramatic evolutionary histories as they build heavier and heavier elements (see Chapter 9).

The evolution of the sun

The evolution of the sun (see Fig. 8.3b) is of particular interest for obvious reasons. It is the one star for which radius, mass, and luminosity are accurately known. Its chemical composition, except for the hydrogen-to-helium ratio, is also well established. Calculations by R. Weymann and R.L. Sears and by J.H. Bahcall and others are plotted in Fig. 8.5, which compares the present-day internal structure of the sun with that existing when it started as a main-sequence star. Since then, the core has contracted, both central density and temperature have risen and the central hydrogen content has been reduced to 48 percent of its initial value. Meanwhile the luminosity has increased by about 33 percent.

The future evolution of the sun has been considered by a number of workers including P. Demarque and R. Larson (1964) and R.K. Ulrich. The radius and luminosity of the sun increase at an ever accelerated pace over the next five thousand million years. Before the solar system has passed its ten-thousand millionth birthday, the surface temperature of the earth will certainly exceed the boiling point of water. Terrestrial life will have been extinguished sooner, for the blanketing effect of steam will produce a severe greenhouse effect. The last life on the earth, like the first, will be very primitive organisms.

Five thousand million years from now, the curtain falls quickly on the inner solar system. Earth's oceans boil away and escape into space; finally the surface rocks melt. Our planet's last seas will be of liquid rock and metal. The envelope of the expanding sun, by then a red giant and perhaps even a Mira variable (see Chapter 9) will engulf successively, Mercury, Venus, and the earth, dragging them inexorably to a fiery death in the hot gases of the rapidly expiring sun!

Between the pressure cooker and the deep freeze

On the broad canvas of stellar evolution, it is easier to predict the future than to reconstruct the past. We can assert confidently that the earth achieved, and more importantly, maintained habitability by the thinnest of margins. There is a continuous, uninterrupted record of steadily evolving life forms over a period exceeding three thousand million years. With this constraint in mind, and our present knowledge of the chemistry of the solar system, atmospheric sciences, and stellar evolution, we can attempt to reconstruct the history of the earth as an abode for life. There are two critical factors for life to be possible: liquid water must be available and the temperature must fall in the right range. It should not exceed 40 °C if the higher life forms are to be preserved, neither should it freeze solid. Factors that control the earth's temperature are:

(1) the solar luminosity (which has increased by roughly 25 percent over the last 4.5 thousand million years);
(2) the cloud and ice cover of the earth that reflects light back into space; and
(3) the greenhouse effect, which depends critically on the earth's atmosphere and how its constituents change with time. (The greenhouse effect leads to the trapping of heat, which makes the surface temperature higher than it would otherwise be.)

If the present-day sun were suddenly replaced by the substantially fainter zero-age-main sequence sun, the earth would immediately be plunged into a permanent ice

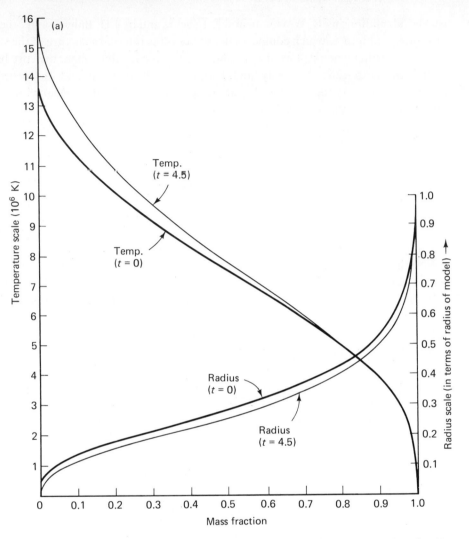

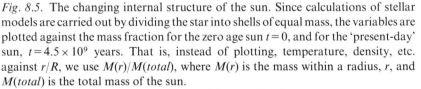

Fig. 8.5. The changing internal structure of the sun. Since calculations of stellar models are carried out by dividing the star into shells of equal mass, the variables are plotted against the mass fraction for the zero age sun $t = 0$, and for the 'present-day' sun, $t = 4.5 \times 10^9$ years. That is, instead of plotting, temperature, density, etc. against r/R, we use $M(r)/M(total)$, where $M(r)$ is the mass within a radius, r, and $M(total)$ is the total mass of the sun.

(a) Shows the temperature in millions of degrees (left-hand ordinate) and radius (right-hand ordinate) in terms of radius of model. Note the rise in temperature in the deep interior and the contraction of the core that is indicated by a decreasing radius for each mass point.

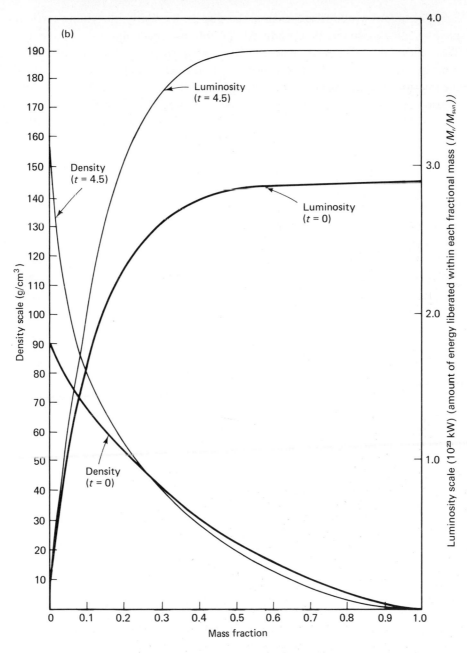

Caption for fig. 8.5 (*cont.*)

(b) Shows the density (left-hand ordinate) and total energy generation within a given mass fraction (right-hand ordinate). Note that essentially all the solar luminosity is produced in the inner half of its mass. Note also the pronounced rise in central density as the sun ages.

The zero-age-main-sequence model is from R. Weymann and R.L. Sears, 1965, *Astrophysical Journal* **142**, p. 174. The present-day model is from J.H. Bahcall, W.F. Heubner, S.H. Lubow, P.D. Parker, and R.K. Ulrich, 1982, *Reviews of Modern Physics* **54**, p. 767.

age. The oceans would freeze, even in the tropics. Evidently, the young earth must have had an atmosphere that produced a greenhouse effect of just the right amount. We believe that the earth lost whatever primordial atmosphere it had. The atmosphere that evolved into our present atmosphere was produced by degassing the interior on a time scale of about eight hundred million years. Nitrogen, carbon dioxide, and steam were among the gases escaping from the rocks. The steam condensed to form lakes and eventually oceans in which the carbon dioxide dissolved to react with silicates to form limestone, while methane and ammonia possibly were abundant in the early atmosphere. On earth, degassing proceeded at just the right rate. On the other hand, Venus lost any oceans it may once have had, and had no means of removing carbon dioxide from its atmosphere, so it suffered an inexorable, runaway greenhouse effect. On the earth the greenhouse effect just sufficed to keep the surface warm and comfortable, in spite of the fact that most of the incoming sunlight may have been simply bounced back into space by clouds.

As the composition of the earth's atmosphere changed, the trapping of solar radiation declined and the surface temperature fell. If the earth had escaped the pressure cooker by a stroke of good luck, it now escaped universal glaciation by an even more harrowing margin. According to an analysis by Michael Hart, the atmospheric properties changed abruptly with the disappearance of methane and other hydrocarbons and the decline of the cloud cover about two thousand million years ago. The surface temperatures fell and the glaciers advanced for a time. They were turned back as temperatures rose as the chemistry of the earth's atmosphere again changed. Free oxygen appears to have been liberated, probably largely by plants as a consequence of photosynthesis.

A number of scenarios were explored by Hart. Place the earth seven million kilometers closer to the sun, and a runaway greenhouse effect would have given the earth a fate very similar to that of Venus, whose surface bakes at 700 K under a pressure of 90 atmospheres. The margin of 'error' in the other direction is even more scary – a couple of million kilometers further out, and global glaciation!

The implications for habitable planets in other solar systems are clear. The zone of habitability is narrow for stars like the sun. For stars cooler than K0 it vanishes, while for more massive stars the stellar lifetime is too short. Although refinements in knowledge will certainly modify the quantitative details of Hart's scenarios, the essential point remains that the earth developed into a habitable planet by an amazing stroke of luck or, as some might suggest, by the Hand of Providence!

Some observational consequences of stellar evolution

In Chapter 6 we saw that star clusters offered us a unique opportunity for studies of stellar evolution. Consider first the galactic clusters whose Hertzsprung–Russell (HR) diagrams are shown in Figs. 6.7 and 6.8 (pp. 120 and 121). The heavy line bordering the left of the main sequence defines the positions occupied by homogeneous stars that have just started burning hydrogen into helium. Thus it defines the zero-age main sequence (ZAMS). As the helium content of the core increases, so does the mean molecular weight. Then, as we saw in Chapter 7, the temperature in the central regions must rise. When this occurs, the star brightens and moves upward and to the right in the HR diagram. The main sequence made up from a sample of stars of varying ages in the neighborhood of the sun has a finite width because it contains objects with cores of different hydrogen to helium ratios. In contrast, most of the stars in a given cluster probably were formed at nearly the same time, though red dwarfs, which condense from the interstellar medium rather slowly, may be younger than other stars. The main sequence in any one cluster tends to be relatively narrow.

The main-sequence lifetime of a star depends on its mass. A star twice as massive as the sun will liberate energy eight or ten times as fast and will exhaust its hydrogen fuel faster than will the sun. A star of ten solar masses uses up its hydrogen fuel even more rapidly. Once the core hydrogen is gone and the star has to rely on hydrogen burning in a thin shell it will evolve to the right in the HR diagram.

Various galactic clusters illustrate these phenomena very well. All stars of NGC 2362 fall on the main sequence. In h and χ Persei a handful of the most luminous stars have left the main sequence to become supergiants, while brighter blue stars are beginning to depart from the main sequence. No Pleiades stars are as bright as those in h and χ Persei, but the main sequence bends conspicuously to the right; there are not yet any giants. The clusters M11 and the Hyades are obviously older. Here the main sequence has been 'rolled back' to stars like Sirius and Procyon, while there are now a number of red giants. In M67 and NGC 188 the main sequence includes only yellow and red dwarf stars, whereas the giant branch is connected to the main sequence by a bridge across the subgiant region. In the other clusters from the Hyades to h and χ Persei, there is a conspicuous gap between the main sequence and red giants; this is often called the Hertzsprung gap.

Thus, of the clusters depicted in Figs. 6.7 and 6.8, NGC 2362 and h and χ Persei are the youngest, M11 and the Pleiades are older, the Hyades and M67 are much older, and NGC 188 is the oldest of all. By noting the turn-off point and applying the results of stellar evolution calculations which give theoretical tracks for stars of different masses, we estimate the age of h and χ Persei to be but a few million years while that of NGC 188 is about five thousand million years.

The thus-determined ages become less accurate the more ancient the cluster. Computed evolutionary tracks are sensitive to assumed chemical composition; even slight variations can cause large changes in predicted ages. Exact comparisons between theory and observation may be difficult for the brightest clusters because all stars may not have been formed at the same time. According to a detailed study by Pierre Demarque, the Pleiades and M11 have an age of the order of sixty million years, the Hyades are about ten times older, and M67 has an age of about four thousand million years.

Although open clusters show a huge spread in ages, ranging up to six or seven thousand million years, globular clusters are all very ancient objects. Some of them have been assigned ages as great as sixteen thousand million years. The theoretical tracks and derived ages depend on the assumed hydrogen-to-helium ratio. Differences between clusters in the observed HR diagrams may also be caused by differing metal-to-hydrogen ratios. The giant-branch stars have only recently evolved from the main sequence; in some clusters they are less massive than the sun and were less luminous on the main sequence. Consider the composite HR diagram for several globular clusters (Fig. 6.10) and the HR diagram for 47 Tucanae and its interpretation (Figs. 6.9 and 8.3c.) Demarque assigns ages of 14 to 16 thousand million years for M15 and M92, M13 with an age of 12–14 thousand million years is younger, while Hesser *et al.* find an age of 13.5 thousand million years for 47 Tucanae.

Galactic or 'open' clusters, such as the Hyades, M67, or NGC 188, probably do not differ greatly from the sun in chemical composition. The situation is different in globular clusters. The metal-to-hydrogen ratio in 47 Tucanae is probably a third the solar value. In M13 this ratio may be down by a factor of 25, while in M15 and M92 it may be down by a factor of 100! Most globular clusters appear to have lower metal-to-hydrogen ratios than galactic clusters. Yet giant stars in globular clusters do attain greater luminosities than giants in open clusters like M67 or NGC 188. Metals in galactic cluster stars are not sufficiently abundant to influence the molecular weight but they do affect the opacity. The material making up globular cluster stars is more transparent and radiation escapes more rapidly from their hot interiors.

Thus, theoretical calculations of evolutionary tracks agree at least semi-quantitatively with empirical color–luminosity diagrams. The giant and supergiant stars are unequivocally understood as ex-main-sequence stars. Discussion of late stellar evolutionary stages are deferred to the next chapter. Let us now examine the engaging stellar evolutionary insights offered by double or binary stars.

The evolution of double stars

In a binary star system the course of stellar evolution will depart from the scenario described for a single star because of the presence of the companion. The details will depend on the masses of the stars involved, their orbital separation, and the eccentricity of the orbits. As a star evolves, it expands in size; if a companion is nearby, the outer layers can be stripped off.

Fig. 8.6 depicts the gravitational situation for two stars, A and B. We can define a two-lobed surface called a 'Roche surface'. If a particle lies within either of the two lobes it moves under the control of one or the other of the two bodies. If a particle slips through the Roche surface inclosing A, for example, it may be captured by B, move in a very perturbed orbit about both bodies, or even escape from the system. Thus stellar evolution in binary systems can become a kind of strip-tease in which first one component and then the other loses its outer envelope.

That eclipsing binaries often contain strange combinations of stars has been known for many years. The most famous of these objects is probably Algol, the prototype of the binary system depicted in Fig. 1.5. The bright, hotter, main-sequence star (spectral class B5–A5) is eclipsed by a larger, cooler, less massive star of about spectral class G. This companion cannot be a main-sequence star because

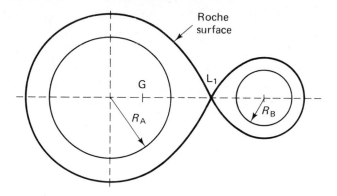

Fig. 8.6. Domains of 'entrapment' in a close double star. Two stars, A and B, of differing mass and radius, R, are held together gravitationally in a close binary system. G is the center of gravity. We can define a three-dimensional surface called a Roche surface whose two-dimensional cross-section is depicted in the figure. This surface has two lobes. If a particle falls in the lobe shown to the left of L_1 it will be trapped by A, while if it falls in the lobe to the right of L_1 it is trapped by B. If it lies outside of the Roche surface, it will move under the influence of both bodies, or it may eventually escape from the system.

on the main sequence size decreases with temperature. It must be a subgiant. Such subgiant components are overluminous for their masses. When their dimensions can be derived, it is found that they fill their Roche lobes.

In some of these binaries, the spectroscopic velocity observations could not be reconciled with the geometry of the eclipses. At times when eclipses occur the stars are moving at right angles to the line of sight so we observe only the velocity of the center of gravity of the system. Half-way between eclipses, the to and fro motions should be maximized as the stars then have their greatest speeds towards or away from the observer. Often this simple picture was not observed and the velocity data made no sense until it was realized that what we were actually observing were streams of gas flowing about the stars. This conclusion was substantiated for some systems by bright emission lines (especially Ca II and Hα) that varied with phase and were particularly strong at primary eclipse.

As we shall see, the pattern of evolution depends critically on the masses of the stars and their separation. Consider first a pair of stars such as VZ Hydrae whose components, a little brighter, larger, and more massive than the sun, are separated by several times their diameter (see Fig. 6.3). We called the initially more massive star the principal component. As the stars evolve, it will be first to leave the main sequence and swell up, but it cannot follow the normal evolutionary path for an isolated star. When its surface expands to the Roche lobe, material will be stripped off and lost to the star. Presumably, much of it will be acquired by the secondary star. Hence the giant phase of evolution will be truncated and strongly modified while the secondary star still continues to shine on the main sequence. Eventually the principal component may lose most of its mass and the core will settle down as a white dwarf. Next, the secondary star starts to evolve off the main sequence and its

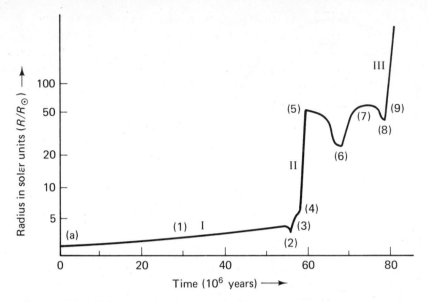

Fig. 8.7. Variation of stellar radius with time for a star of five solar masses. Following Kippenhahn, Thomas, and Weigert, the star's radius (in solar units) is plotted against age in millions of years. The Arabic numbers and letter (a) refer to the evolutionary stages in Fig. 8.3. In the core-burning phase, denoted as I, the star evolves from a radius of 2.7 to 4.3 solar units over a period of about 55 million years. In the next stage, denoted as II, it develops an extended outer convection zone and the expansion halts at a radius of 55 solar units at the red giant tip. During subsequent phases of helium burning the radius actually decreases part of the time. When a helium-burning shell is formed around a thermally unstable core that consists mostly of carbon, a convective envelope develops. This envelope expands as the carbon core shrinks and then carbon burns in the core. Stage III refers to an advanced evolutionary stage where the radius and luminosity increase rapidly as carbon is burned into heavier elements.

outer envelope eventually reaches its own Roche lobe. If by this time the principal component has become a white dwarf, very interesting things can now happen. The secondary star sheds material that falls toward the white dwarf to form a thin plate of spinning gas called an accretion disk. The impact of the material on the accretion disk and its eventual cascade on to the white dwarf with its high surface gravity can produce instabilities such as ultraviolet radiation, X-ray emission, and even nova outbursts.

With this broad-brush picture as our general guide let us look at this problem in more detail. Consider the evolution of a single star of five solar masses as depicted in Fig. 8.3a. Following Kippenhahn *et al.* let us plot the radius as a function of time (see Fig. 8.7). Note that the star does not expand at a uniform rate. For about fifty-five million years the radius increases smoothly. Then there is an abrupt increase as the star develops a convective envelope. The advent of helium burning in a dense core causes the radius to decline for a time, but when the helium shell source is developed, the radius once more increases. In a binary system a star's evolution is modified in a manner dependent on the distance of the companion. Effects of rotation and tidal distortion are minor compared to the fact that in a binary system each component has only a limited volume available for its long-term expansion. Eventually the Roche lobe guillotine will fall; when it does depends on the separation of the stars and their mass ratio. Close binaries are going to be severely affected even in early stages of their evolution. If a star evolves to a vastly extended cool red giant stage or as a supergiant, even visual binary systems suffer significant effects.

The evolution of binary stars has been studied by Kippenhahn and Weigert, Paczynski, M. Plavec, and more recently by R.F. Webbink, Iben and A.V. Tutukov, C. De Loore and J.P. De Greve, and many others. There are many ways in which binary systems can evolve, depending on the initial conditions. Here we will describe only some 'basic scenarios' as outlined by Plavec. He lists three broad cases.

A. The radius of the primary companion reaches the Roche lobe before core burning has ceased (epoch I in Fig. 8.7).
B. The star has exhausted hydrogen in the core and is producing energy in a thin, hydrogen-burning shell when the expanding envelope reaches the Roche lobe (epoch II in Fig. 8.7).
C. The star reaches the supergiant phase before it gets to its Roche limit (denoted by III in Fig. 8.7).

It must be emphasized that the consequences of the peeling off of the outer layers of a star has a traumatic effect on the star's own evolution and eventually on that of its companion. The initially more massive principal component expands until it fills its Roche lobe. Material is then torn off at what is called the first Lagrangian point, L_1 on Fig. 8.6, and most of it falls on to the companion. Mechanical stability is quickly restored in the star but a longer time interval is needed for the star's luminosity to exactly balance its nuclear energy production. During this phase we say that the star is not in thermal equilibrium. If one calculates a new model for the star which has lost an amount of mass, ΔM, we find that the size of the Roche lobe has decreased. The star gets hit by a double blow. As it strives to become a giant, the outer layers get stripped off but, as this happens, the cutting edge of the Roche lobe moves inward and more material is taken away. Consequently, the mass-loss process is accelerated since the condemned star cannot attain an equilibrium state. It is caught in the trap of a vicious circle.

An important point is that two stars may be nearly identical on the main sequence with masses differing by only a few percent, but in the course of evolution the principal component may have become a giant before its companion has departed very far from the main sequence. The two components of AR Aurigae have 2.57 and 2.29 solar masses, respectively. Using Iben's theoretical models, Plavec found that the principal component would reach the top of the red giant branch some fifty million years before the secondary star would have ended its main-sequence evolution. The evolutionary scenario depends strongly on the mass of the principal component, the mass ratio, and on the separation.

To illustrate some of the vast number of possibilities two examples of binary-star evolution calculated by Plavec and his group are described. In Fig. 8.8a an example of their case A is considered.

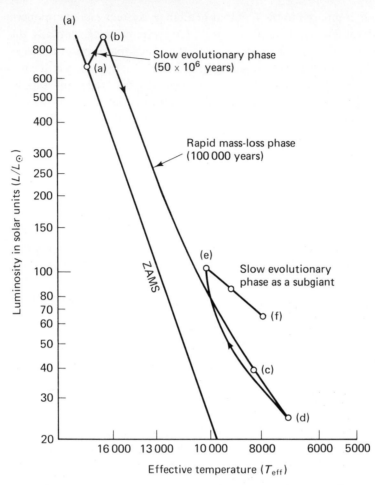

Fig. 8.8. Evolutionary tracks in binary systems.
(a) Evolution of an Algol-type system. In this system (Plavec's case A), the principal
component expands to the Roche surface before all the hydrogen has been burned in
the core. Note the speed with which the evolving star fades parallel to the main
sequence before the luminosity rises again and the star evolves into a subgiant.
(After M. Plavec.)

The two components of the binary initially have masses of five and three suns.
Their separation is 8.8 times the radius of the sun. The principal component is a B5
star which reaches the Roche limit with a radius of 3.74 solar radii in 53.7 million
years. When mass loss starts, the core temperature and rate of energy generation fall
and much of the energy produced is absorbed in expanding the outer envelope of the
star against gravity. Thus the luminosity of the principal component plummets and
its effective temperature falls. The maximum mass loss rate is about 2×10^{-5} solar
masses per year. It requires only 110 000 years for the luminosity of the principal
component to reach its lowest luminosity at point (d). Then the mass-loss rate falls
and the star slowly brightens. Some 700 000 years after it left (b) the star reaches (e)
where it has become an A1 subgiant of 2.6 solar masses. If the star were undisturbed

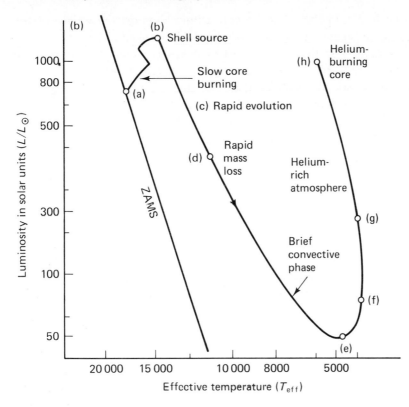

Caption for fig. 8.8 (*cont.*)

(b) Evolutionary track for Plavec's Case B. In solar units the mass and radii of the principal component are 5 and 5.5, respectively, while the companion has a mass of 4. The stars are initially separated by 13.8 solar radii and the orbital period is 2 days. At (b), where the core hydrogen is exhausted, the principal component encounters the Roche lobe and starts to evolve rapidly. The maximum mass-loss rate, 2×10^{-5} solar masses per year, is reached at (d) but the star is faintest at (e) when the opacity of the expanding layers is large. The star temporarily develops a convective zone, and reaches spectral class G8III as the effective temperature reaches a minimum. Envelope expansion ceases at (h) when the star collapses to a compact helium-burning object. (After S. Krisz.)

it would move from (e) to (f) in 65 million years, lose an additional 0.6 solar masses, and become an A8 subgiant. Meanwhile, the companion has increased its mass to 5.4 solar masses and what had been a star of spectral class B9V is now a B4V star. It will last perhaps 40 million years before its expanding surface penetrates the Roche lobe. The observer is most likely to catch the system during this phase. He or she will see what resembles a B4 main-sequence star accompanied possibly by a thin gas ring from the principal component which is now an A-type subgiant that is overluminous for its mass by about one magnitude. We obtain an Algol-like system (see Table 6.3) very similar to the eclipsing binary Z Vulpeculae. Note that what happens is that the roles of the two stars are switched. The initially more massive and luminous principal component has become the fainter and less massive member of the pair.

In Plavec's case B, hydrogen has already been exhausted in the core and energy is being produced in a thin shell, but the core cannot support the weight of the overlying layers. An isolated star would evolve quickly across the Hertzsprung gap and become a giant. For the star trapped in a binary system of appropriate dimensions, the outer layers are in rapid expansion when the surface reaches the Roche limit and the star becomes unstable. It cannot attain any semblance of equilibrium until helium burning is ignited in the core.

An evolutionary track is shown in Fig. 8.8b. As a consequence of rapid mass loss the star shrinks in size and luminosity. At point (e) the luminosity is a minimum and the principal component, now of two solar masses is a G5 subgiant star. About 420 000 years after the onset of instability, mass loss ceases at point (g). The thermal instability in the core still persists as the star strives vainly to achieve equililbrium for about 660 000 years as the star continues to bloat up. Finally, helium burning is ignited in the core, the expansion stops and the star shrinks.

The principal component is now brighter than it was on the main sequence, but it corresponds to a late F-type giant of 27 solar radii; it is quite unlike any normally evolved star. The vast bulk of the hydrogen-rich material of the star has been lost, leaving what was once the convective core of the star when it was just evolving on the main sequence. Up to point (h), energy is still being supplied by burning in a thin shell of less than a solar radius. Most of the volume of the star is at this stage occupied by a helium-rich mixture. When helium burning starts, core contraction ceases and the outer envelope collapses within a few hundred thousand years, yielding a compact helium-burning star of less than a solar radius. It has a mass of 0.7 solar masses. No single star of such mass proceeding through normal evolutionary channels would ever burn helium in its core, but this is a very abnormal object. The core has an exceedingly high density and temperature appropriate to a single star of much greater mass. Hence helium can be burned into carbon. The atmosphere of this star is now helium-rich, resembling υ Sagittarii or HD 30353.

What happens next? Helium burning will continue until the fuel is exhausted and then the dense cinder will settle down as a white dwarf. But what of the companion star? It has been evolving quietly, slowly departing from the main sequence during the trauma of its companion. Its serene evolution is disturbed, however, by the debris poured onto it from the other star. Thus its own evolution will be accelerated and eventually its envelope will begin to spill over onto its dense white-dwarf partner. Helium-rich material is now being accreted by a hydrogen-deficient star with an enormous surface gravity. Free fall of material upon a white dwarf would produce spectacular effects, including intense ultraviolet and X-ray emission with maybe dramatic outbursts such as are shown by novae. Current theories indicate that the infalling gas does not cascade directly to the surface of the white dwarf but collects in a broad thick ring (accretion disk) around the white dwarf (see Fig. 8.9).

Half a century ago, long before we knew about stellar evolution, studies of main-sequence eclipsing systems gave an important check on one prediction of stellar models, namely that the density increases very rapidly toward the center of a star. The orbits of some eclipsing binaries are not circular but elliptical, so that stars travel at different speeds at different times in their orbits. Consequently, the

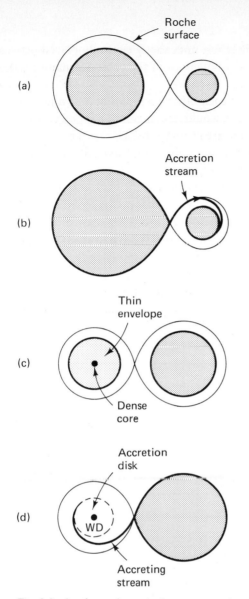

Fig. 8.9. A schematic evolutionary scenario of a close binary star.

(a) Initially both stars are on the main sequence. The situation is as depicted in Fig. 8.6.

(b) The principal component has now expanded to fill its Roche lobe through which material is now flowing. Much of it moves in a stream that impinges on the secondary star. Some of the material (not shown) may be lost from the system.

(c) The principal component, having lost most of its mass, now consists of a dense core and a thin envelope. The compànion, whose mass has been enhanced by accretion, has begun to swell up as it starts to evolve away from the main sequence.

(d) The principal component has lost all material outside a dense core which has become a white dwarf (WD). The core is surrounded by a thin accretion disk fed by material from the companion which is now losing material through its Roche surface. The final stage (not shown) is when both stars have lost their outer envelopes; there exists now only a pair of white dwarfs. (After M. Plavec.)

secondary minimum shown in Fig. 1.6 is not exactly centrally placed between the two primary minima unless the long dimension of the orbit (its major axis) points towards the observer.

Two spherical stars would attract each other like point masses and, unless there was a third star in the system, the orbit would remain pointed for all time in the same direction in space. In reality the stars are distorted by their own rotations and tidal effects on one other. They no longer mutually attract like spheres, and the orbit rotates in space. Thus if the light curve is observed over a period of years, the position of the secondary minimum will be seen to shift back and forth as the orbit turns in space. The rate of turning of the orbit depends on the separation of the stars, the period, and the concentration of density toward the stellar center. The greater the rise of density toward the center, the slower will the orbit turn. The observed low rate of orbital turning (apsidal motion) is in harmony with the high density concentrations predicted by stellar structure theory.

In the next chapter we will turn to the late stages of the evolution of an ordinary star, how the outer envelope is stripped off by a combination of dusty winds and pulsations, how the outermost material is lost to the interstellar medium, and how the core settles down to a white dwarf. More massive stars can die more spectacularly, as is discussed in Chapter 11.

9

Wind, dust, and pulsations:
a star's last Hurrah!

In the last chapter we traced the evolution of a star as it exhausted the hydrogen in its core, developed a shell source, and evolved into a giant or, in the case of a massive star, into a supergiant. We mentioned the 'burning' of helium into carbon and alluded to the building up of even heavier elements. The ultimate fate of most stars is that the envelope escapes into space while the dense core becomes a white dwarf. Important details remain to be discussed.

(a)   The evolution of the hot core, especially in massive stars and the building up of heavy elements therein.
(b)   The role of stellar winds in affecting observable properties of stars and influencing stellar evolution.
(c)   Stellar pulsations and oscillations which provide some of the most important clues we have pertaining to stellar interiors.

Composition differences between stars

In Fig. 5.17 we compared the chemical composition of the sun with that of carbonaceous chondritic meteorities to obtain an estimate of the initial composition of the solar system. We emphasized that this pattern was not a product of chance but a 'fossil' record of element-building events.

To properly attack the problem of nucleogenesis we investigate other stars whose origins, ages, and histories differ from that of the sun. Not only elemental abundances but also isotope ratios can differ from object to object. Stellar composition differences may be attributed to either (a) element building within the core of the star itself with subsequent mixing of processed material to the surface or stripping away of outer layers as in highly evolved close binaries, or (b) the initial chemical composition of the material destined to form the star differing from the sun or similar stars.

In Chapter 4 we described the splitting of the spectral sequence between oxygen-rich, M-type giant stars and the carbon-rich C giant stars. The 'heavy-metal' S-type stars present an even more engaging problem. Here bands of zirconium oxide (ZrO) rather than titanium oxide (TiO) dominate the spectrum suggesting an enhanced abundance of zirconium. Since the formation of TiO and ZrO depend on density and temperature in slightly different ways, some argued that in S stars conditions were just right for ZrO formation to be favored. If that were true, atomic Zr lines ought to be systematically weaker in S stars because Zr atoms were used up in forming ZrO, but they were actually stronger. More remarkably, not only were Zr lines strengthened but so also were lines of neighboring elements, e.g. molybdenum (Mo), niobium (Nb), and ruthenium (Ru). The clinching discovery was the detection of the man-made element technetium (Tc), with an atomic number of 43, which is unstable against beta decay. Even more remarkable was the behavior of FG Sagittae in which a normal-looking B-type spectrum evolved into one with lines of the heavy-metal group. All C and S stars are highly evolved giants or supergiants; no C or S dwarf has ever been found.

In Chapter 4 we also mentioned the bright Wolf–Rayet stars which appear to be massive, highly evolved objects whose spectra clearly indicate profound composition differences. Other examples appear in spectroscopic binaries such as υ Sagittarii, while white dwarfs often appear to have unusual chemical compositions.

A quite different type of abundance difference is exhibited by stars such as HD 19445 or HD 140283 where the metal-to-hydrogen ratio is about a hundred times lower than in the sun. The spectra of these stars show strong hydrogen lines and weak metal lines, superficially resembling A-type, but the relative intensities of iron lines, the appearance of molecular bands of the hydroxyl ion (OH), and the spectral energy distributions all showed them to have effective temperatures near that of the sun.

These apparent spectral anomalies can be easily understood. The continuous absorptivity which causes the general fogginess or opacity of a stellar atmosphere is produced in the sun by the negative hydrogen ion. At a given temperature it depends on the number of electrons which are supplied by the metallic atoms, but when the metal-to-hydrogen ratio is low few metal atoms are available to supply the necessary electrons. The transparency of the atmosphere is greatly enhanced. Instead of looking through about 2 grams of material per square centimeter, we look through about 200 grams! The hydrogen lines build up in strength, the metal lines are still weak, because although we see to greater depths we see fewer metal atoms per gram of material.

Stars of low metal-to-hydrogen ratio are invariably old. It is no surprise that globular clusters such as M15 show much lower metal-to-hydrogen ratios than do younger galactic clusters. We cannot draw a smooth curve relating metal-to-hydrogen ratio with age. Element building proceeds at different rates in different galaxies and even at different rates at different places in the same galaxy. One utilizes chemical analyses of gaseous nebulae as well as of stars (see Chapter 10). In the spiral galaxies, M33 and M101, the abundance ratios N/O, Ne/O, S/O, and Ar/O remain essentially the same as we go out from the center, but the O/H ratio declines as shown in Table 9.1. Evidently the element-building factory runs faster near the core of a spiral galaxy than in the outskirts, and it has run faster in our Galaxy than in these other spiral systems.

Table 9.1 *Variation of oxygen (× 1000)/ hydrogen ratio with distance from the center of spiral galaxies*

Distance from center $(R/R_0)^a$	M33	M101	Our Galaxy
0.0	0.44	1.75	2.14
0.5	0.23	0.416	0.71
1.0	0.131	0.14	0.234

[a] R is the radius being considered and R_0 is the effective radius of the galaxy as defined by G.de Vaucouleurs.

How did the chemical elements originate?

Our basic postulate is that the abundance distribution of elements is not a result of mere chance but contains the clues to mechanisms whereby different kinds of atoms were created. Two theories were proposed. In the 1940s, S. Chandrasekhar and particularly G. Gamow and H. Alpher suggested that all the elements were created in an early stage of the evolution of the universe. In the 1950s, G. and M. Burbidge, W.A. Fowler and F. Hoyle noted that more elaborate processes were necessary. They proposed that only hydrogen and helium were created initially; other elements were made from them by processes occurring in stars, in stellar envelopes, or in the interstellar medium.

In the 'Big Bang' theory of the origin of the universe all material was compressed to such incredibly high density and temperature that at one stage all matter consisted essentially only of neutrons. As the universe expanded, the neutrons broke down into protons and electrons. Protons captured neutrons to form heavy hydrogen (deuterium) which in turn was quickly built into helium. Elements heavier than helium cannot be built by a continuation of this process since stable nuclei of mass 5 or 8 do not exist in nature.

The most impressive argument for the Big Bang is an omnipresent 2.73 K background radiation. If this is interpreted as what we now see of the emission from the great fireball at the dawn of creation (Big Bang) and if we make the best guess we can about the density of matter in the present observable universe, we can estimate the conditions for element building in the first few minutes or hours after 'the beginning'. Calculations based on models of the Big Bang universe produce abundances of hydrogen (H), deuterium (D), helium (He) and lithium-7 (Li) that fit our best estimates of the composition of primordial matter. In fact, determinations of the initial D/H and He/H ratios are of great importance for fitting some of the initial conditions for the expansion of the universe.

Except for He and H and the relatively rare elements, Li, beryllium (Be), and boron (B), with their fragile nuclei, all chemical elements are built in stars, and are made almost entirely in the dense, high-temperature cores. Some of this material is returned to the interstellar medium from which new stars are eventually formed. The processes are not simple. The many peculiarities of the abundance curve are such that no one routine or set of conditions seems capable of explaining them. In fact M. and G. Burbidge, W.A. Fowler, F. Hoyle, A.G.W. Cameron and their associates and successors have identified no fewer than eight different types of scenarios required to explain the observed heavy-elemental abundances. The principal mechanisms are sketched briefly below.

Hydrogen is converted into helium in the course of normal stellar energy production. In Chapter 7 we described how, in the dense cores of many stars, helium may be converted into carbon by the triple-alpha process (suggested by E. Opik and worked out in more detail by E. Salpeter and by Hoyle) wherein three alpha particles are jammed together almost simultaneously. The brief but finite (10^{-14} second) lifetime of the ^{8}Be nucleus formed when two alpha particles collide permits another alpha particle to be captured to produce carbon if the density and temperature are high enough (temperature $T > 100\,000\,000$ K; density > 6000 g/cm^3).

$$^4\text{He} + {}^4\text{He} \rightarrow {}^8\text{Be}^*; \quad {}^4\text{He} + {}^8\text{Be} \rightarrow {}^{12}\text{C}$$

Once ^{12}C appears, oxygen of atomic weight 16 is produced by the reaction

$$^{12}\text{C} + {}^4\text{He} \rightarrow {}^{16}\text{O} + \gamma$$

Then, subsequently, neon of atomic weight 20 could be produced by $^{16}\text{O} + {}^4\text{He} \rightarrow {}^{20}\text{Ne}$, but another possibility is spallation or the shattering of a nucleus by collision with another particle, e.g. $^{16}\text{O} + {}^4\text{He} \rightarrow 5{}^4\text{He}$.

At temperatures in the neighborhood of a thousand million degrees, which can occur in cores of highly evolved massive stars, even some carbon nuclei have energies sufficient to overcome the strong electrical repulsion between one nucleus and another. Then 'carbon burning' can occur in reactions like

$$^{12}\text{C} + {}^{12}\text{C} \rightarrow {}^{20}\text{Ne} + {}^4\text{He}, \text{ or } \rightarrow {}^{23}\text{Na} + {}^1\text{H}, \text{ or} \rightarrow {}^{23}\text{Mg} + \text{neutron},$$
$$\text{or } \rightarrow {}^{24}\text{Mg} + \gamma.$$

If the star is so massive that further compression and temperature rise can occur, 'oxygen burning' may take place with reactions like

$$^{16}\text{O} + {}^{16}\text{O} \rightarrow {}^{28}\text{Si} + {}^4\text{He}.$$

What happens as the temperature and density are raised to higher and higher values? Reactions will occur at a faster and faster rate; some nuclei will be built up, others will be shattered. E. Bodansky, W. Fowler, and D. Clayton concluded that at temperatures above $3\,000\,000\,000$ K an equilibrium is established involving reactions like

$$^{28}\text{Si} + {}^4\text{He} \rightleftarrows {}^{32}\text{S} + \gamma;$$

that is, silicon captures an alpha particle to form sulfur with an atomic weight of 32 with emission of a gamma ray; conversely, a ^{32}S nucleus absorbs a gamma ray and undergoes distintegration to $^{28}\text{Si} + {}^4\text{He}$. They were able to reproduce the solar abundances by invoking an equilibrium situation in which silicon is 'burned'.

In such an equilibrium situation, the most stable nuclei, that is, the nuclei that hold their constituent particles most tightly, will be favored over those that hold them less tightly. In the interval between calcium (atomic weight = 40) and nickel (atomic weight = 60), iron will be the most favored element, followed by chromium, nickel, and other metals of the iron group.

Thus we can easily understand the high abundances of elements like neon, magnesium, silicon, and sulfur, which are presumably produced by carbon burning, oxygen burning, and so forth, in the dense cores of highly evolved stars. Some of this material must have escaped to the interstellar medium from which a later generation of stars is born. The 'iron peak' requires very high temperatures and densities, which can occur only in the cores of extremely massive stars. The problem of getting this material dispersed into the interstellar medium without modifying it is an extremely difficult one. It is generally supposed that such material is actually supplied by the explosions of supernovae. Good observational evidence supports this conclusion. Remnants of the Cassiopeia A supernova appear to be ejecta from an explosion that occurred in the middle of the seventeenth century. Studies by M. Peimbert, S. van den Bergh, and particularly by R. Chevalier and R. Kirshner, show that different fragments have different chemical compositions, which can be identified with various strata in the pre-supernova, where oxygen burning, silicon burning, etc, occurs.

Although the abundances decline at first rapidly beyond iron, the distribution flattens out; beyond germanium and tin the numbers fall off irregularly. These elements cannot be created by simply raising the density and temperature in the nuclear furnace. In fact, elements such as mercury, gold, or bismuth should not appear at all. On the contrary, raising the temperature and density would simply shatter the iron nuclei into alpha particles, with considerable absorption of energy from the surroundings. Ultimately, if the temperature and density were increased indefinitely, there would be nothing left but neutrons. In fact we would approach the conditions of the Big Bang.

The clue to the origin of these heavier elements was found by reassessing Gamow's hypothesis that all nuclei were built up by successive neutron captures. The problem was, if such elements were built in stars, where did the neutrons come from? A possible answer to this question was given by Cameron and by the Burbidges, Fowler, and Hoyle, who pointed out that the actual number of neutrons required, compared to the number of atoms of carbon, oxygen, and neon, for example, was very small, and that neutrons (n) could be produced by reactions like $^{21}Ne + \alpha \rightarrow ^{24}Mg + n$ or $^{13}C + \alpha \rightarrow ^{16}O + n$. Once formed, these neutrons could be captured by nuclei of iron and other elements to build up successively heavier nuclei. A neutron carries no charge; hence there is no electrical repulsion and it can penetrate a nucleus with ease. It is believed that in violent events such as supernova explosions large numbers of neutrons may be produced suddenly. In the more orderly evolution of massive stars, neutrons can be produced more slowly.

Different types of nuclei are built up depending on whether the atoms are subjected to a high or a low density of neutrons. Suppose a nucleus of atomic weight A and charge (atomic number) Z captures a neutron. It becomes a nuclide of atomic weight

$A + 1$, still with a charge Z. The ratio of neutrons to protons in the nucleus is shifted in favor of the neutrons. The nucleus may eject an electron and become a nuclide of atomic weight $A + 1$ and atomic number $Z + 1$. Thus we write the reaction as

$$(A, Z) + n \rightarrow (A + 1, Z); (A + 1, Z) \rightarrow (A + 1, Z + 1) + \epsilon^-,$$

where ϵ^- is a beta particle (see Chapter 3), that is, an electron ejected from the nucleus. Speaking very generally, nuclei tend to maintain roughly comparable numbers of protons and neutrons. Among heavier stable nuclei, the number of neutrons exceeds the numbers of protons; for example, in ^{56}Fe, $Z = 26$ and $A - Z =$ number of neutrons $= 30$. One cannot go on adding neutrons indefinitely without allowing the nuclei to emit beta particles in an effort to restore some balance, otherwise the nucleus becomes unstable or simply will not accept the neutron.

If the density of neutrons is low, the time interval between the capture of one neutron and then another by the same nucleus is long enough for the ejection of a beta particle to occur. Suppose the density of neutrons is so very high that a nucleus captures a second neutron before it has had a chance to eject a beta particle, that is, $(A, Z) + 2n \rightarrow (A + 2, Z)$ or even $(A, Z) + 3n \rightarrow (A + 3, Z)$. Such nuclei will eventually decay by ejection of beta particles, but note that the resultant nuclei (decay products) will consist of combinations (A', Z') that cannot be produced by slow, leisurely addition of neutrons. In fact, nearly all the nuclei of heavier elements are produced by one (sometimes both) of these processes. A few exceptions are the so-called proton-rich nuclei, which appear to have been produced by the ejection of neutrons from nuclei during the violence of a supernova explosion.

To summarize, heavy nuclei can be produced within the turmoil of a supernova explosion, wherein large quantities of neutrons are produced, or in the cores of massive stars, where neutrons are liberated less copiously.

We now have a plausible explanation for the heavy-metal Class S stars and the carbon stars. The S stars show an enhanced supply of metals of the zirconium group, including technetium (atomic number $= 43$) which has been produced by humans but does not otherwise exist on the earth. Some of these stars also show great quantities of barium and other heavy elements. The *melange* of elements with enhanced abundances in S stars could only have been produced by slow neutron capture in the stellar core. Isotope abundances are of the utmost importance since some isotopes are produced by rapid (r) and others by slow (s) neutron capture processes. Yet other isotopes result from both r and s processes. Stellar isotope ratios generally can be measured only for those elements that are contained in molecules.

Products of nuclear processing can be dredged up from cores of stars by penetration of turbulent plasma from the convection zone into this element-building region. Possible scenarios have been described by J.M. Scalo and R.K. Ulrich, by I.J. Sackmann and R.F. Christy, and by I. Iben and Renzini. One is shown in Fig. 9.1. Here the star is in an advanced evolutionary state. The core hydrogen is long since gone, the hydrogen-burning shell has moved far outward, a helium core has been developed and exhausted. At least part of the time there now exists a helium shell source. Now it zigzags between phases when hydrogen is being burned in an outer shell, helium is burned to carbon in an inner shell, and the convective envelope occasionally dips down to dredge up the products of nuclear reactions. The essential point is that in the hydrogen-burning shell source, all the carbon and oxygen are converted into nitrogen. The situation is unstable. As temperature and density increase, vigorous helium burning in a turbulent layer converts the helium to carbon, energy generation in the hydrogen-burning zone is extinguished, and nuclides like ^{22}Ne and ^{25}Mg are built out of nitrogen by successive alpha-particle capture. An important consequence is that neutrons are produced. Some of them are captured by iron nuclei and build up s-process elements. At this stage equilibrium is disturbed so convective plumes can dredge up material rich in s-product elements and carbon, eventually conveying it to the surface. After a bit of time, things settle down, the hydrogen-burning shell is re-established and the cycle repeats.

Details will be revised as our understanding improves, but these scenarios do give us ideas on specific processes of stellar nuclear alchemy. Three elements, lithium, beryllium, and boron cannot be manufactured in the stars. Their fragile nuclei are quickly broken down to alpha particles and protons. They appear in cosmic rays, however, and the general belief is that these nuclei are produced by breaking down nuclei of heavier elements, such as oxygen or neon, in the interstellar medium through impacts with very-high-energy particles. Some of these metals may have been formed in the solar system as a consequence of flare activity in the youthful sun. In any event their production rate is extremely slow.

The dusty winds of doom; the beginning of the end

Thus are normal elements built in stars, but what happens to the stars themselves while this is going on? The dramatic events in the core are accompanied by drastic changes in the now rapidly expanding outer envelope. As the star moves into the giant region along the 'asymptotic giant branch' (AGB) it undergoes spectacular changes which are brief in terms of the life span of a star.

This evolutionary stage involves three new phenomena: mass loss by wind, pulsation of the outer layers, and formation of solid grains (dust) that may ultimately hide the star from optical view. Interactions occur. Mass loss is affected by pulsations and by dust.

In fact, mass loss now becomes a primary influence in stellar evolution as it can exceed by a factor of hundreds the rate at which nuclear ash is added to the core. So, ultimately the star dies from the hemorrhage of its life-giving fuel as winds carry it away into space. The mass-loss rate varies from 10^{-8} to 10^{-4} solar masses per year, the rate increasing as the star evolves. It is clear that a star of one to three solar masses cannot survive very long at a mass-loss rate of 0.1 the mass of the sun per thousand years.

As the star moves up the AGB, substantial pulsations in radius, driven by instabilities near the base of the vast envelope, can occur. These pulsations

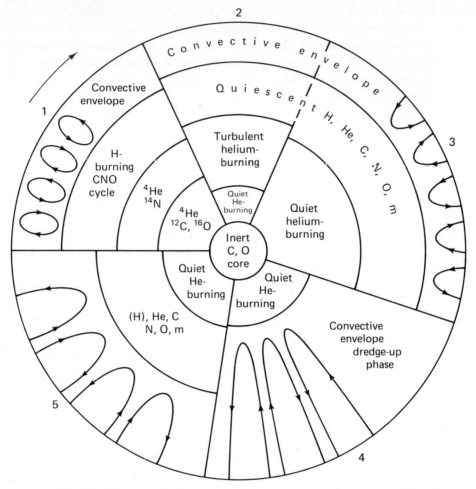

Fig. 9.1. Nucleosynthesis and dredge-up cycle: a complex scenario. This diagram, due to I. Iben, illustrates a scenario during an advanced evolutionary phase when core helium(He) has been exhausted leaving an inert central region of carbon(C) and oxygen(O). It is not drawn to scale either in steps of space or time. In phase (1) hydrogen(H) is being burned to He by the CNO cycle in a shell source with the result that C, N, and O are predominantly reduced to nitrogen(N). Recall that ^{14}N was the bottleneck in the CNO cycle. A zone of ^{4}He and ^{14}N 'ash' is deposited below the H-burning shell. Below this stratum lies a zone in which He has been burned to C and O in the previous cycle. The accumulation of reaction products drives up temperature and density so He is ignited in a vigorous 'combustion' that involves the layers between the core and the H-burning shell (phase 2). The region expands, so the hydrogen-burning shell is driven upwards and soon extinguished, temporarily. In addition to the energy production, the destruction of ^{14}N by the cycle: $^{14}\text{N} + {}^4\text{He} \rightarrow {}^{18}\text{F} + \gamma$; $^{18}\text{F} \rightarrow {}^{18}\text{O} + \epsilon^+ + \gamma$; $^{18}\text{O} + {}^4\text{He} \rightarrow {}^{22}\text{Ne} + \gamma$; $^{22}\text{Ne} + {}^4\text{He} \rightarrow {}^{25}\text{Mg} + \text{n}$, produces a supply of neutrons (n) which are captured by local nuclei, including heavy metals, denoted by m. Iron plays a particularly important role in the production of nuclides by slow neutron captures. In phase (3), convection ceases, He burns quietly, but the equilibrium is destroyed in such a way that the convective zone (here depicted by long convection plumes) reaches into

apparently enhance mass-loss rates and hasten the demise of the star. At this stage the star has become a Mira-type or long-period variable. These variables have been patiently observed for many years. Most of them have periods between 200 and 400 days, luminosities of 2000 to 6000 that of the sun, and radii of 200 to 300 that of the sun, or even greater! Note that when it reaches the Mira stage the sun will engulf the earth. Long-period variables show visual brightness variations of five magnitudes but the range in true luminosity is much smaller since they are red, cool, stars. Most of their energy emerges in the near infrared where the variations are significantly smaller. At Mount Wilson Observatory, many years ago, Pettit and Nicholson used a thermocouple to measure minimum–maximum temperatures of 1900°–2600° and 1600°–2200° for Mira and χCygni, respectively. It is difficult to say what these measurements actually mean since these stars do not radiate as black bodies. Furthermore, we see through huge depths, millions of kilometers, in the radiating layers so the usual concept of effective temperature which require knowledge of both luminosity and radius is not very helpful. We cannot specify a unique radius for a Mira star.

Interpretations of these evolving stars is enormously complicated by the presence of dust, which can completely enshroud the optical source. Thus, a star may evolve to yet higher luminosities than a Mira but become observable only in the infrared (IR) and radio-frequency ranges. Oxygen-rich stars of this type show a peculiar enhancement 'masering' of the 18-cm (1612 MHz) radio-frequency line emission from the OH (hydroxyl) molecule (see masers pp. 236–40). They are called OH–IR stars and some are brighter than the most luminous Miras. Their regular, large amplitude 'light' curves as measured actually in the infrared reveal periods up to 2000 days, i.e. longer than those of Miras. Mass-loss rates up to 10^{-4} the mass of the sun per year show this stage to be short-lived (i.e. less than a million years).

What has happened? As the star ascends the AGB (see Fig. 9.2), the envelope expands, cools, and breaks into pulsations. Mass-loss rate rises and dust hides the star. Then, the shell may escape from the star to form a cool gas cloud of dimensions ten to a hundred times larger than the solar system. This stage is hard to study as it is veiled in secrecy behind a dense dust envelope of its own making. A clue is provided perhaps by some very red OH–IR stars, where pulsations have ceased. Apparently, the envelope is now detached from the star; the observed radiation comes from the hot, dusty shell. This final phase of mass loss may be completed in 100–1000 years. What had once been a star is now mostly a hollow shell. The envelope has become a low-density cloud escaping forever from the gravitational field of the hot core at a rate of ten or more kilometers per second. As its nuclear energy sources are nearly gone, the core settles down gradually to become a white dwarf. The hollow shell is

Caption for fig. 9.1 (*cont.*)

layers where products of slow neutron capture have been deposited (phase 4). In phase (5) He-burning continues for a time. The amount of H involved is very small as it would disappear violently at the edge of the He zone. The mass but not necessarily the size of the core increases. The edge of the convective envelope recedes; finally stage (1) is re-established. The cycle repeats.

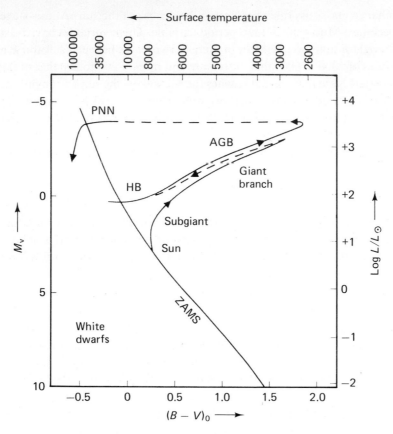

Fig. 9.2. Schematic tracks for late stages of stellar evolution. Luminosities are given in terms of that of the sun and as absolute visual magnitudes M_v. The abscissa is the $B-V$ color of the star corrected for interstellar extinction. The temperature scale at the top of the diagram corresponds to that valid for giants. ZAMS indicates the zero-age main sequence, HB the horizontal branch, and AGB the asymptotic giant branch. Dashed lines denote portions of evolutionary tracks where developments go very fast (as following the first ascent of the giant branch, or the development of the core to a planetary nebula nucleus (PNN), following the detachment of the circumstellar envelope). Artistic license is invoked at the top of the AGB, where the evolving star is hidden by a dust cloud, so neither B nor V magnitudes are measurable; this stage occurs when a Mira or long-period variable evolves into an OH–IR star or a carbon-rich IR star, such as IRC + 10216 (Table 9.2), and the core starts on its way to becoming a white dwarf.

filled by a hot, tenuous gas. S. Kwok, C. Purton, and M. Fitzgerald suggested that at this stage a high-velocity (1400–5000 km/s) wind emerges from the core. Although the mass loss in this swift wind is small (perhaps 10^{-9} to 10^{-7} solar masses per year), it rushes outward, scouring out all the gas in the 'hollow' zone, and hits the sluggishly moving, previously detached circumstellar shell. The momentum speeds up the velocity of the shell from about 15 km/s to 30–50 km/s. Shock waves are set up, traveling both inwardly and outwardly.

The core consists of the inert, dead ash of nuclear reactions plus a small amount of inner envelope material that was unable to escape. The surface temperature of the core steadily rises. For a core mass near $0.6\,m$ solar masses, the temperature rises to about 30 000 K in about 1500 years. This high-temperature radiation now impinges on the dusty shell and, since it is rich in ultraviolet quanta, it ionizes the shell of gas and causes it to glow in visible light as a planetary nebula (PN) (see fig. 9.3).

The core mass largely determines the scenario. If m (core) $< 0.57\,m$(sun), the core evolves so slowly that the circumstellar envelope all escapes into space before the temperature of the core reaches 30 000 K. If m(core) $> 0.7\,m$(sun), its subsequent evolution may be so rapid that the power source disappears before the ionization of the shell is complete.

We defer further discussion of planetary nebulae to Chapter 10 which describes the interstellar medium (ISM), into which these objects gradually fade away. We now look more closely at stellar winds, which play so important a role in stellar evolution, and at stellar pulsations, which give insights into the structure not only of highly evolved stars, but also of the sun. Finally, we want to trace the fate of the cooling core as it gradually expires as a white dwarf.

Stellar winds

Except for a few obvious examples, such as novae, supernovae, or Wolf-Rayet stars, the importance of mass loss by stellar winds was not appreciated until fairly recently. We can measure the solar wind directly by space probes; it amounts to about 10^{-13} of the sun's mass per year, which means that mass loss now has a negligible effect on the evolution of the sun. When the sun reaches the AGB, however, it will lose nearly half its present mass by winds as the core settles to a 0.56 solar mass white dwarf. Stars of greater mass lose a larger proportion of their substance in winds. Those of up to about 6 or 7 solar masses evolve down to white dwarfs usually of less than 1.4 solar masses. Stars exceeding 12 solar masses may lose mass all their lives. Many of these never evolve into giants or supergiants at all; they eventually become neutron stars, black holes, or supernovae.

Mass loss is important, not only because it insures that most of the stars in the sky will evolve into white dwarfs, but also because much nuclear processed material is returned to the interstellar medium. From this material a new generation of stars is spawned, so that the metal-to-hydrogen ratio in the Galaxy gradually rises. Material locked in white dwarfs is entombed forever.

Methods of studying winds and mass loss

(1)   *P-Cygni phenomena.* Broad emission-line profiles with absorption on their violet edges are indicative of mass loss from a star (see Fig. 9.4 a, b, c). Such profiles are seen in catastrophic ejections as occur in novae (Chapter 11) and related stars, but also in the ultraviolet spectra of 'normal' early-type stars such as τ Scorpii, indicating that the atmospheric layers that we see

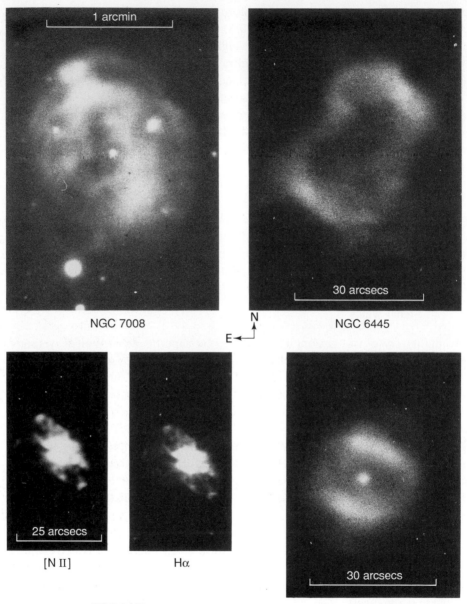

NGC 7008   NGC 6445

[N II]   Hα

NGC 2440   NGC 6058

Fig. 9.3. Examples of some planetary nebulae. Images of NGC 7008, NGC 6445, and NGC 6058 were obtained with a narrow band-pass Hα filter. For NGC 2440, we compare [NII] and Hα images. The NGC 6445 and NGC 2440 nuclei are so hot that nearly all their energy is emitted in the far ultraviolet; in the optical region the starlight is overwhelmed by the bright nebular emission. These photographs were obtained with an image tube at the Cassegrain focus of the Mount Wilson 1.5 meter reflector with the collaboration of S.J. Czyzak.

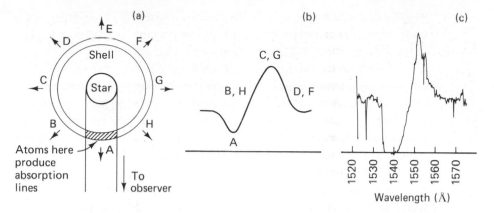

Fig. 9.4. Mass loss in an expanding shell.
(a) Schematic diagram of an expanding shell about a star. The arrows indicate the directions in which the various parts of the shell are moving. Only the atoms in the shaded part of the shell (part A) lie between the star and the observer and hence produce an absorption line.
(b) Schematic 'P-Cygni'-type profile from an expanding shell. Atoms in region A of Fig. 9.4a produce the absorption line; those in region E are occulted.
(c) Shows an observed line shape, namely the profiles of C IV 1548, 1550 Å in the O4 supergiant, ζ Puppis. (*International Ultraviolet Explorer Atlas of O-type Spectra, 1200–1900Å*, N.R. Walborn, J. Nichols-Bohlin, and R.J. Panek, 1985 NASA ref. publ. 1155.)

may not be firmly bound to the star gravitationally but may be to some extent escaping into space!

(2) *Winds from supergiants.* Although the shapes and velocity shifts of atomic and molecular lines in the spectra of supergiants often strongly suggest a wind component, the first unambiguous detection of such a wind was made in the binary α Herculis. A.J. Deutsch noticed that the spectrum of the G-type component shows sharp, weak lines characteristic of its M-type companion and correctly concluded that the light of the hotter star was shining through a cloud of cool gas escaping from its companion.

(3) *The technique of interferometry*, using optical, infrared, and radio-frequency radiation allows us to obtain high spatial resolution. Sometimes it is even possible to construct an image of the source. From such very detailed observations we can get the wind velocity, and data on the atmospheric structure and uniformity of wind outflow. For example, the size of Betelgeuse, measured in the red hydrogen line, Hα, is 20 percent greater than its size measured in photospheric light. By comparing optical observations with infrared measurements (which give data on dust and molecules) and radio-frequency measurements (which give data on molecules), we can find how far from the star the bulk of the dust and gas are concentrated.

(4) *Ultraviolet spectral observations* give evidence of extensive envelopes in cool stars, chromospheres, and hot coronas (which appear to be heated by shock waves in a magnetized gas).

(5) *Profiles and velocity shifts of molecular lines*, especially the hydroxy (OH) and carbon monoxide (CO) molecules are important for studies of the late evolutionary stages, supersonic flows, etc. That is, we can study spatial distributions, velocities, and time variabilities of different species of molecules that exist under regimes of different temperatures and pressures. For example, in Mira variables we can monitor a shock wave running through the atmosphere. The excitation temperature varies with the phase of the oscillation, the gas cooling when it expands. Most of the outflowing gas is molecular hydrogen which cannot be detected directly. We can get the mass-loss rate for the constituents, CO, OH, and H_2O; then, if we are clever enough in doing the chemistry, we can evaluate, for example, the ratio of CO (the most stable molecule) to molecular hydrogen, H_2, and thus evaluate the total mass-loss rate. The chemistry of these expanding circumstellar envelopes poses some engaging problems as fairly complex and exotic molecules can form.

What actually causes the winds to blow? In stars like the sun, which is surrounded by a tenuous corona whose temperature of 2 000 000 K is maintained by the dissipation of magnetic energy, the small mass loss can be explained by a theory given by Eugene Parker. In cool giants and supergiants the driving mechanism seems to be radiation pressure acting primarily on dust grains. These particles are efficient interceptors of radiation and, once they form, they will dominate the motion of the wind. The gas is dragged along with the grains and the entire mass of material is swept away from the star.

In the environs of hot stars, dust cannot easily form and here radiation pressure acts primarily on atoms in their ground states via strong 'resonance' lines such as Lyman α in the case of hydrogen. The P-Cygni-type profiles (see Fig. 9.4) of very hot stars indicate fast winds (velocity = 1000–3000 km/s) and mass-loss rates of 10^{-7} to 10^{-8} solar masses per year. Often the level of excitation in the wind is higher than in the atmosphere of the star. For example, lines of [O VI] (five times ionised oxygen) appear in the spectrum of τ Scorpii, but such highly ionized oxygen does not occur in the normal atmosphere of such a B0V star. Presumably the rapid wind is heated by the dissipation of mechanical or magnetic energy via shock waves.

We must emphasize that mass loss is not confined to evolved and dying stars. Material seems to be ejected from very young stars at the epoch of their formation. For example, T Tauri stars, which are but recently formed from dense clouds of the interstellar medium, appear to eject copious amounts of gas. Material may also be falling in; the behavior of circumstellar envelopes in such objects can be complex.

Dust in circumstellar envelopes

Since the pioneering surveys of G. Neugebauer and R.B. Leighton, much progress has been made in studies of dust in circumstellar envelopes. In many instances the infrared flux exceeds the photospheric flux; most of the starlight is absorbed by the grains and re-emitted by them as heat radiation.

Of what material are the dust grains composed? What are their sizes? How far from the parent stars and under what conditions are grains formed? Infrared spectroscopic and interferometric measurements are especially valuable although optical, radio-frequency and even ultraviolet data can be helpful.

In addition to thermal radiation from the dust, a number of discrete, broad spectral lines are observed. Silicate grains (revealed by 9.7 micrometer (μm) radiation) are found especially in circumstellar envelopes of oxygen-rich giant stars later than M3. This feature, discovered by N. Woolf and E. Ney (1969), appears as an emission line when the dust shell is thin, but as an absorption line when the dust shell is thick. Some oxygen-rich stars show a 3.1 μm feature due to water. In envelopes of some carbon-rich stars, R.R. Treffers and M. Cohen found the broad 11.3 μm line of SiC. Other diffuse lines are known in both the infrared and visual spectral regions. They may arise in complicated molecules, such as polycyclic aromatic hydrocarbons (PAHs – see Chapter 10).

Alternatively, some truly remarkable carbon structures are possible. Particularly striking is the hollow cage network containing 60 carbon atoms primarily arranged in distorted hexagons and pentagons, closely resembling the pattern employed by Buchminster Fuller in his design of the US pavillion at Expo 67 in Montreal. Harold Kroto of the University of Sussex has shown how such a molecular structure can be very stable.

The gas-to-dust ratio appears to be about 100:1. Consider oxygen-rich stars where we can assume solar abundances to be a good guess. There, most of the refractory material that could be condensed into grains actually is! For example, the low abundance of gaseous SiO indicates that most silicon is held in solid grains. Alternatively, in the circumstellar envelopes of carbon stars, where oxygen is depleted, amorphous carbon, soot, or Kroto's structures, rather than tiny graphite crystals, appear to prevail.

Grain sizes are often deduced from their optical properties. Because these particles strongly redden starlight we infer that most of them must be smaller than about 1 μm in radius. Large chunks, like snowballs, would simply block starlight without changing its color. If the particles were of the size of ordinary molecules, about 10^{-8} centimeters in diameter, they would be very potent reddeners. We have a daily example of this phenomenon in the blueness of the sky and in the redness of the rising and setting sun. The sky is blue only because rays of sunlight passing through the atmosphere are deflected or scattered sideways from their original paths by air molecules. Blue rays are scattered more readily by molecules than are red rays. Consequently, most of the scattered sunlight is blue; this diffused blue light from the sun provides the beauty and glory of clear sky. Manifestly, the scattering process removes blue rays from the original solar beam so the sun appears redder than it would be in the absence of the atmosphere. Also the reddening of the sun is most pronounced near the horizon, when its rays traverse a long column of blue-eliminating atmosphere (see Chapter 10).

Particles much smaller than the wavelength of light show strongly wavelength-dependent scattering (Rayleigh-type scattering) which varies inversely as the fourth power of the wavelength of light that falls upon them. That is, they scatter ultraviolet light (3300 Å) sixteen times as efficiently as red light (6600 Å), whereas actual observations of stellar colors show that the circumstellar grains cut down the intensity of ultraviolet light by a much smaller factor, only about a factor of two.

Other clues to grain size are provided by the polarization the grains produce in the infrared and optical spectral regions. Many of the grains have sizes of about 0.1 μm, but they show a considerable spread in diameters. The smallest grains polarize and scatter inefficiently. The grain sizes appear to be similar to those of the interstellar medium.

The details of grain formation in stellar envelopes are not too well understood. Carbon-rich grains can form at 1700 K, oxygen-rich at 1200 K, but local temperatures in zones where solid particles actually condense appear to be much less, perhaps 1000 K and 600 K, respectively. With interferometric techniques, we can measure the distances of the condensing grain clouds from the parent stars. For Betelgeuse they appear at about 35 times the radius of the star; for other stars circumstellar-envelope distances of 5 to 40 R(star) are found. At present, accurate values are hard to establish. Thus most grains do not condense in the actual atmosphere of a giant or supergiant, but rather at a considerable distance from the photosphere.

Circumstellar grains play a dominant role in assisting the ejection of stellar nucleogenesis products into space. They provide catalytic surfaces for complex chemical reactions, they greatly complicate optical studies of late stellar evolutionary stages, and, by collecting in huge clouds which eventually collapse under their own gravity, they are intimately involved in the birth of new stars.

Stellar pulsations

Stellar pulsation is a normal stage of stellar evolution. Fig. 9.5 and Table 9.2 summarize some of the basic data. Pulsations occur in a wide variety of stars, in dense, very hot ($T = 100\,000$ K) objects on their way to becoming white dwarfs, and in stars so cool ($T = 1000$ K) as to be found only by their infrared radiation; in very massive stars of 4–15 solar masses, such as β Canis Majoris variables or in classical Cepheids; in old novae about 0.55 the mass of the sun, in tenuous long-period or Mira variables and OH–IR stars with densities below a millionth that of the sun; and in white dwarfs 50 000 times as dense as the sun. The periods range from ∼200 seconds for a while dwarf to about 6 years for an OH–IR star. Stellar pulsations elucidate many phenomena not otherwise comprehensible. They explain rapid mass loss in long-period variables (LPVs), OH–IR stars, and mass-losing carbon stars that produce most planetary nebulae (see Chapter 10). They also provide a useful tool for checking stellar interior models, not only for late evolutionary stages, but also for the sun. Pulsations are not confined to the final stages of a star's life, but are found in earlier phases of evolution such as β Canis Majoris stars, Cepheids, and RR Lyraes.

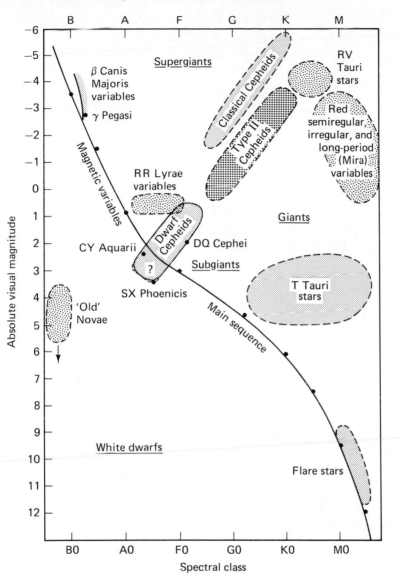

Fig. 9.5. Schematic position of variable stars in the Hertzsprung–Russell diagram. The main sequence is denoted by the solid line: positions of giants, subgiants, supergiants, and white dwarfs are indicated, and the domains of different kinds of variables are shown. The RV Tauri stars and semiregular red variables probably owe their variations to pulsations. The T Tauri stars are probably stars in the process of formation; the red flare stars are probably very young main-sequence stars. The positions of several representative type stars are indicated. The dwarf-Cepheid sequence actually may not cross the main sequence. Stars of Populations I and II are built on different models, so the overlapping of two different kinds of variables is not surprising. The internal structures of the stars differ from one type of star to another, but within each group they are probably similar.

Table 9.2 *Data for pulsating variables*[a]

Type of star	Range of period (days)	Population type	Spectral class	Absolute magnitude (visual)	Mass (sun)	Typical star
RR Lyrae variables	0.06–0.8	II	A2–F6	0 to +0.5	0.6–0.7	RR Lyr
Classical Cepheids	1–50	I	F6–K2	−0.5 to −6	3.5–14	η Aql
Type II Cepheids	2–45	II	F2–G6	0 to −3	<1	W Vir
RV Tauri stars	20–150	II	G, K	−3	<1	RV Tau
Semiregular[b] variables	100–200	I, II	M, S, C	−1 to −3	∼1	BQ Ori
Long-period (Mira) variables	100–700	I, II	M, S, C	−5 to −6 (M_{bol})	1–2	Mira
OH–IR stars	1000–2000	I, II	'M'	−5 (M_{bol})	1–5	AFGL 230
Mass-losing C stars	∼600	I	'N'	−8 (M_{bol})	1–5	IRC+10216
β Canis Majoris variables	0.17–0.25	I	B1–B3	−3.5 to −4.5	7	β Cephei
Dwarf Cepheids	0.06–0.13	I	A2–F5	+2 to +4	2	δ Scuti
Possible white dwarf pulsators[c]	0.002–0.014	I, II	WD	>+5	0.6	ZZ Ceti

Notes:

[a] R Coronae Borealis stars are sometimes included as pulsating variables. Novae (which sometimes show oscillations) are discussed in Chapter 11.

[b] Semiregular stars often switch from large to small amplitudes, with regular variations superposed on long-term changes.

[c] Among white dwarfs, pulsations may be confined to the hottest objects.

The Cepheid variables

We begin with the most orderly and intensively studied type of pulsating variable, the Cepheids, named after the prototype, δ Cephei, discovered by J. Goodricke in 1794. Actually, there are two groups of Cepheids. Classical Cepheids belong to population type I. They are massive stars much younger than the sun, have periods in excess of 1 day, and their light curves repeat faithfully cycle after cycle. They have evolved apparently from main-sequence O and B stars. Population type II Cepheids have evolved from stars of a solar mass or less. The subgroup of type II with periods less than 0.8 day and amplitudes of about 1 magnitude are called RR Lyrae stars. They are numerous in globular clusters and in the galactic bulge. Those of longer period include BL Herculis, W Virginis and RV Tauri type variables. They tend to be fainter than classical Cepheids of the same period.

Classical Cepheids show conspicuous changes in brightness, color, spectral class, and velocity during their cycle of variation. Fig. 9.6 compares the light variation of η Aquilae with its radial velocity curve and the corresponding change in radius. As the star steeply rises to maximum light it becomes bluer and the spectral class becomes 'earlier'. After maximum, this star fades more slowly but shows a hump in its light curve. The spectral lines also undergo displacements with the same period as the light variation. The radial velocity curve is plotted 'upside down' in the sense that velocities of approach (radial velocities) increase upward numerically. The bottom curve shows the changes in the radius of the star. Maximum light occurs before the star has reached maximum radius. With many Cepheids, maximum light occurs simultaneously with, or slightly earlier than, the greatest velocity of approach.

Mean spectral classes, colors, surface temperatures and luminosity (*L*), are smoothly correlated with period (*P*); the longer the period, the cooler the star. In 1910 Miss Henrietta S. Leavitt at Harvard noted that the classical Cepheids in the Small Magellanic Cloud showed a well-defined correlation between period and luminosity in the sense that the longer the period, the brighter the star. Since, with sufficient accuracy, all stars in this nearby galaxy can be regarded as being at essentially the same distance from us, the period–luminosity relation represents an intrinsic property of these stars. If the classical Cepheids in the Magellanic Clouds are typical of population I Cepheids everywhere, this period–luminosity correlation makes it possible to estimate the absolute magnitude and therefore the distance of any Cepheid whose period and apparent magnitude are known. Thus, this empirical relation has important application to the establishment of the distance of any stellar system where such stars can be observed. In order to use this relationship we must know the intrinsic luminosity of at least one Cepheid or, in technical jargon, we must establish the zero-point of the period–luminosity (*P–L*) law. In practice this step turns out to be difficult since good trigonometric parallaxes cannot be measured and statistical methods are inaccurate.

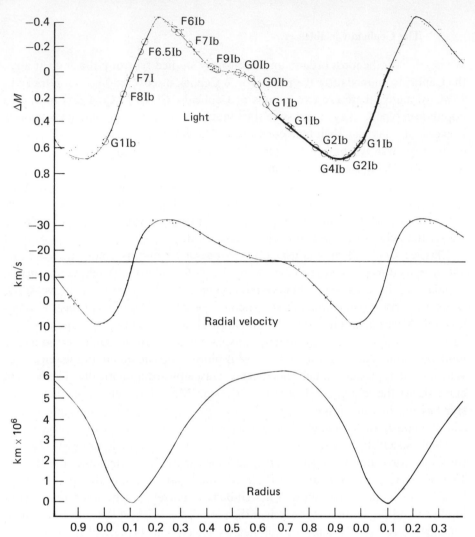

Fig. 9.6. The light, radial velocity, and radius variations of the classical Cepheid, η Aquilae. All quantities are plotted against phase or fraction of period (1 unit = 7.177 days). The top curve shows the light variations ΔM, in magnitudes; the spectral class corresponding to each phase is indicated. Notice that the spectral classes differ at points corresponding to the same brightness on the rising and falling branches of the light curve. The earliest spectral class comes *after* maximum light. Notice the bump on the descending branch of the curve, which is also mirrored in the curve of radial velocity (in kilometers per second). The lowest curve shows the change in radius (in millions of kilometers) as a function of time. Notice that maximum radius comes just before minimum light and minimum radius occurs on the rising branch of the curve. (Courtesy Arthur Code.)

Pulsation theory

The light and spectral variations in Cepheids cannot be explained by eclipses. Furthermore, if the radial velocity variations were attributed to motions of the components of a spectroscopic binary, the orbits would have to be smaller than the stellar dimensions. Impressive observational and theoretical evidence was accumulated years ago to support the pulsation theory proposed by H. Shapley in 1914. Actually, the possibility of stellar pulsations had been demonstrated as early as 1879 by A. Ritter who used sound mathematical and physical arguments, but astronomers then knew little about the interpretation of light curves and nothing about the variable radial velocities of the Cepheids.

Just what are the conditions within a star that would enable it to pulsate? Why do some stars pulsate and others do not? As we saw in Chapter 7, at any point within a normal star, gas pressure plus radiation pressure just suffices to support the weight of the overlying stellar layers. Let us now suppose that the delicate balance between gravity and pressure is upset by the slight adjustments the star has to make in its internal structure as it converts hydrogen into helium and slowly evolves across the HR diagram. When the pressure exerted upon a gas is lessened, the gas expands, just as bubbles rising from the depths of a pond grow larger as they approach the surface. As the heated stellar gases expand, they force the overlying layers upward, but only for a time because expanding gas cools and thus exerts less and less pressure as the speeds of its atoms slacken. When the pressure falls sufficiently, gravity reasserts control and the expanded layers fall back. Does the star then regain its stability? No, for the momentum of the downward-moving gas is usually enough to cause it to overshoot the stable position. Again the gases become compressed and heated enough to overcompensate gravity and the sequence of events is repeated.

The situation recalls that which prevails when a weight is suspended from a coiled spring. The spring stretches and its tension increases until it just compensates for the weight. If we disturb the equilibrium by pulling slightly on the weight and then releasing it, the increased tension in the spring overpowers the force of gravity. The weight shoots up beyond its original position, until the force of gravity pulls it back. The oscillation that is set up persists until it is gradually damped out by friction.

The ionization and recombination of hydrogen and helium may play an important role in the pulsation of a Mira or Cepheid variable. Consider the ionization of hydrogen in a small parcel of gas. Energy is taken up from the gas to pull the electrons and protons apart, and the temperature falls. When electrons and protons recombine, energy is liberated and the temperature rises. Thus occurs a see-sawing between neutral and ionized states. The same kind of flip-flop can occur between neutral and ionized or between singly and doubly ionized helium. The hydrogen ionization flip-flop seems important for the Mira-type variables, helium ionization plays the critical role in Cepheids.

A. Eddington worked out the first mathematical theory of Cepheids. He proved that large stars should pulsate more slowly than smaller, denser, ones. In mathematical terms, the product of the period and the square root of the average density should be nearly constant, a prediction that seemed to be in harmony with observation. The period is measured directly but the density must be found from the radius of the star and its mass. The radius follows from the absolute magnitude and surface temperature, since the latter determines the radiation rate per unit area and the absolute magnitude gives the total luminosity, which is the surface area times the emission per unit area. The mass of the star is found indirectly. One calculates the evolutionary tracks for stars for different masses and identifies the one that gives the observed luminosity and surface temperature.

Eddington's theory immediately encountered difficulties. First, the calculations showed that the oscillation energy should gradually decay, just as the oscillation of a weight attached to a spring dies out. A normal Cepheid should stop pulsating in a few thousand years. Clearly, the mechanical energy dissipated must somehow be replenished from within the star.

Also the theory predicted that the star should be brightest when it was smallest because the temperature rise would more than offset the effects of a decreased surface area. Correspondingly, it should be faintest when largest and at both maximum and minimum the radial velocity should be zero. Reference to Fig. 9.6, however, shows that at maximum and at minimum light, the velocity differs considerably from zero.

The standing wave set up in the deep interior evolves into a running wave further out in the star in a region of much lower density commonly called the envelope. The most familiar examples of running waves are wind-blown waves on a lake or ocean which produce breakers as they hit the beach. Oscillations in the visible layers need not be in phase with the oscillation in the deep interior. In 1938, M. Schwarzschild showed that this straightforward modification of the original theory removed a fundamental difficulty.

Nevertheless, detailed quantitative agreement between theory and observation required an exact knowledge of the properties of matter at the temperatures and densities encountered in stellar interiors and an exact physical theory for handling them. Many features of a variable star's behavior depend on near cancellation of opposing effects, so accurate calculations are needed. Consider first the nature of energy flow in a pulsating star. The predominant flux is a steady stream outward; only in the outer 25 percent or so of the radius will there arise a significant ripple from the alternate damming up and release of energy in pulsations. When a volume of gas is compressed, its temperature is raised. Ordinarily, the opacity of the material will be decreased so that energy will escape more easily. Under these circumstances, the pulsations will tend to damp out. In certain outer regions of the star where ionized helium prevails, compression may actually increase the opacity of the material. More energy is trapped during the compression phase and when it is released during the expansion phase it gives an extra boost to the outer layers, thus tending to build up the pulsations. There are also effects arising from the influence of changes in the molecular weight and from the geometrical property that contraction must always tend to compress the gases. Energy tends to be dissipated in deeper layers, so the amplitude of the pulsation adjusts itself so there is no long-term storage or depletion of energy in the pulsating layers.

Whether or not a given star will pulsate depends on the details of its structure and evolutionary history. John Cox and Arthur Cox (no relation) showed that if one calculated a static model for a star in an 'unstable' region of the HR diagram, it would start to oscillate and the amplitude would build up until storage and dissipation balanced. As evolution carries the star out of the unstable region, the oscillations die away and the star ceases to pulsate. Thus pulsation occurs in certain zones or strips in the HR diagram and evolution may carry a star more than once through an unstable strip.

Detailed calculations with precise models, taking into consideration the large amplitudes and the storing and dissipation of energy in the outer layers, account for the observed behavior of the velocity and light curves. In agreement with observations, the time of mean rising light is predicted to come near minimum radius. Christy was, in fact, able to explain exactly the light curves of many classical and cluster-type Cepheids. In particular, he accounted for the bumps on many curves as follows. A large acceleration of material in the outer layer where helium is becoming ionized sends a pressure pulse inward which bounces off the core and is reflected back to the surface. In one model, the secondary bump is an echo of a primary bump 1.4 periods earlier. Besides such echoes, Christy found resonance effects in that ingoing and outgoing waves may mutually interfere; sometimes they reinforce each other and sometimes they cancel. Hence one can understand the complex light and velocity curves observed. In all models, a change in radius of only a few percent suffices to account for the observed light variations.

The efforts of the Coxes were devoted mainly to an interpretation of classical Cepheids, whereas Christy's analysis gives us an understanding of RR Lyrae stars and population type II Cepheids. Further refinements are required, especially to take into account atmospheric phenomena and show how they are related to the internal structures of the stars.

Eddington pointed out, too, that stars might pulsate in overtones, as well as in their fundamental frequencies, just as a musical string emits notes one, two, or three octaves apart, depending on how it is plucked. One example is the so-called 'dwarf Cepheid' star δ Scuti, observed by E.A. Fath, which may pulsate not only in its fundamental period, but also in overtones.

Some of the best examples of overtone pulsations are found among RR Lyrae stars. Christy's theoretical studies indicate that stars of high luminosity-to-mass ratio tend to pulsate in the fundamental mode while those of low luminosity-to-mass ratio prefer the first overtone. At the dividing line between the two types of behavior, the state of pulsation depends on the past history of the system.

The success of theory in interpreting pulsating variable stars constitutes one of the most impressive advances in modern astrophysics. Miss E. Hofmeister obtained an excellent agreement between the theoretical period–luminosity relation and the empirical one determined by Robert Kraft. Christy's calculations indicate that it may be possible to determine mass, luminosity, radius, and helium content for these stars from observational data on light and velocity curves alone. The luminosities so determined appear to agree with data from other sources. For example, for β Doradus, whose period $L_\odot$ is 9.84 days, Christy finds the luminosity as 3.7×10^3 solar luminosities ($L_\odot$), or the bolometric magnitude as -4.2, which fits the period–luminosity relation. Application of the theory (in this instance the transition period at which RR Lyrae stars switched from fundamental to overtone pulsations) gave $L = 46\ L_\odot$ for these stars in ω Centauri and $L = 37\ L_\odot$ for those in M3. On the other hand, the masses seem to be lower than those found from evolutionary arguments both for RR Lyrae stars and for classical Cepheids. Do these stars lose mass during late stages of their evolution?

A curious feature of RR Lyrae stars is that the prototype RR Lyrae has a magnetic field that was found to vary between -1580 and $+1170$ gauss with the secondary period of 41 days. What role do these magnetic fields play in pulsations?

The long-period variable stars (Mira stars)

Another important group of variable stars, those of long period (often called Mira stars), present even more engaging and puzzling problems than do the Cepheids. The stars in this group are all cool red giants and supergiants. They include oxygen-rich M stars, carbon-rich R and N stars, and 'heavy-metal' S stars. Their periods range mostly from 200 to 500 days, and their visual fluctuations in brightness amount to about 5 magnitudes (i.e. a hundredfold range).

Their light curves are not regular, like those of Cepheids, but show fluctuations in maximum and minimum brightness, shape of light curves, and periods. Most appear to originate from main-sequence stars of 1 or 2 solar masses; their luminosities range from about 2000 to 6000 times that of the sun. As we shall see, their vast sizes, 200–350 times that of the sun, are due to their pulsations. Their radii cannot be defined in any satisfactory manner. Hence effective temperatures are difficult to specify. The local temperature shows a huge variation throughout the readily observable layers.

The spectra of long-period variables

The spectra of Mira-type stars are dominated by molecular bands. With high dispersion these bands may be resolved into individual lines. In class M variable stars, only the violet region is free from the obscuring bands of titanium oxide, TiO, which produces great gaps in the spectrum from 4600 to 6400 Å. Fig. 9.7 shows the spectrum of Mira itself. In cool carbon stars the blue and violet regions are largely blotted out by bands of carbon, and in the S stars these regions are obliterated by absorption by zirconium oxide, ZrO. The absorption is strongest at the band heads where individual molecular lines are tightly packed and then gradually lessens as we move away from this wavelength because the separation of the individual lines becomes greater and greater.

Long-period variable spectra are also rich in atomic absorption lines characteristic of low temperatures, except when these lines happen to fall in regions of band absorption. Thus the sodium D lines, although quite strong in S stars, are smothered by the bands of TiO in M stars. As the variable star fades and its temperature falls, the dark lines change as we would expect them to do from the theory of ionization (Chapter 4). The H and K lines of ionized calcium fade; easily excited lines such as that of neutral calcium at 4226 Å strengthen. Also, at the lower temperatures, additional compounds form and all bands become intensified.

A striking phenomenon in red variables is the appearance before maximum light, of strong, bright lines, especially in hydrogen (Fig. 9.7). The lines appear generally in red variable stars and their intensity range often far exceeds the range of light variation. They reach maximum intensity about one-sixth of the period after maximum light. Bright lines in stellar spectra are not unusual, although most stars show exclusively dark lines. These emission lines are often found in very hot stars with extended envelopes, but Mira-stars are red, cool objects. Observations with the International Ultraviolet Explorer (IUE) reveal many cool stars with extended outer envelopes that show emission lines. The phenomenon is common in binary systems, but many single stars also show extended chromospheres or coronas that appear to be excited by the dissipation of mechanical or magnetic energy.

In the long-period variables, however, the bright hydrogen lines are not radiated by the outermost portions of the atmosphere, but at levels below those in which the molecules are absorbing. The evidence for this remarkable behavior comes from a close examination of the intensities of the bright hydrogen lines in Mira and similar M-type variables. In both laboratory and celestial sources, such as the solar chromosphere or gaseous nebulae, the intensities of the bright lines of the Balmer series fall off regularly from $H\alpha$ to $H\beta$, etc. But in the Mira variables, when a H line falls within a TiO band, it is greatly weakened. Thus, $H\alpha$ and $H\beta$ are much fainter than $H\gamma$, which in turn is not as strong as $H\delta$. The $H\epsilon$ line is weakened by its close proximity to the 'H' line of Ca II. We can only conclude that the bright-line radiations are absorbed by overlying layers of TiO and even ionized calcium. This conclusion is supported by high dispersion observations that reveal individual absorption lines within the H emission feature. Emission lines excited by fluorescence also occur (see Chapter 10). Quanta of bright hydrogen lines are absorbed by atoms which excite them to higher levels from which they can cascade downward with the emission of energy.

The Doppler shifts of visual region absorption lines are small, complex, and impossible to reconcile with the type of atmospheric motion found in classical Cepheids. Presumably they are formed very high in the atmosphere. A.H. Joy found that the differences between various cycles of Miras exceeds the mean velocity change observed in a single cycle. Velocities at bright maxima appreciably exceed those observed at the fainter maxima. The bright lines always show an outward displacement, suggesting a hot layer rushing upward from some point of origin far below the layers where the TiO bands are formed.

Spectroscopic observations (largely secured by K. Hinkle in the near infrared at

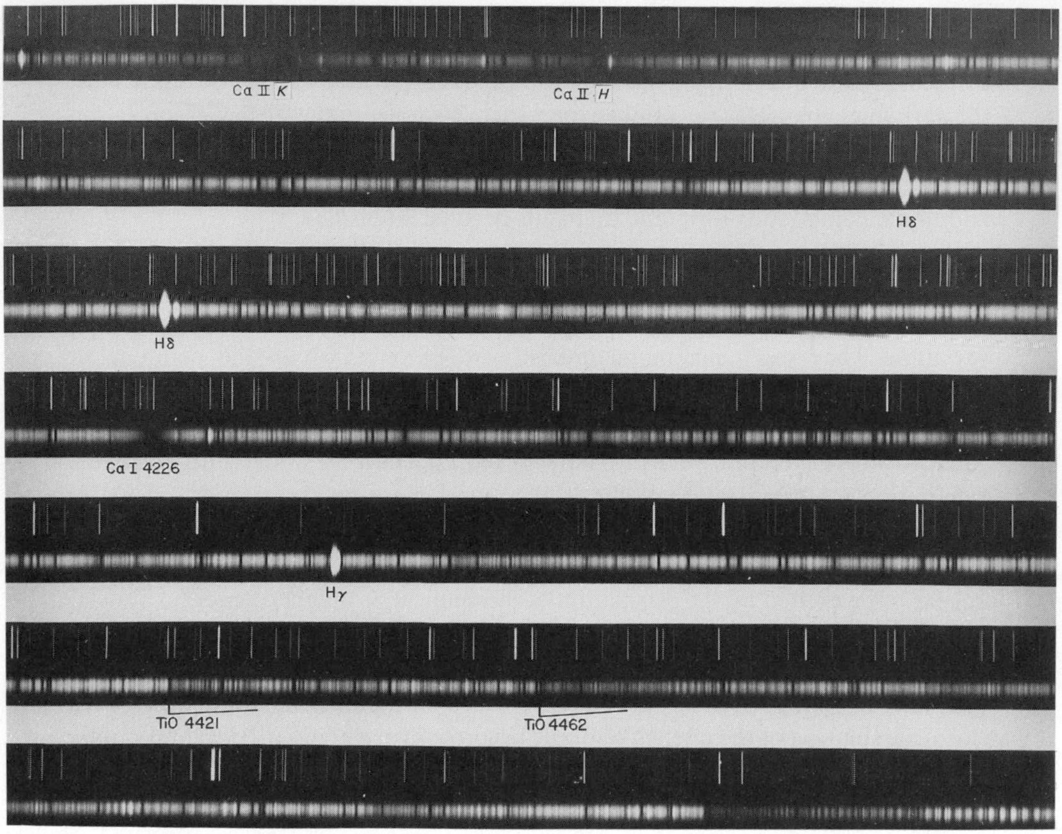

Fig. 9.7. A portion of the spectrum of Mira (o Ceti) as photograped by George Herbig with a dispersion of 2 Å/mm with the coude spectrograph of the 120-inch reflector at the Lick Observatory, University of California. Notice the strong Hγ and Hδ emission lines of hydrogen and the weakness of the hydrogen Hε (3970 Å) line near the Ca II *H* line. The fine structure of the TiO bands and the rich metallic spectrum are well exhibited in this spectrogram.

Kitt Peak Observatory) revealed that at 2 micrometers we see large-amplitude relatively regular velocity variations corresponding to motions at rather deep layers in the atmosphere.

Thus, at first sight, the spectroscopic data for long-period variables seem to be gibberish. Clearly, the final emergent spectrum must be contributed by a host of layers at widely different temperatures, densities, and conditions of excitation, everything being time dependent.

The puzzle was solved by George Bowen and Lee Anne Willson at Iowa State University in Ames. The first important fact to realise is that pulsations, and pulsations alone, act to vastly extend the photosphere of a star. Fig. 9.8 compares the overall pattern of the atmosphere of a normal supergiant with that of a long-period variable. This effect is due to mechanical consequences of the pulsation, not to temperature effects. The 'scale height' of the atmosphere (measured by the

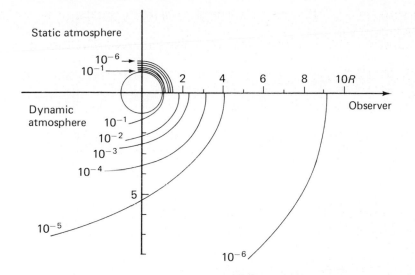

Fig. 9.8. Distention of the atmosphere of a pulsating Mira star. The contour lines denote a constant column density as summed up along lines of sight at various distances from the center of the stellar disk. All column densities are expressed as fractions of the column density at the static photosphere on a line of sight through the center of the star. All distances are expressed as multiples of the static, photospheric radius (R). The upper part of the diagram shows the column density contours of the static atmosphere; the lower part shows corresponding column density contours for a typical pulsating model – a Mira pulsating in the fundamental mode with a mass of 1–2 solar masses, an effective temperature of 2955 K, a luminosity 5000 times that of the sun, and a period of 350 days. Strong spectral lines are formed between a thousandth and a millionth of the photospheric column density; hence the sizes of these stars are difficult to determine. (Courtesy George Bowen and Lee Anne Willson, 1984.)

distance over which the density falls by a factor 2.7) is a significant fraction of the stellar radius in Mira stars. At large distances from the stellar photosphere, where the radiation is attenuated, relatively high densities still occur. Under these circumstances, solid grains can form possibly more easily than near a non-pulsating giant star.

Pulsational waves are not smooth events and where they have large amplitudes shock waves can occur. Here the velocity of mass motion of a parcel of matter exceeds the velocity of sound in the gas. Shock waves can occur, for example, when a traveling wave propagates into a medium where the density is falling off rapidly. Under these circumstances, even waves of relatively small amplitude can develop into shocks.

In a shock wave two masses of gas collide violently with one another and the kinetic energy of this relative motion is converted into heating the material. The consequent sudden temperature rise can cause the dissociation of molecules, the ionization and excitation of atoms, and the emission of radiation. The gas may recombine, radiate energy, and thus cool. Thus shock waves can explain, for

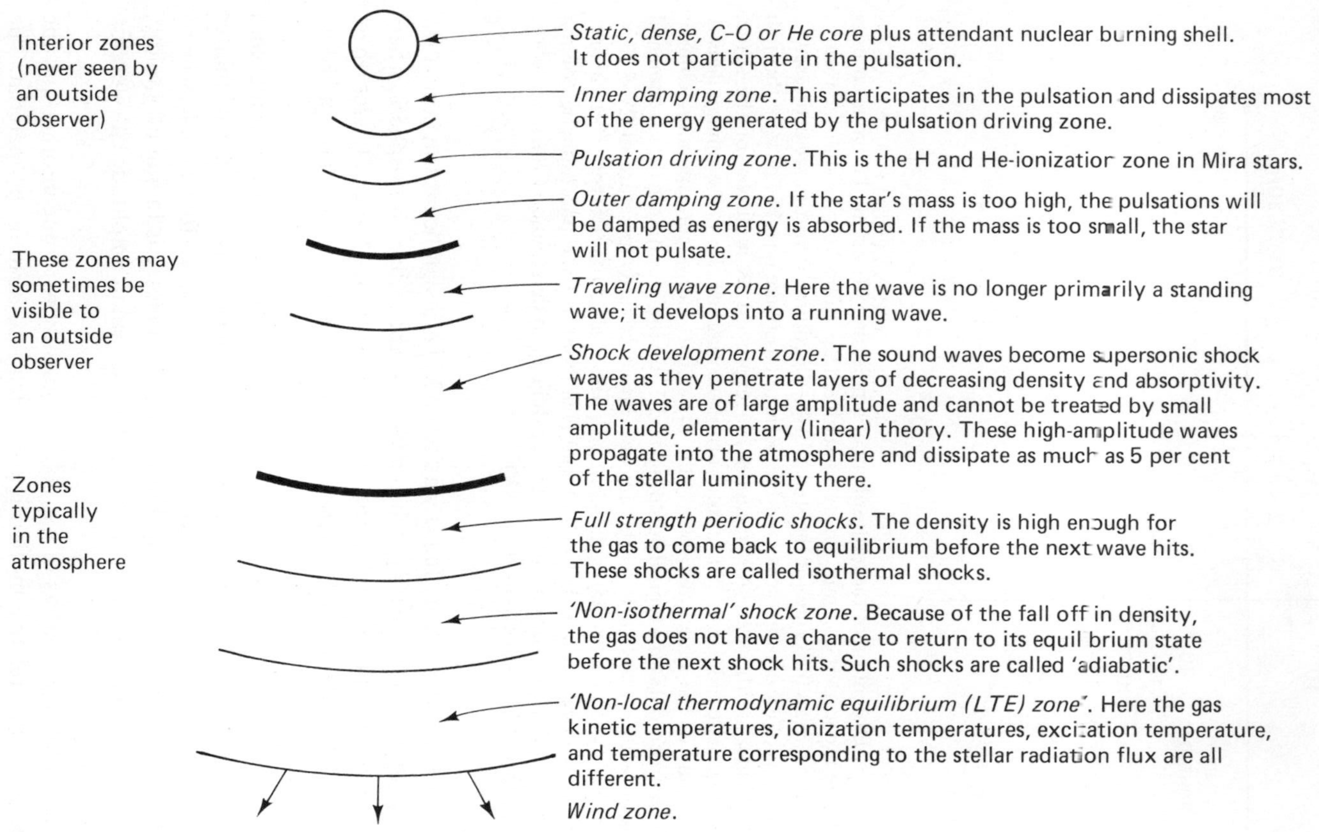

Interior zones (never seen by an outside observer)

Static, dense, C–O or He core plus attendant nuclear burning shell. It does not participate in the pulsation.

Inner damping zone. This participates in the pulsation and dissipates most of the energy generated by the pulsation driving zone.

Pulsation driving zone. This is the H and He-ionization zone in Mira stars.

These zones may sometimes be visible to an outside observer

Outer damping zone. If the star's mass is too high, the pulsations will be damped as energy is absorbed. If the mass is too small, the star will not pulsate.

Traveling wave zone. Here the wave is no longer primarily a standing wave; it develops into a running wave.

Shock development zone. The sound waves become supersonic shock waves as they penetrate layers of decreasing density and absorptivity. The waves are of large amplitude and cannot be treated by small amplitude, elementary (linear) theory. These high-amplitude waves propagate into the atmosphere and dissipate as much as 5 per cent of the stellar luminosity there.

Zones typically in the atmosphere

Full strength periodic shocks. The density is high enough for the gas to come back to equilibrium before the next wave hits. These shocks are called isothermal shocks.

'Non-isothermal' shock zone. Because of the fall off in density, the gas does not have a chance to return to its equilibrium state before the next shock hits. Such shocks are called 'adiabatic'.

'Non-local thermodynamic equilibrium (LTE) zone'. Here the gas kinetic temperatures, ionization temperatures, excitation temperature, and temperature corresponding to the stellar radiation flux are all different.

Wind zone.

Fig. 9.9. Model of a Mira star. (After George Bowen and Lee Anne Willson.)

example, the temporary bright hydrogen lines that are produced below the strata responsible for titanium oxide absorption in Mira stars. They can also account for the strange behavior of the velocities as measured by Joy, and for the asymmetrical light curves with their steep rise and slow decline.

At higher densities where recombination of ions and electrons is rapid, the gas may cool off and come to some kind of equilibrium between the successive shocks caused by the pulsation of the star. On the other hand, at lower densities the cooling time required for the gas to reach some kind of a steady state may be longer than the time interval between successive shocks.

Fig. 9.9 depicts schematically the Bowen and Willson model of a Mira variable. Their model invokes ten distinct zones corresponding to the dominance of different physical processes. The zones amenable to direct observation include those in which shocks become fully developed, zones where the gas readjusts between successive bashings, a zone where the gas does not have a chance to cool off before the next shock comes along, a zone in which the ordinary formulations of ionization and excitation cannot be handled quantitatively by familiar equations, and finally the outward-blowing wind zone.

Meanwhile, material between the shocks is expanding and doing work through-out the cycle as it also absorbs radiant energy flowing outward from the star. The rate of energy dissipation by shock waves varies greatly and can amount to as much as a hundred solar luminosities in a strongly driven Mira wind. This energy dissipation can strongly damp the pulsations.

Are shock waves important in other types of pulsating stars? They certainly are, but in RR Lyrae stars, the shock energy may be radiated away below the visible layers and modify the internal stellar structure with no easily noticeable effect on layers amenable to direct observation. In classical Cepheids, shock waves do not seem to play as important a role as in Miras. We discuss shock waves further in Chapter 10.

Pulsations seem to play a central role in mass-loss processes in Miras. As the star ascends the asymptotic giant branch (AGB), pulsation may become more pro-nounced, the luminosity increases, the surface temperature falls, and the rate of mass loss increases. In the OH–IR stars and mass-losing C stars, which represent a more advanced evolutionary phase than do the Miras, continual mass loss with attendant particle formation may cause the circumstellar envelope to hide the star. For most such stars, mass loss is heaviest during epochs of large-amplitude pulsation.

The mechanism of mass loss in some Miras and all OH–IR sources seems to be an interplay of pulsations that lift the cooling gas out to a point where grains can condense, and radiation pressure that then acts on the dust. Since gas and dust tend to be dragged along together, the whole mixture is expelled. If the mass loss is greater than 10^{-6} solar masses per year, shocks and radiation pressure are both required; shocks alone might suffice at lower mass-loss rates.

Pulsations enhance mass-loss rates by orders of magnitude over what would otherwise be the case and suffice to end AGB evolution in Miras and OH–IR stars in about a million years, terminating a stellar lifetime that may have lasted as long as ten thousand million years!

Although details remain to be worked out, for example, the influence of grain formation on pulsation patterns and the final uncoupling of the expanding envelope from the core, the Mira scenario seems well defined. The mysteries of the intricate spectrum and the shapes of the light curve seem resolved; as George Bowen expressed it: 'Simplicity emerges from awesome chaos.'

Faint, short-period variables (dwarf Cepheids)

Stellar pulsations may occur in stars that are not much brighter than the sun, as investigations by T. Walraven, Harlan J. Smith, O. Struve, and others have shown. SX Phoenicis is a pulsating subdwarf star (spectral class A5) for which Smith found an absolute magnitude of $+3.9$ from its trigonometric parallax. CY Aquarii, with a period of 90 minutes, is a similar star of absolute magnitude 2.5, as perhaps are several other stars including δ Scuti, VZ Cancri, DQ Cephei, and AI Velorum. Smith calls such objects 'dwarf Cepheids'. Their periods are less than 0.2 day, they are about 2 magnitudes fainter than RR Lyrae stars, and they show a definite period–luminosity relation. The existence of such a relation among classical Cepheids, dwarf Cepheids, and β Canis Majoris stars indicates that, within each group, the period (P) multiplied by the square root of the density (ρ) is a constant (K), $P\sqrt{\rho} = K$, but that K changes from one group to another.

Some complexities of the pulsation process

We have previously alluded to the subject of overtone vibrations. A vibrating string or organ pipe may oscillate not only at its lowest or fundamental frequency (of largest wavelength) but also at overtones or multiples of the basic frequency. In more complex systems, such as a drum head, bell or star, the overtone frequencies are not simple multiples (2, 3, etc.) of the fundamental frequencies but are related to it by well-defined but not integral ratios. This property gives a drum a 'hollow', booming sound. If we could listen to the song of a pulsating star, it might resemble that of a drum more closely than an organ pipe. Classical Cepheids and Mira variables pulsate in the fundamental mode.

RR Lyrae has two periods, a fundamental around 0.5 day and a harmonic or overtone of about 0.25 day; these two periods interfere to produce a harmonic with a period of about 41 days. Although some dwarf Cepheids have light curves that indicate a single fundamental period of pulsations, others have two or more periods with resulting beats. SX Phoenicis, which has the shortest primary period known, 79 minutes, has a beat period of about 280 minutes. AI Velorum is an even more remarkable star. In the course of a single day, its light curve runs through nine cycles of different shape, which T. Walraven found to be the consequence of six different superposed sinusoidal oscillations.

Up to this point we have discussed radial pulsations but deformation oscillations, in which the mean radius does not change but the shape of the star varies, may also occur. P. Ledoux suggested such oscillations may occur in a class of early-type (B1–B3) called β Cephei or β Canis Majoris stars. D. McNamara found these stars to fall above the main sequence and to exhibit a period–spectrum relationship. Although their light variations are small, the line profiles often showed marked changes. β Canis Majoris was intensively studied by W.F. Meyer and by O. Struve. Its velocity curve consists of two interfering harmonics, one with a period of almost exactly 6 hours and the other with a period of 6 hours 2 minutes. The line shape changes are correlated with the second period.

In discussing vibrations of a string, an organ pipe, or the purely radial pulsations of a star, we need only one set of numbers to identify the fundamental and the various overtones. But a star is really a three-dimensional object, the description of whose vibrational pattern needs three distinct sets of overtone numbers, since it can vibrate in longitude and latitude as well as radially. Thus we employ an overtone number, n, which is concerned with radial pulsations, while two additional numbers, l and m, are used to denote non-radial or deformation oscillations.

Solar seismology

In the mid-1950s, observers at the McMath–Hulbert Observatory of the University of Michigan noticed that if a slit was placed across a solar image under conditions of good seeing, spectral lines tended to show a wiggly appearance, indicating that material in closely neighboring areas would be rising or falling in some time-dependent pattern. The whole solar surface seemed to be trembling up and down. From moving pictures of the phenomenon, R. Leighton noticed that the wiggles seemed to be dominated by an oscillation period of about 5 minutes. Convection currents thumping against the base of visible layers of the solar photosphere were invoked to explain this phenomenon. R.K. Ulrich (1970), and J.W. Leibacher and R.F. Stein (1971), concluded that these 5-minute oscillations actually revealed large-scale, high-overtone, non-radial pulsations of the sun. The waves running through the superficial layers of the sun were basically acoustical (i.e. sound) waves. Ulrich had the further insight that the depth of penetration depended on the overtone degree number, l, the larger the value of l the shallower the depth of penetration. Each wave of overtone number l corresponded to a length, L, of the wave on the surface (and an associated horizontal wavenumber $k^*(=1/L)$). High values of l thus correspond to short wavelengths, L, and high values of k^*.

Clearly, if from the data we could unscramble the different inherent frequencies corresponding to different ls or k^*s, we would have a tool for exploring the interior of the sun, i.e. doing solar seismology. The observed frequencies will depend on the internal structure of the sun, but their interpretation requires a knowledge of the physics involved.

Two types of waves seem to be involved: (a) acoustical waves trapped below the photosphere, and (b) gravity or bouyancy waves (analogous to water waves) which occur in regions below the convection zones. The influence of both types of oscillations can be measured in the surface layers. Actually, about ten million acoustic wave modes with amplitudes of 10–50 centimeters per second are involved simultaneously. Their combined effects, acting through constructive and destructive interference, produce the observed patterns.

A high-overtone, large-l sound wave will be reflected in the lower photosphere where the density decreases rapidly outwards. When the wave starts to move downward, the deeper part of the crest moves more rapidly since sound speed increases as temperature rises with depth. Thus in Fig. 9.10a the ray ABC is refracted until it is eventually turned backwards to the photosphere at point C. The greater the l-value, the less the depth of penetration.

Fig. 9.11 shows a plot of the frequency against the overtone number, l. The brightness at each point reveals the power in the wave at that particular combination of frequency and l. Note that the observed frequencies fall along well-defined ridges corresponding to different values of the radial order number n! The positions of the ridges can be calculated by theory and compared with observations. The existence of such characteristic ridges was first shown by observations by F.L. Deubner. More detailed information was secured by Deubner, Ulrich, E. Rhodes, G. Simon, and others. Further theoretical studies by D. Gough, J. Toomre, J.W. Leibacher, Ulrich, and J. Christensen-Dalsgaard have greatly extended our understanding of the complex problem of solar oscillations. Observations extending over long time intervals are needed to assess the long-period oscillations associated with low overtones and deep penetration in the sun.

With a precise theory of the structure of the sun we can calculate the frequency–overtone relationships. By comparisons with observations we can improve solar models, much as seismic wave data are used to construct and improve models of the interior of the earth. What are some of the possibilities?

(a)   We can obtain a better estimate of the original solar helium content.
(b)   In particular, from the gravity waves, we may deduce the solar structure near the center where energy is generated, and perhaps gain insights on the enduring neutrino problem.
(c)   We can obtain the internal rotation pattern of the sun, the depth of the convection zone, test stellar convection theory, and ascertain the possible presence of a strong magnetic field.
(d)   We can get a better idea of how mass is distributed within the sun and perhaps provide a test on the general theory of relativity.

The great power of solar seismology depends on the possibility of observing velocity patterns over the entire, well-resolved surface of the sun. Other stars supply less information, as we can observe only the radiation from the unresolved disk, but even so enough data should be obtainable to evaluate the structures of individual field stars and calculate their ages.

The white dwarfs, a sad tale of the death of stars

The longest established and most securely anchored fact in stellar evolution studies is that the last visible stage in the life of most stars is a white dwarf. The first of these objects to be discovered was the companion of Sirius. Its existence had been

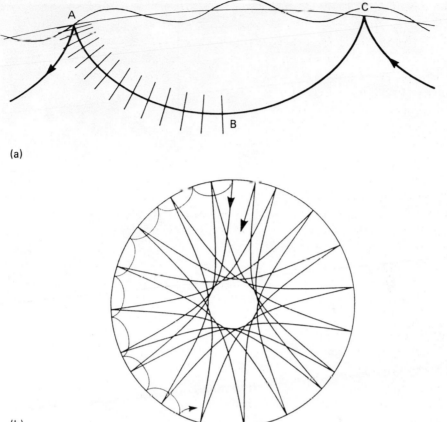

(a)

(b)

Fig. 9.10. Movement of acoustical (sound) waves in the sun.
(a) The segment, CBA denotes the path of a relatively shallow acoustical wave; the ticks perpendicular to the ray indicate the wave front and represent projections onto the page of surfaces of equal phase. The separation between the ticks corresponds to some constant phase difference which we are free to choose, e.g. $\pi/8, \pi/4, \pi/2, \pi$, or 2π. Since the sound speed is less near the stellar surface where the temperature is less, the wave crests fall closer and closer together as the surface of the star is approached. The wave becomes refracted because the portion at the bottom of the tick travels faster than the top portion since temperature increases with depth and sound speed increases as the temperature rises. At point B the wave is running horizontally. Thereafter it is refracted upwards.

The projection of the tick lines of constant phase interval intersect the stellar surface at regular distances. If the phase separation of the individual ticks in the diagram is $\pi/8$, the resultant surface wave would be as shown. If there is a node at A, there is neither a node nor a crest at C. The ray path is not closed; in other words, surface waves are not directly analogous to waves in a string. For example, the surface wave pattern must join up with itself after going all around a great circle of the solar or stellar sphere, even though the ray path does not.
(b) Here are illustrated waves of both low and high l-values. Waves of large n-value and low l-value (e.g. $n = 23, l = 6$) oriented to descend at a steep angle may penetrate to a great depth, thus probing the deep solar or stellar interior. Waves of large l, such as depicted above in (a) penetrate only to relatively shallow layers. (Courtesy, Douglas Gough.)

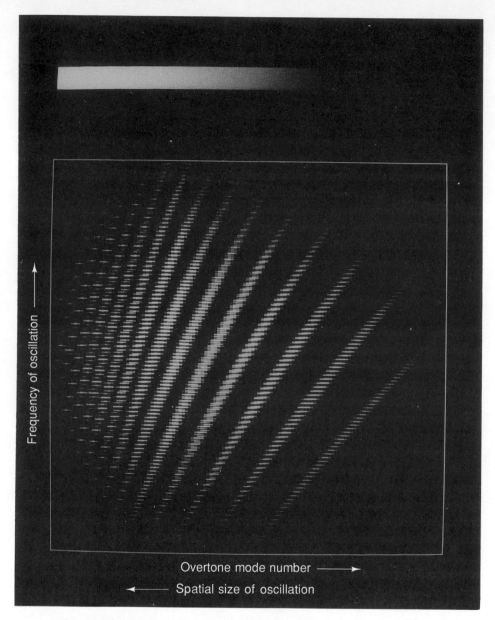

Fig. 9.11. Solar overtone oscillations. The figure shows the relationship between the frequency, v, of the oscillations and the overtone mode number, l, which is related to the surface (spatial) wavelength (L) of the oscillation by $L = \text{radius (sun)}/\sqrt{l(l+1)}$. The frequency, from 2.57 to 3.74 MHz, is plotted vertically, while the overtone mode, l, is plotted horizontally. l varies from 5 to 132, that is, from dimensions, $L = 127\,000$ to $53\,000$ kilometers. The data were obtained at the 60-foot tower telescope of the Mount Wilson Observatory by S. Tomczyk, in collaboration with A. Cacciani, E.J. Rhodes, and R.K. Ulrich.

predicted by F.W. Bessel from its gravitational effects, which produced a smooth sinusoidal variation in the angular (proper) motion of Sirius across the sky. It was actually discovered in 1862 as an extremely dim object about 400 times fainter than the sun. Since it had about the same mass as the sun, one would expect it to be a cool red star. Measurements of its color indicated it to be really a blue star with a temperature of about 8200 K. Hence, since each unit area radiates four times as much energy as the sun, the ratio of areas must be about 1600. Therefore the diameter is about 1/40th that of the sun and the volume $40^{-3} = 1.56 \times 10^{-5}$ that of the sun. The mean density must be about 64 000 times that of the sun, i.e. about 90 000 times that of water!

Under such conditions, matter has very peculiar properties. Electrons are all or nearly all stripped from the atoms and so the mass can be compressed to a very high density. The electrons do not obey the ordinary gas laws at all, but are said to constitute a 'degenerate' gas. Such material has very high electrical and thermal conductivity. Any excess energy could quickly flow from one region to another by simple heat conduction. The temperature no longer appears in the gas law; the pressure depends solely on the density. Once a star evolves into this state it cannot get out of it. The star now shines very feebly, radiating away its last resources of heat energy, the nuclear sources being long since exhausted. The process may be slow, requiring many millions of years, as the object gradually fades as a burned-out cinder.

Although they are called 'white dwarfs' after the prototype, Sirius B, these virtually defunct stars actually show a considerable spread in color from blue to red. Their spectra, which have been studied in greatest detail by J.L. Greenstein, show some remarkable properties (Fig. 9.12). Spectral lines are broadened because of the high densities. Most white dwarfs show H lines; of the remainder the majority show He lines or what appears to be a continuous spectrum. The spectra are what we would anticipate for the residual cores of defunct giant stars. Probably most are predominantly carbon or a mixture of carbon and oxygen. Carbon-rich stars, which show pressure-broadened bands of molecular carbon, are also known.

A quantum of light escaping from the surface of a dense white dwarf has its frequency lowered; that is, the corresponding spectral line is shifted redward. Measured 'gravity shifts' in white dwarf stars of known velocity in binary systems give direct information on their masses. Other data are supplied by evolutionary models.

Although Sirius B has a mass of 1.02 solar masses (m(sun)), the average mass of a white dwarf is in the neighborhood of 0.57 m(sun). A main-sequence star up to 5 solar masses may evolve to a 1.0 m(sun) white dwarf, while stars up to ~ 8 m(sun) may evolve to the strict limiting mass of 1.4 m(sun). S. Chandrasekhar showed that the radius of a white dwarf depends on its mass. The more massive it is, the stronger does gravity pull the material together, the smaller the radius, and the greater the value of the gravity, g. As the mass approached 1.4 m(sun), the radius would shrink to zero. Actually, at this point the electrons would be forced back into the nucleus and a neutron star would result.

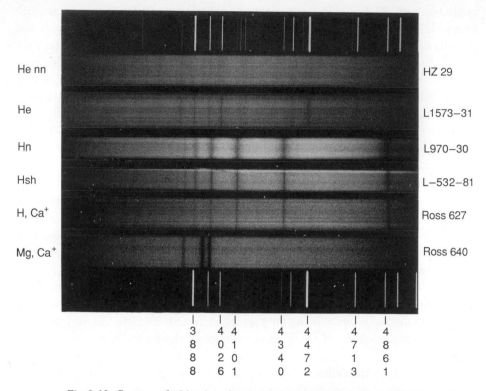

Fig. 9.12. Spectra of white dwarf stars. Helium-rich stars such as HZ 29 have fuzzy-lined spectra; stars with sharp-lined spectra, such as L1573–31, L970–30, and L532–81, show prominent hydrogen lines. Ross 627 shows weak lines of Ca II, while Ross 640 is characterized by strong lines of Mg and Ca II, with no hydrogen. The numbers along the bottom of the figure are ångström units. (Courtesy J.L. Greenstein.)

　　Much progress has been made in recent years, both theoretically and observationally. In 1970, J.C. Kemp, J.B. Swedlund, J.D. Landstreet, and R. Angel found that light from the faint white dwarf, GRW +70° 8247 was circularly polarized, suggesting a magnetic field of a hundred million gauss. Corroborative spectroscopic evidence for this star was obtained and interpreted by Angel, R. O'Connell, Greenstein, and G. Wunner to show that the surface magnetic field was between two hundred and six hundred million gauss, i.e. about a thousand million times as strong as the earth's field! Other white dwarf stars with fields in the neighborhood of ten million to a thousand million gauss have also been found. How are such fields possible? A clue may be found in the fact that in an ionized gas, magnetic lines of force are locked into the material, so if a parcel is compressed the magnetic field is correspondingly enhanced. Possibly these stars evolved from strongly magnetized A-type stars (commonly called Ap stars) which have magnetic fields of thousands of gauss.

　　Another curious type of white dwarf star are the pulsating stars of the ZZ Ceti type, which have periods between 200 and 1200 seconds, surface temperatures of about 12 000 K, and fall on an extension of the Cepheid instability strip to low luminosity. Any oscillations would appear to be of the deformation type rather than the radial type, and temperature fluctuations may occur. There are also pulsating white dwarfs with surface temperatures of about 20 000 K and helium-rich atmospheres, and a group of variables with temperatures near 100 000 K, presumably of the type found as nuclei of planetary nebulae, i.e. objects that have only recently evolved from red giants and are now starting the long descent to stellar oblivion.

In spite of progress in both white dwarf observation and theory, many problems remain. The chemical compositions of their atmospheres, although mostly understandable qualitatively, present puzzling problems. Many years ago E. Schatzman pointed out that in the intense gravitational field of a white dwarf, hydrogen would float on top – like an oil slick on water. Heavier elements would sink down. Puzzling exceptions to this rule exist. There are stars where the carbon-to-silicon ratio is less than the expected interstellar medium value although gravitational separations should remove silicon faster than carbon. In some stars, G. Michaud suggested that a tiny amount of hydrogen would diffuse into layers so hot and dense that it would be instantly consumed in the CNO cycle, thereby supplying a small amount of energy to the star and explaining the depletion of white dwarf atmospheric hydrogen with age.

Rates of cooling of white dwarfs should be low. The only heat energy available is that provided by nuclei of He, C, N, O, etc., since the energy of the electrons cannot be radiated away. The speeds of the electrons in the degenerate gas are fixed by the density which in turn is determined by the mass and radius of the white dwarf. The higher the density the greater the average velocity. Energy cannot be extracted by reducing the speeds of the electrons as then the radius of the star would have to be increased, which would require the input of energy. The heavy nuclei gradually become entrapped in a crystalline structure and cannot release any more energy, so the star fades, now on a time scale of five to fifteen thousand million years.

There are no cool white dwarfs (temperature ~ 5000 K) of extremely great age and it seems that the oldest objects appeared shortly after the formation of the galactic disk. A defunct star takes a long time to turn cold!

Thus a single star like the sun passes into oblivion quietly and without fanfare. The material locked within the core is removed from the cycle of stellar death and rebirth forever. White dwarfs in binaries, however, sometimes experience a more interesting fate, as we shall explore in Chapter 11. We now turn our attention to the interstellar medium, into which merge the envelopes of dying stars and from which new stars are eventually formed.

10

The interstellar medium and gaseous nebulae

A glance at any ordinary photograph of the Milky Way shows that the space between the stars is not empty. The bright patches of nebulosity, the giant rift stretching from Cygnus to Sagittarius as well as the inkiness of the Coal Sack, all testify to the reality of giant clouds of occulting matter that fill up 'empty' space. Readers may judge for themselves by inspecting the famous nebula around η Carinae, shown in Fig. 10.1. One sees that dark lanes are not holes through which we look into empty space, but rather clouds of some material, probably fine dust and gas, that dims and obscures the light of distant stars.

One is struck by the extreme patchiness of the material. Not only do dust and gas tend to concentrate in spiral arms, but within the arms themselves, the material collects in features ranging from evanescent, gravitationally unstable globules a few times the size of the solar system to vast molecule-containing clouds, sometimes a hundred light-years across. The density within these clouds can range from ten to a thousand times that in intervening space. Within the dark clouds, far away from bright, hot stars, the gas can get as cold as 10 K. The vast bulk of the interstellar medium (ISM) is gaseous, mostly hydrogen.

When the dusty clouds chance to occur near one or more bright stars, however, they are illuminated and can appear as diffuse nebulae. We may visualize the stars as giant spotlights shining on a mixture of fluorescent material (atoms) and reflecting particles (dust). The continuous spectra of emission nebulae appear to be produced, not only by reflection from grains, but also by emission processes involving mostly hydrogen. When the spotlight is rich in ultraviolet radiation, the gas fluoresces and bright lines appear, but when only cool stars are present, there is no rich supply of ultraviolet radiation for the atoms to absorb and re-emit as visible bright lines. Starlight falling upon the nebula is then scattered by dust particles; the nebular spectrum is simply a reflection of the stellar one.

Nuisance or treasure trove?

Astronomers who wanted to explore the Galaxy regarded the dusty interstellar clouds as a pest, since they often hid exciting regions like the galactic center. Early-twentieth-century optical observations suggested a quiescent gaseous stratum

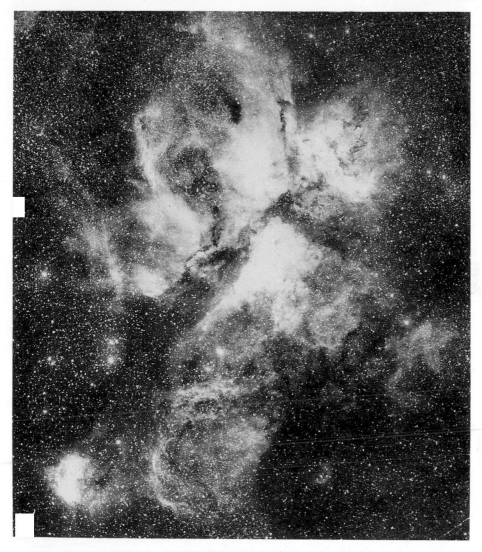

Fig. 10.1. The η Carina Nebula, photographed at Mount Stromlo Observatory.

stretching through the spiral arms of our Galaxy with a sprinkling of dust grains, that were sometimes concentrated into clouds.

The opening of new windows in the spectrum, those provided by radio waves, ultraviolet, and X-rays, and the advent of new insights has changed that picture dramatically and transformed the interstellar medium from the *bête noire* of astronomy to a jewel box, laden with clues to the formation of stars and the evolution of galaxies. Although the behavior of some parts of the ISM seems to suggest steady-state conditions, most of it appears as a dynamic milieu beset by storms engendered by the winds and often violent explosions of massive stars. Much of it is an incredibly hot, attenuated gas feebly emitting X-rays. Some of it contains dark shielded pockets of cold smog where remarkably complex molecules formed.

Table 10.1 *Mass budget of the interstellar medium*

Contributions to enrichment of the ISM (solar masses/year)	
Winds from hot O and B stars	$\sim$0.08–0.5
Material ejected by Wolf–Rayet stars	$\sim$0.06
Ejecta of novae and cataclysmic variables	$\sim$0.005
Planetary nebulae and AGB stars	$\sim$0.5
Supernovae	$\sim$0.03–0.05
Infall from intergalactic space, etc.	$<$1.4
Material removed by star formation[a]	3–10

Note:
[a] The estimate of the rate of material used in star formation is due to M. Jura.

Here the primordial chaos that existed before the creation of the solar system and the earth still prevails.

That stars must continuously be in the process of formation is indicated by the existence of objects such as Canopus, Rigel, ζ Puppis, and γ Velorum. These luminous stars pour out energy at such a rate that they must have lifetimes measured in mere millions of years; some of them are younger than the race of man! We have a clue to the formation of such luminaries; distinct groups of very luminous stars, known as stellar associations, were first recognized by V. Ambarzumian. If the space velocities of these stars are carefully measured, it is found that all of them within each group appear to have emerged but a few million years ago from a small volume of space (still observed to be gas and dust laden).

We can think of the interstellar medium as a reservoir of material from which new stars form, and which in turn is fed from the debris of dying stars, stellar winds, and perhaps even from infall of material from remote regions of the Galaxy or even intergalactic space (Table 10.1). Rates of such processes are hard to evaluate.

The material returned to the interstellar medium is enriched in carbon, nitrogen, oxygen, and heavier elements so that the later a star is formed, the greater will be its metal-to-hydrogen ratio. In particular, although the mass returned by supernovae is small, the contribution consists largely of highly processed material.

Fig. 10.2 depicts the cycling. A large part, presumably most of the material of dying stars, is returned to the interstellar medium, but a certain fraction is forever lost from this cycle of death and reincarnation by entombment in white dwarfs, neutron stars, and black holes. Unless there is infall of fresh gas and/or dust from remote galactic or intergalactic regions, the interstellar medium will simply run down, in a few thousand million years. Since the interstellar medium is still with us, some of it must have been protected from star formation and released slowly or relatively recently brought in from great distances.

In spite of the high activity level of the interstellar medium, much of it now seems to be in pressure equilibrium which would require high densities in cold regions and low densities in hot regions. Why? An even greater mystery is the apparent equality

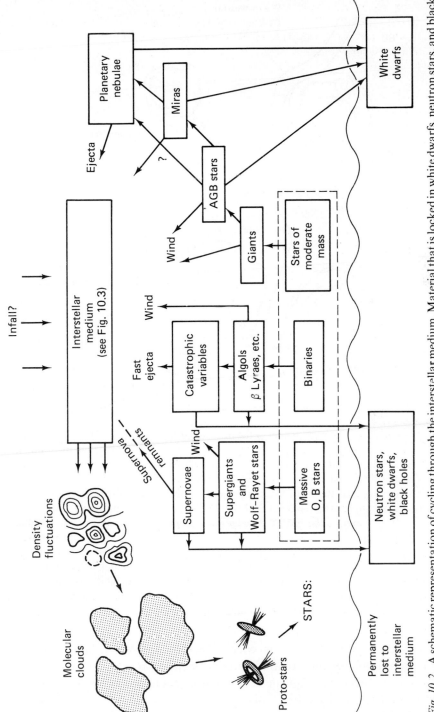

Fig. 10.2. A schematic representation of cycling through the interstellar medium. Material that is locked in white dwarfs, neutron stars, and black holes is lost forever to the depicted cycling process, thus constituting a drain on the mass of the interstellar medium. Much dust and gas is ejected to the interstellar medium by stars in the asymptotic giant branch (AGB), by winds from stars, by stellar pulsations, and by the formation of planetary nebulae.

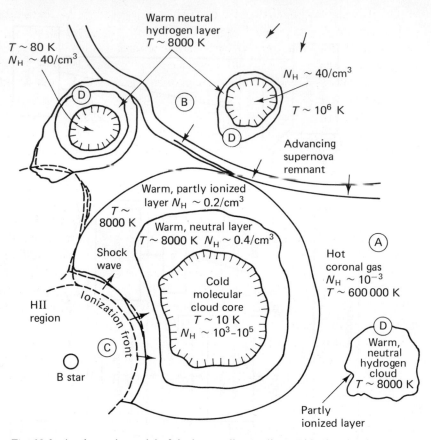

Fig. 10.3. A schematic model of the interstellar medium. This sketch is intended to display some of the complexity of the structure of the interstellar medium. It is essentially the McKee–Ostriker model in which the space between the clouds is occupied by a hot, very rarefied gas (A). At the upper right, a supernova remnant (B) is moving into the local medium, parts of which have already been heated by previous supernova blasts. At the lower left (C), radiation from a recently formed O or B star ionizes the gas at an edge of a cloud, the ionization front gradually advancing until recombinations of ions and electrons balance the number of ionizations. This is the situation existing in the Orion Nebula. It is not certain that hot coronal gas is as extensive as in the McKee–Ostriker model; in fact much of the space between discrete cloudlets may be occupied by relatively cool neutral hydrogen which is revealed by the 21-centimeter radiation. (D) denotes a 'buffer zone' between cool, dense regions and hot gas.

of different kinds of energy. Thermal energy, i.e. the energy of gas kinetic motion is about 1 electron volt per cubic centimeter, as is the energy density of starlight, that of cosmic rays, and even the magnetic field if it amounts to about 2 microgauss!

Let us start with a broad-brush picture of the interstellar medium (*c.* 1988), as attempted in Fig. 10.3. Basically there exist clouds of varying mass, density and size, and an intercloud medium. In the McKee and Ostriker model (1977) a gas of extremely low density ~ 0.001–0.0001 atoms per cubic centimeter occupies 20–60 percent of available space in the galactic disk. This gas is fully ionized and heated to a temperature of the order of 600 000 to 3 000 000 K by successive supernova explosions. Throughout our Galaxy and neighboring ones such as M33 we recognize supernova remnants (SNRs) which show the interaction of exploding supernovae with gas clouds of the local interstellar medium. Behind the shock front, temperatures can rise to as high as 10 000 000 K. This highly rarified gas cools somewhat but never drops below $\sim 600\,000$ K. Such regions are revealed by their X-ray emissions. There is some question as to whether this very hot, low-density gas fills as large a fraction of the Galaxy as this model suggests; in fact much of the space may be occupied by neutral hydrogen (H I) gas.

In the plane of the Milky Way itself, the interstellar medium has a clumpy structure (see Fig. 10.3). There are core clouds of neutral hydrogen of density N_H about 40 atoms per cubic centimeter and temperature (T) roughly 80 K. Sometimes these cool clouds form great aggregates of molecular clouds with densities typically of the order of 500 atoms/cm^3, although occasionally they may equal or exceed 10 000 atoms/cm^3. In these regions complex molecules appear and stars are made. They occupy about 5 percent of space. There also exist relatively thin, warm, neutral hydrogen clouds with densities of about 0.4 atoms/cm^3 at a temperature near 8000 K, filling about 25 percent of space. In addition, there is a warm, ionized component with a density ~ 0.2 atoms/cm^3 at $T \sim 10\,000$ K. Of course, in regions where bright, hot stars have recently been 'turned on' as in the Orion Nebula (Figs. 2.10 and 10.21, pp. 24 and 246–7) or Trifid nebulae, the density may be $\sim 10\,000$ atoms/cm^3. Note that we can have structures with cold clouds in the center, a layer of neutral hydrogen, and a warm H II volume in contact with a very-low-density region that has been heated by a supernova explosion. The interstellar medium model depicted in Fig. 10.3 seems to be terribly complicated. How were astronomers ever led to such a seemingly messy, untidy model of an ubiquitous mixture of dust and gas?

Tools to unlock the mysteries of the interstellar medium

The opening of new windows across the electromagnetic spectrum, and steady improvements in radiation detectors, have changed our perceptions profoundly. Observations in the optical region revealed many important facts; the extinction of starlight by dust grains, the existence of large clouds of hot, glowing gas, such as the Orion, η Carinae, and 30 Doradus nebulae, and the presence of a low-density gas that produced absorption lines of Ca II, Na I, and a few molecular fragments. Radio astronomy allowed us to look through clouds of smog and gas and detect the neutral hydrogen (H I) 21-centimeter emission, the glow of ionized gas, and eventually the tell-tale lines of molecules in the millimeter range. We were able to construct maps of large areas of the sky, such as the Orion complex, in H I or H II, free from dust obscuration. Radio data and measurement of the polarization of light from distant stars also gave clues to the presence of interstellar magnetic fields.

Advances in infrared detectors permitted the astronomer to measure atomic and molecular radiation from the heart of obscured cool clouds, and even to study cloud chemistry. Ultraviolet data extending down to ~ 1000 Å revealed the nature of the extinction law for interstellar smog in our own Galaxy and the Magellanic Clouds. In hot, diffuse nebulae, emission lines of carbon ions, C III and C IV, reveal an element whose abundance in interstellar space is not otherwise easily obtainable. High-dispersion spectra of hot stars, obtained particularly with the Copernicus satellite and the International Ultraviolet Explorer, disclosed hosts of metals and other elements in the interstellar medium and gave the motions and properties of individual clouds. Molecular hydrogen is an important constituent of the interstellar gas. We find that many common elements arc depleted on to grains. A startling discovery was that of the O VI 1023, 1038 Å lines indicating interstellar medium regions at temperatures of $\sim 600\,000$ K. X-ray observations revealed even more extreme conditions: gas heated by shock waves to a temperature of 10 000 000 K. Finally, gamma rays showed that there exist 'supra-thermal' cosmic rays of energies 10^7 to 10^{20} electron volts (see Chapter 12). These can produce heating even in centers of dense clouds.

Keys to understanding the radiation from gaseous nebulae

Hot gaseous nebulae probably constitute the simplest component of the interstellar medium, but diffuse nebulae such as Orion are complicated, irregular structures wherein gas and dust are intimately mixed. To understand how the astronomer analyzes spectra of gaseous nebulae, it is better to start with the seemingly simpler planetary nebulae which are usually roughly symmetrical and excited by a single hot imbedded star. The name comes from the fact that they often appear as small disks, similar in telescopic appearance to Uranus or Neptune. But other than that, they have nothing in common with the planets of our solar system. We mentioned these objects in Chapter 9, where they were identified with the last stages in the life of a star, as its remnant core evolves into a white dwarf after the outer layers are ejected. The central star of a planetary nebula thus differs fundamentally from bright blue stars like the Orion Trapezium, which were but recently formed from the interstellar medium.

We now want to look at planetary nebulae from a different point of view. Indeed, they are important suppliers of material to the interstellar medium, but they also serve as celestial laboratories where we can study radiation processes under conditions unattainable in any terrestial environment. The insights are applicable to diffuse nebulae in galaxies, to supernova remnants, to envelopes of novae, and even to exotic sources like radio galaxies and quasars. Many are aesthetically beautiful objects, works of art, worthy of study in their own right.

An essential point is that the hot central stars or nuclei of planetary nebulae emit much ultraviolet radiation, which after absorption by the nebular gases produces the observed emission by processes we shall describe. Another important feature is the extremely low density of gaseous nebulae. Events important in stellar atmospheres may play no role here. Some idea of the extreme tenuity of a planetary nebula may be gained from the following illustration. Imagine a drinking glass filled with hydrogen at room temperature and pressure. Now add a few thimblefulls of helium, a half thimblefull of ordinary air, and a sprinkling of dust particles. The chemical composition of the contents of our drinking glass will now roughly resemble that of a planetary nebula. Now seal the glass and expand it until it is as tall as Mt Everest and about 3 kilometers in diameter. The gas inside our hypothetically enlarged vessel will still be about a hundred times as dense as the gas in a typical planetary nebula! Diffuse galactic nebulae are often ten to a hundred or more times more rarefied; it is only because these objects are so large that they can be seen at all!

Most of the radiation of planetary nebulae is concentrated in emission lines, principally those of hydrogen and ions of helium, carbon, nitrogen, oxygen, and neon. Since different ions may have quite different patterns of spatial distribution, an ordinary photograph which involves input from many ions is of limited usefulness. It is better to employ narrow band-pass filters and charge-coupled devices. In addition, radio-frequency imaging with 1.3 to 2 centimeter wavelengths with the Very Large Array (or VLA) in New Mexico gives a resolution of 0.1 vs ~ 1.0 arcsec, which is comparable to that obtained in the very best optical observations. Furthermore, these radio data are free from the influence of dust obscuration.

Many planetary nebulae appear to be symmetrical, but they show bilateral symmetry rather than spherical symmetry. Their forms tell us much about ejection processes in dying stars, and particularly about the role of fast winds. Recall the scenario from Chapter 9 (p. 175). Most of the envelope of a post-asymptotic-giant-branch star, Mira, OH–IR, or carbon star is ejected at a velocity of a few kilometers per second and wafted into space.

The core evolves quickly towards a white dwarf state. S. Kwok postulated that at this time the dying stellar core spews forth a rapid wind with a velocity of about 2000 km/s involving a mass-loss rate between 3×10^{-9} and 3×10^{-7} solar masses per year. This wind impinges upon the red giant envelope and compresses it. Shock waves are produced since the velocity of wind greatly exceeds the velocity of sound in the gas (see p. 243 and Fig. 10.20). These waves move into the outer shell and also towards the central star, heating the cavity to 100 000 K or more. Bruce Balick has shown how the structures of planetary nebulae can be interpreted in terms of patterns shaped by brisk winds blowing into sluggishly moving masses of gas (Fig. 10.4).

A quantitative understanding of the geometry and morphologies of planetary nebulae now becomes possible because images can be recorded with charge-coupled devices. These detectors have two crucial advantages: the response is strictly proportional to the intensity, and a huge brightness range can be accommodated. Extensive observations have been made by D.C. Jewitt, G.E. Danielson, P.N. Kupferman, G. Jacoby, Y.H. Chu, and their associates, and particularly by B. Balick and collaborators who systematically observed more than 50 planetaries in emission line radiations of Hα, and ions of oxygen, nitrogen, and helium, and sometimes neutral oxygen. They found that the planetaries fell into three broad structural classes, round, elliptical, and butterfly, although there are a few objects like NGC 7008 (Fig. 9.3) and NGC 6543 which are definitely 'peculiar'.

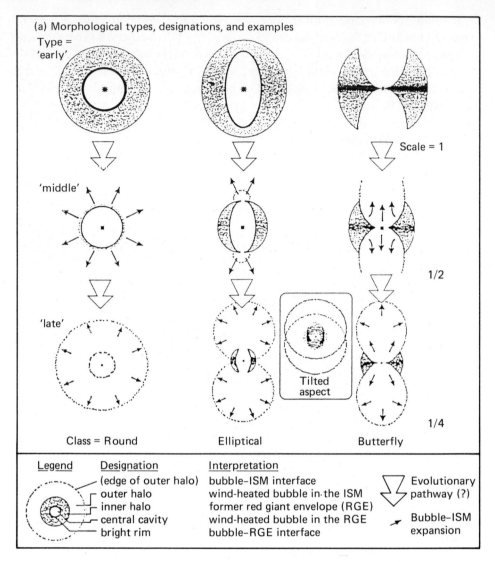

Fig. 10.4. (a) Balick's interpretation of the shapes and evolution of planetary nebulae. These schematic diagrams depict two-dimensional cross-sections through the centers of three types of planetary nebulae; uniform shells, shells with an enhanced density near the equatorial plane, and shells with an extreme density enhancement near the equator. These classes represent round, elliptical, and butterfly structures. The shaded areas denote remaining regions of the red giant envelope, the tone being related to the total emission along the line of sight (emission measure). Heavy lines indicate bright inner rings, while light broken lines indicate edges of haloes in advanced evolutionary stages. As the nebula expands, the hot gas bursts forth into the interstellar medium (ISM) and produces wind-heated bubbles. The central cavity is raised to a high temperature by the shocks produced by the fast wind. (Courtesy B. Balick, 1987, *Astronomical Journal* **94**, p. 671.)

Abell 30   NGC 3242

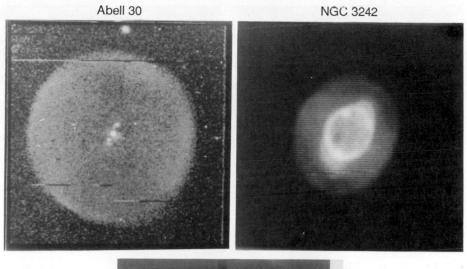

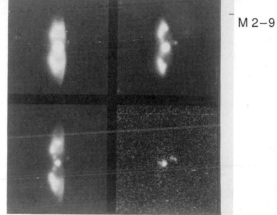

M 2−9

Fig. 10.4. (b) Example of round, elliptical and butterfly-type planetaries. The Hα images are shown for the round planetary nebula, Abell 30, and the elliptical planetary NGC 3242. Abell 30 is a very old object with low surface brightness; NGC 3242 is relatively young.

For the 'original' butterfly nebula, M2–9, the Hα image is at the top left, the [O III] image is at the top right and the [N II] image is at the bottom left. There is essentially no He II 4686 Å in this low excitation nebula, so we see only a star (bottom right). (Courtesy B. Balick, 1987, *Astronomical Journal* **94**, p. 671.)

The essential point is that the envelope of the red giant star is usually not ejected as a spherically symmetrical shell. Rather, the density and shell thickness may decrease with latitude, the density in the equatorial region being often five or more times that in the polar region. Then the snow-plow shock of the fast wind pushes much more quickly through the polar than through the equatorial region. In fact, for thick, dense equatorial regions, the gas may remain cold and neutral throughout most of the development of the nebula. Modest enhancements of material in the

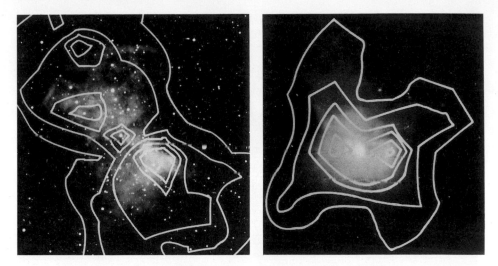

Fig. 10.5. The planetary nebulae, NGC 6853 and NGC 2346. Contours of molecular hydrogen emission are superposed on optical region images of two highly evolved planetary nebulae. The hydrogen emission delineates the neutral molecular envelope surrounding a zone of ionized gas that glows in visible light. Notice the vast extent of the neutral gas that surrounds NGC 6853 (left). NGC 2346 (right) is excited by a spectroscopic binary star which is still girdled by a dusty cloud, so dense as to produce occasional eclipses of the star. The molecular hydrogen maps are very important for assessing the ratios of neutral to ionized gas in planetaries. (Courtesy Benjamin Zuckerman, Bruce Ballick, and Ian Gatley, *Sky and Telescope*, Feb. 1987, p. 129.)

equatorial zone lead to planetaries of the elliptical type, while pronounced equatorial concentrations give objects of the butterfly class.

When the shell is punctured, high-density gas moving with the local sound speed rushes outward and cools, producing two lobes of expanding gas which appear as bubbles along the major axis. In the extreme case of a butterfly planetary nebula, a bipolar nebula can result.

The wind plays a dominant role in shaping planetary nebulae; does the dissipation of mechanical energy contribute significantly to the production of the observed nebular emission! Wind energies may be 3 to 30 times the power output of the sun, but the radiant energy from the central star may be a hundred or a thousand times larger, so it dominates the overall nebular energy output. Nevertheless, shock effects can be important, sometimes at the edge of the red giant envelope where observations show that a huge range of excitation can occur over a strip smaller than 0.01 parsec. In some nebulae, as the shock waves penetrate regions which are still neutral, they may dissipate energy in collisional excitation of molecular hydrogen (see Fig. 10.5).

The bipolar interacting wind theory appears to explain the shapes and evolution of the majority of planetaries, but we must not expect a tidy mathematical model to represent in detail the ejecta of real stars. Red giant envelopes are almost certainly not smooth, symmetrical, homogeneous structures. Convection patterns, larger in size than the sun, are likely to be present. Persistence of these structures into the ejected envelope may provide the basis for the irregular, mottled, knotty appearance of many planetary nebulae.

Distances of individual planetary nebulae are poorly known. One method is to use direct images to measure the rate of angular growth in arcsecs per year, and also the radial velocity of expansion. The difficulty of this method, which has been applied by W. Liller, is that part of the observed angular expansion actually may arise from an increase of the ionized zone of the nebula, and is not simply due to the outward motion of a fixed parcel of gas. C.R. Masson used the VLA radio telescope to find the annual angular expansion of the planetary nebula NGC 7027. By combining this datum with the radial velocity of expansion and allowing for the increase in the size of the ionized zone, he was able to derive a distance of 940 parsecs. Another method, due to R. Kudritzki and his associates, involves determining the spectroscopic parallax of the central star. Alas, both methods can be applied to only a few objects. Statistical methods are often used. The best of these depend on the association of planetary nebulae with objects at a known, but – alas – great distance, such as the galactic bulge, the Magellanic Clouds, or other galaxies. The maximum luminosity attainable by a planetary nebula and the average luminosities in certain samples such as the Magellanic Clouds appear to be moderately well known, but it is difficult to use such data to get a reliable distance for any specific object.

The mystery of nebulium

Although gaseous nebulae display the well-known Balmer lines of hydrogen and familiar lines of helium, their strongest spectral lines, in the green at 4959 and 5007 Å and ultraviolet at 3726 and 3729 Å, have never been observed in any laboratory. Believing themselves well acquainted with the spectra of all common elements, astronomers at first were inclined to ascribe the unexplained nebular lines to a mysterious element unknown on the earth, which was called nebulium. Advances in physics and chemistry left no room for such a hypothetical element, and the aforementioned strong lines turned out to be due to oxygen. Other unidentified lines were shown to come from nitrogen, neon, argon, sulfur, and other elements shining under physical conditions not readily attainable on earth. Let us see how these lines were identified.

As we found in Chapter 3, spectral lines arise when an electron jumps from one energy level in an atom to another. The transitions that are most probable, that is, easiest for the electron to accomplish, normally give rise to strong lines; those that are highly improbable, that is, difficult, result in weak lines. The rules governing transitions are relatively simple, and are so restrictive that the number of spectral lines is far less than the number of possible combinations of pairs of energy levels. When a physicist begins to analyze a spectrum, he or she knows only the wavelengths, and hence the frequencies, of the spectral lines, and that these frequencies result from differences between atomic energy levels. The process of

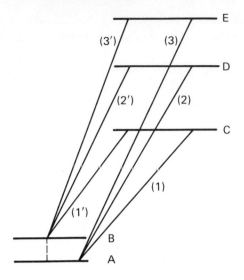

Fig. 10.6. How atomic energy levels may be identified. We observe lines (1), (2), and (3), corresponding to jumps from levels C, D, and E to level A, and likewise lines (1'), (2'), and (3') corresponding to jumps from levels C, D, and E to level B. From these observations the existence of energy levels C, D, E and also B is inferred, even though the jump from B to A (which corresponds to a 'forbidden' line, see below) is not observed.

deducing the energy levels from the observed frequencies, which is somewhat like solving a jigsaw puzzle, is illustrated in Fig. 10.6. The observed lines are (1), (2), (3), (1'), (2'), and (3'). From the fact that the differences in frequency between the lines (1) and (1'), (2) and (2'), and (3) and (3') are constant, we infer the existence of the pair of energy levels A and B with the same frequency difference. Likewise, we find the levels C, D, and E, always bearing in mind that the final pattern of levels must be consistent with the frequencies of all the observed lines. In spite of the fact that the line AB is not observed, because the rules of the game demand that this transition be highly improbable, we can still discover the levels A and B. Once an atom gets into level B, perhaps by collision with an electron, it may adopt a circuitous route to return to level A. It may, for example, absorb energy of the right wavelength to take it up to C, D, or E, after which it can come back to A by radiating the lines (1), (2), or (3). A direct jump from B to A has a low probability.

By similar reasoning, physicists were able to deduce that the lowest energy levels of doubly ionized oxygen, O III, formed the pattern shown in Fig. 10.7, even though transitions between these levels had never been observed in the laboratory. In pondering the origin of nebulium, I. S. Bowen noticed, in 1927, that the energy differences between D and P_1 and P_2 correspond exactly to the frequencies of the pair of intense green nebular lines at wavelengths 4959 and 5007 Å. Also, the difference between S and D agreed with the wavelength of another nebular line at 4363 Å.

Bowen's discovery revealed the remarkable nature of physical conditions in

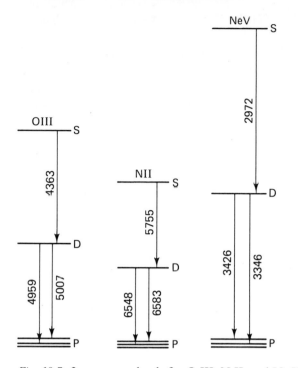

Fig. 10.7. Low energy levels for O III, N II, and Ne V. The lowest energy levels consist of a close group of three levels P_0, P_1, and P_2, collectively denoted as P, and the two higher single levels, D and S. Transitions between D and P in O III give the strong green nebular lines at 4959 Å and 5007 Å; in N II they give a pair of red lines 6583 Å and 6548 Å, straddling Hα at 6563 Å. The S–D transitions are called 'auroral type' because in the earth's aurora the strongest line usually observed is [O I], 5577 Å, which is of this type. Intensity ratios of 5755 Å and the 6548, 6583 [N II] lines and of 4363 Å and the 4959, 5007 Å [O III] pair are used to get electron temperatures in nebulae. The nebular type [Ne V] lines are seen in the near ultraviolet spectra of high excitation planetaries but the λ 2972 auroral type line falls in the far ultraviolet and is properly observable only with instruments such as the International Ultraviolet Explorer.

gaseous nebulae, as the following arguments show. According to modern theory, an O III atom may jump from level D to level P, or from S to D, but its chances of doing so are exceedingly small – about a hundred million times less than the chance that a hydrogen atom will emit a line of the Balmer series. Another way of putting it is that, although an atom will linger only a hundred-millionth of a second in an ordinary level, it will remain in levels like D or S, so-called metastable levels, for seconds or minutes before returning to the ground level. For this reason, transitions of the type SD or DP_2 have been called 'forbidden', although actually they are only highly improbable. Forbidden lines are generally indicated by brackets around the symbol of the ion. Thus the forbidden violet line of doubly ionized oxygen at 4363 Å is denoted by 4363 [O III].

Why, then, do the forbidden lines dominate the spectra of planetary and many other gaseous nebulae? The answer is that the normal, or so-called permitted, lines are very hard to produce under the conditions existing in these objects, while the forbidden lines are not.

In the discharge tube, atoms are excited and de-excited by collisions with fast-moving electrons. An oxygen atom that happens to land in level D, for example, would have about one chance in a hundred of emitting a forbidden line in a second. Since collisions with other atoms and electrons occur at the rate of several million per second, only a tiny fraction of the total number of oxygen atoms excited to level D will produce a forbidden line.

In a typical gaseous nebula, the electrons are not moving fast enough to excite atoms to the normal levels (whose excitation potentials are often as high as 10 or 20 volts). On the other hand, the free electrons are moving sufficiently fast to excite atoms from the ground level to one or the other of the metastable levels that are close to the ground level. The nebular density is so low that once an oxygen atom is in one of these metastable levels it is almost certain to radiate a forbidden line. Once a quantum of forbidden radiation is created, it is sure to escape from the nebula, since the probability that it will be reabsorbed is negligible. In the laboratory or in the nebula, the rate of emission of forbidden line quanta per unit volume and time will always be the number of atoms in the upper level multiplied by the transition probability expressed in reciprocal seconds, for example, [O III] λ4363 $1.6N(S)$, and λ5007 $0.021N(D)$. In the laboratory, even with O^{2+} ions present, this radiation would be overwhelmed by permitted line and continuum radiation and possibly light from the discharge-tube walls. Although the forbidden radiation per atom per second is actually somewhat weaker in the nebula than in the discharge tube, the nebula is so vast (of radius about 10^{12} kilometers or greater) that the forbidden lines produced there may attain a considerable intensity. The emission per unit volume in normal lines is weakened enormously and that in the forbidden lines relatively little in going from a vacuum tube to a gaseous nebula.

In doubly ionized oxygen, the D level lies about 2.48 electron volts (eV) above the P levels, while the S level is about 5.3 eV above the lowest P level. Under nebular conditions, virtually all ions reach the S level by direct electronic collisional excitation from the P level. At densities less than about 10 000 electrons per cubic centimeter, ions are excited from P to D and D to S by collisions: few escape from D or S to P by collisions. It is possible to calculate the A-values or 'transition probabilities' for radiative escapes from levels S and D, and also the 'target areas' for direct collisional excitation from P to D or S. These are obviously zero unless the electron has sufficient energy, $E(D) - E(P)$ or $E(S) - E(P)$, to lift the ion to the appropriate energy.

Measurement of the intensity ratio $I(4363)/[I(5007) + I(4959)]$ enables us to compare the numbers of ions excited to levels S and D respectively, i.e. $N(S)/N(D)$. Now $I(4363)$ is proportional to $N(S)$ multiplied by $A(SD)$, where $A(SD)$ is the transition probability for a jump from S to D with emission of a quantum of radiation, and $h\nu(SD)$ where $\nu(SD)$ is the frequency of 4363 Å and h is Planck's constant. Likewise, $I(5007) + I(4959) = N(D) \times A(DP) \times h\nu(D - P)$. From the intensity ratio $I(4363)/[I(5007) + I(4959)]$, we can compare $N(S)/N(D)$, and thus the

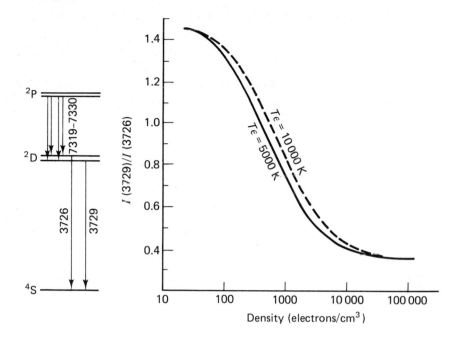

Fig.10.8. The use of [O II] lines to get electron densities.
(a) This shows the lowest energy levels of ionized oxygen. The [O II] P–D transitions fall in the deep red. Jumps from the close pair of ^{2}D levels to the single ground level, S, produce the 3726 and 3729 Å lines, whose intensity ratio depends primarily on density.
(b) Variation of the [O II] 3729/3726 ratio with electron density. Notice that the ratio depends strongly on density, and that while the dependence on electron temperature T_e due to radiation and collisions involving the P levels is small, it increases as T_e increases.

relative rates of excitation from P to S and P to D by electron collision. This ratio depends only on the collisional target areas and the temperature, since only electrons moving with energies greater than 2.48 and 5.3 eV can excite ions to the D and S levels, respectively. Thus, from measurements of this and similar ratios in [N II] and [S III], for example, we can find the electron temperatures in gaseous nebulae and extended stellar envelopes of very low density.

Another class of forbidden-line transitions are represented by lines of [O II], [S II], and (Ne IV), which are illustrated in Fig. 10.8a. Note that there is a single lower level and two pairs of closely spaced 'doublet' levels labelled ^{2}D and ^{2}P. (Don't get distracted or worried about the notation; there is a good technical reason for using these designations, but we need not be concerned here.)

Note that the close '3726' [O II] pair at 3726.1 and 3728.8 Å is of special interest since atoms can remain in the ^{2}D level for hours before emitting radiation. This pair is often strong in low-density diffuse nebulae in the Galaxy, as well as in some planetary nebulae. The 3726/3729 ratio is a useful indicator of electron density. Let us see why this is so.

In the first approximation, we can ignore the influence of higher ^{2}P levels. Ions can be excited to them and drop back to ^{2}D but the effect can be easily handled. At high densities, such as occur in novae, the population of the two ^{2}D levels is determined only by collisions; the ratio of atoms in the two doublet ^{2}D levels is actually the ratio of the number of Zeeman states into which each of them is split by a magnetic field, a constant of nature. Far more atoms escape by collisions than by radiation, so the $I(3726)/I(3729)$ ratio is now determined by the ratio of the A-values, since the number of radiative jumps is the number in the upper level multiplied by the A-value. At the low-density extreme, all ions must escape by emitting at either 3726 or 3729 Å (depending on which doublet ^{2}D level an ion occupies). The ratio now depends only on the target areas; the target area ratio for the two doublet levels is not the same as the A-value ratio. Fig. 10.8b shows this dependence. Similar curves may be calculated for lines of [S II], [Cl III], [Ar IV], and [Ne V]. Refined work takes into account the trickle down from the doublet ^{2}P levels.

Progress in this field depends not only on excellent observed data but also on accurate theoretical calculations of collisional effects, A-values, etc. Such calculations are being carried out by M.J. Seaton and his associates, including A.K. Pradhan, C. Mendoza, and C.J. Zeippen, by K.L. Baluja, P.G. Burke, and A.E. Kingston in Belfast, by S.J. Czyzak, by H. Nussbaumer, and by others. In the spectral region from 1200 to 3000 Å, a number of strong permitted lines of C III, C IV, N III, N IV, N V, O III, O IV, Si III, and Si IV fall; these have been studied in planetary nebulae and other objects with the International Ultraviolet Explorer (IUE). As with the forbidden lines, the upper levels of these transitions are predominantly excited by electron collisions although for some of them recombination from the next higher ionization stage may play a role.

Fluorescence in gaseous nebulae

Fluorescent rocks are among the most fascinating of all minerals displayed in museums. We enter a windowless room and see a case filled with specimens, most of which shine dully by reflected white light, but, on pressing a button, we work an almost miraculous transformation. The white light vanishes, and suddenly the rocks glow in a sparkling array of colors. What has happened? When we extinguish the white light, we also turn on a source of ultraviolet radiation. Although invisible to the eye, the ultraviolet light is absorbed by the rocks and reradiated in visible colors; each ultraviolet quantum is split up into two or more quanta of longer wavelength.

A similar process is at work in gaseous nebulae. In spite of the fact that the nebulosity must derive all of its energy from the illuminating stars, the total amount of visible light radiated by a planetary nebula may be as much as 40 or 50 times greater than that emitted by the central star itself. The explanation may be traced to the fact that the central star is so hot that most of its energy is given out in the form of invisible ultraviolet light, as illustrated in Fig. 10.9. After the invisible light energy has been absorbed by the nebular atoms, it is re-emitted in visible form. Hydrogen, by far the most abundant constituent of the stars and nebulae, is the atom that is mainly responsible for the transformation of the unseen into the seen.

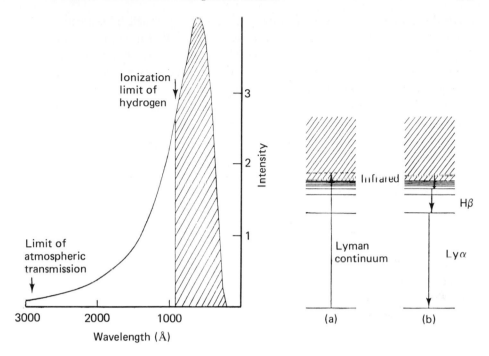

Wavelength (Å)

Fig. 10.9. Distribution of radiant energy in the spectrum of a star that radiates as a black body whose temperature is 50 000 K, plotted for the region shortward of 3000 Å. Note that the energy maximum lies in the far ultraviolet. The hatched area indicates the fraction of total energy available for the ionization of hydrogen in the nebular shell. Virtually all energy observed in planetary nebulae comes ultimately from the far ultraviolet. More sophisticated calculations based on stellar atmospheric models give a somewhat modified curve whose effects should be considered in refined work. The qualitative character of the energy curve is unchanged. New observations with the International Ultraviolet Explorer (IUE) enable us to reach down to nearly 1200 Å, so some checks on model atmosphere predictions are possible.

Fig. 10.10. The origin of hydrogen emission lines in the spectra of gaseous nebulae.
(a) An atom in the ground level of hydrogen absorbs a far ultraviolet quantum of energy and the electron is ejected from the atom.
(b) In the scenario depicted, the electron is recaptured on the fourth level with the emission of an infrared quantum. Then it falls to the second level with the emission of Hβ, which is easily observed, and finally to the ground level with the emission of Lyα, which is unobservable. This is only one of many possible electronic recapture and cascade routes. Other examples are given in the text.

Let us see what happens when the ultraviolet radiation of the star falls upon a shell of hydrogen gas. A quantum of wavelength less than 912 Å possesses sufficient energy to tear the electron away from a hydrogen atom. Such an electron, detached from an atom, may wander about in space until it is recaptured by a proton. Although the electron was torn away from the smallest orbit, it may, when it is recaptured by some other proton, land in any of the orbits, the higher ones as well as the lower ones (Fig. 10.10). If the free electron is captured in the lowest orbit, a quantum similar to the original quantum from the star will be reborn, and this quantum, escaping from the atom, may ionize yet another hydrogen atom.

Now an electron captured in one of the higher orbits may jump to any one of the lower levels, radiating as it jumps, or in principle it may absorb another quantum of starlight and leap to a still higher orbit. But conditions in gaseous nebulae do not favor this latter process. The nebula is so enormous compared with the star that the starlight is spread over a vast area and its intensity at any point is very low. When radiation is so 'diluted', an excited hydrogen atom has little chance of absorbing another quantum, because it remains excited for only a hundred-millionth of second, whereas it might have to wait 20 years for a quantum of just the right frequency to come along. Also, as we have already seen, the density is so very low that the prospect of a collision with another particle is remote. Hence the electron has no real alternative but to return to the ground level, either in one jump or by cascading down by stages.

Each electronic jump is of course accompanied by the radiation of a light quantum. Thus if the electron is captured in the second level, the atom will radiate energy in the near ultraviolet, beyond the limit of the Balmer series. The exact wavelength is determined by the energy of motion of the free electron. From the second level the electron drops to the lowest level and the atom emits the first line of the Lyman series. Many of the electrons caught in high levels, however, will fall to the second level and thereby produce the bright Balmer lines so prominent in the spectra of planetary nebulae. Eventually, all of the stellar radiations of wavelength shorter than 912 Å are converted into light of lower frequencies, a large percentage of which falls in the visible region of the spectrum.

We saw in Chapter 3 that captures of free electrons in the second hydrogen level produce a continuous range of emission that starts at the Balmer limit and decreases in intensity toward the ultraviolet. A similar continuum is observed at the head of the Paschen series. In addition, there is also a curious continuum produced by atoms escaping from the second hydrogen level to the ground level. Normally we would expect the atom to radiate the Lyman-α line. Actually the $n=2$ level is split into two groups of energy states of very nearly the same energy. Jumps from one group to the ground level are permitted; jumps from the other group are forbidden. The atom caught in one of the second group of states may jump to a virtual (fictitious) level between the $n=2$ level and the ground level and then from the virtual level to the ground level. Since the virtual level may lie anywhere between the $n=2$ level and the ground level, the result of the effects of many atoms will be a continuous spectrum filling the normally observable range. This two-photon continuum may produce most of the visual continuous spectra of many diffuse nebulae, and play an important role in planetaries as well.

There is also another type of continuous radiation that is produced by an ionized gas, the so-called 'free–free' emission often denoted by the formidable-sounding German word: 'Bremsstrahlung'. An electron passing near an ion is accelerated and radiates energy by a process that usually involves very small energy changes. Much of this type of radiation is emitted in the radio-frequency range. At shorter radio wavelengths (higher frequencies), e.g. 4 centimeters, all gaseous nebulae seem to be transparent. At longer wavelengths (c.g. 83 cm) some of the more massive nebulae become opaque. In particular, for the Orion Nebula, measurements at several wavelengths enable one to determine the total amount of material along the line of sight and also the gas temperature. Thus B.Y. Mills and P.A. Shaver found an electron temperature of 8000 K for Orion, a result in harmony with M. Peimbert's [O III] measurements.

Ordinary laboratory lines of carbon, nitrogen, neon and oxygen in several stages of ionization are observed in high-density nebulae. They involve high energy levels, typically 10–30 eV above the ground level, and must originate by recapture of electrons and cascade.

The observed radiations of neutral and ionized helium are produced, like those of hydrogen, by ionization followed by recapture, but the quanta necessary to remove one or two electrons from helium lie far in the ultraviolet, beyond 506 Å and 228 Å, respectively. The nuclei of planetary nebulae that show nebular lines of He II are among the hottest stars known, with temperatures often in excess of 100 000 K. H. Zanstra showed that if all the stellar radiation emitted beyond the Lyman limit of hydrogen were absorbed by hydrogen atoms in the nebula, one could evaluate the temperature of the central star. If the nebula is so thick that there are a great number of absorptions and re-emissions, each quantum of ultraviolet energy eventually becomes broken down into a quantum of Lyman α radiation and one of Balmer radiation. The latter escapes at once from the nebula; the former is repeatedly absorbed and re-emitted until it escapes. By measuring the nebular radiation flux in, say, Hβ, we can calculate the total number of quanta in all Balmer lines as their relative intensities are accurately known. Similarly we can calculate the number of Balmer continuum quanta. We can also measure the stellar energy flux within the optical region of the spectrum and get likewise the corresponding number of visual region photons. Since the number of Balmer quanta equals the number of stellar quanta beyond the Lyman limit, we can find what proportion of energy is radiated by the star in the far ultraviolet as compared with that radiated in the optical range, and hence determine the temperature from the radiation laws (see Chapter 4). If the star does not radiate as a black body, we can use energy distributions given by model atmosphere theory.

Independent evidence for very high temperatures of exciting stars of planetary nebulae is offered by their spectra. Some display Class O spectra with weak emission lines, some show absorption lines and some are almost featureless in the optical region. Yet other planetaries have nuclei of the Wolf-Rayet type with diffuse emission lines superposed on a continuum (see Fig. 10.11). Spectra obtained with the International Ultraviolet Explorer have been especially useful. Stars with virtually featureless optical region spectra often show strong P-Cygni-type profiles, revealing very swift ($\sim$ 1000–2000 km/s) winds.

The thermostatic action of nebular lines

An interesting sidelight on the forbidden lines of ions of oxygen, nitrogen, neon, and sulfur, and collisionally excited ionic lines of ions of carbon, nitrogen and oxygen, is

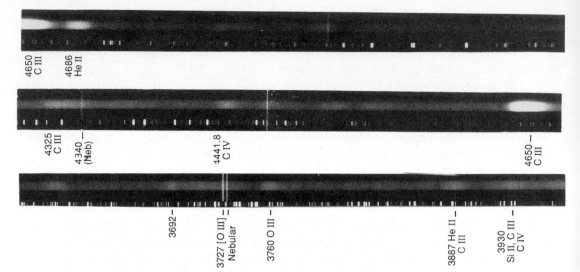

Fig. 10.11. The spectrum of the nucleus of NGC 40. This central star of the fairly bright planetary nebula NGC 40 has a spectrum of the Wolf–Rayet type with strong broad emission lines of ions of helium, carbon, and oxygen. Other planetary nebulae nuclei have Wolf–Rayet spectra of an unusual type, showing strong O VI lines. Yet other central stars have weak emission lines of carbon and nitrogen ions; some show P-Cygni-type lines (particularly in the ultraviolet). Many show extremely faint, narrow, absorption lines. (Courtesy Lick Observatory, University of California.)

that their production acts as a thermostat to regulate the temperature of the nebular gas. Let us consider first a nebula that is composed entirely of hydrogen. The temperature of the gas is measured by the speeds or kinetic energy of its atoms and electrons, which in turn depends on the energy flux from the central star. If the star is hot and therefore rich in high-frequency radiation, electrons will be torn away from hydrogen atoms with high speeds. These electrons dash about and if they collide with neutral hydrogen atoms they bounce away without loss of energy unless they are moving very fast indeed, about 200 km/s (i.e. have energies in excess of 10.2 eV). At this and higher energies, they are able to excite hydrogen atoms from the ground level up to the first excited level. In a pure hydrogen nebula, the gas kinetic temperature of the electrons would depend strongly on that of the star, until the gas temperature rose above about 25 000 K, when excitation of the second and higher levels of hydrogen would become important. The atoms excited to these levels must return to the ground level by radiation, so the energy fed to the atoms by collisions is all radiated away and most of it is lost to the nebula. Any further temperature rise would be inhibited.

Suppose that we introduce small amounts of C, N, O, Ne, and S into the hydrogen nebula. Their ions have energy levels typically 1.5 to 5 eV above the ground level. Now the energies of the free electrons which had almost all been produced by the ionization of hydrogen become dissipated in exciting these 'foreign' atoms to low-

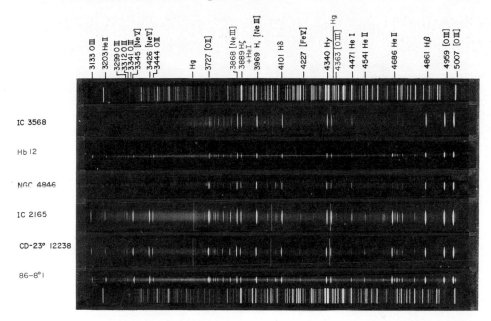

Fig. 10.12. Slit spectra of planetary nebulae. The principal nebular lines are marked. Note the high intensity of the [O III] 4959 Å and 5007 Å lines and the continuous spectrum at the head of the Balmer series of hydrogen, just shortward of [O II] 3727 Å. It is particularly strong in IC 2165. The long lines marked Hg, including the strong one nearly blended with 4363 [O III], are due to mercury vapor lamps in the sprawling metropolis of San Jose. (Photographed with the nebular spectrograph on the Shane 3-meter telescope at Lick Observatory.)

lying energy levels. The radiation from the ultraviolet-permitted lines of C III, C IV, etc., many infrared lines, and forbidden lines arise at the expense of energies of motion of the free electrons. Virtually all of this energy is forever lost to the nebula, because the forbidden radiations cannot be reabsorbed. The C IV 1548,1550 Å radiation is scattered until it escapes from the nebula or heats the dust. Thus the nebular gases are effectively cooled; although central star temperatures may range from 30 000 K to greater than 100 000 K, the nebular gas almost never gets much hotter than 16 000 K to 20 000 K. In planetary nebulae radiation mostly in ultraviolet C III and C IV lines, in the forbidden [O III], [N II], and [S III] lines, and in a few infrared transitions drains away the bulk of the energy. In diffuse nebulae an analogous role is played by molecular lines and the C^+ ion.

On the importance of coincidences

In addition to the forbidden lines of oxygen and nitrogen, which originate from metastable levels, some lines from normal levels have also been observed. Their origin is a fine illustration of the strange whims of nature. Fig. 10.12 shows the strong lines in the ultraviolet spectrum of several high-excitation planetary nebulae.

One line, at 3203 Å, is due to ionized helium, and a pair arises from [Ne V], but our

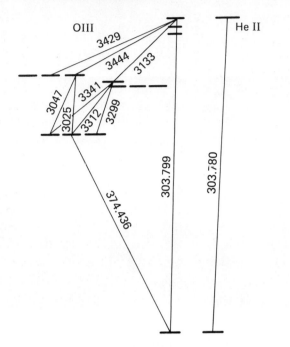

Fig. 10.13. Schematic diagram of the Bowen fluorescent mechanism. The 303.78 Å line of ionized helium is absorbed by the O III 303.799 Å line, which is raised to an upper level, whence it cascades downwards to produce the observed lines marked in the diagram. A similar cycle occurs in N III, triggered by O III!

interest lies chiefly in the other strong lines, which are radiated by O III. These lines are perfectly normal in that they are commonly observed in the laboratory. The puzzling circumstance of their appearance is that other equally intense laboratory lines are absent.

The apparent favoritism has been demonstrated by I.S. Bowen to result from a remarkable coincidence. Bowen noticed that all the observed ordinary lines of O III in planetary nebulae originate from electrons cascading down from a single excited level. Furthermore, the wavelength of the radiation required to excite this level is 304 Å, which coincides almost exactly with the strongest line of ionized helium. The 304 Å line corresponds in ionized helium with the first line of the Lyman series in hydrogen. When a doubly ionized helium atom captures an electron, becoming singly ionized, the final stage in the cascading process is often the transition from level 2 to level 1. Hence this He II line, although invisible, probably attains great strength in some planetary nebulae. As shown in Fig. 10.13, the O III ions in the lower level absorb the plentiful radiation of He II, are excited to the high level, and then return by successive stages with the emission of the observed lines. The missing lines are not observed because they originate from other levels that do not have fortuitous sources of energy. Even more remarkable is the circumstance that as the O III ion finally returns to the lowest energy level it emits radiation at 374 Å, which is of just the right wavelength to generate a similar cycle in producing certain observed lines of doubly ionized nitrogen, N III, near 4640 and 4100 Å. Bowen's explanation is supported by the fact that the permitted lines of O III appear only in those portions of the nebula where the observable lines of ionized helium are strong, and therefore where the 304 Å line also is presumably intense.

Since the observed lines all arise from a trickling down of atoms from a single upper level, it is possible to predict their relative intensities. Theoretical studies by Seaton and H.E. Saraph, by Nussbaumer, by R. Weymann and R.E. Williams, by J.P. Harrington, and by R. McCray and T. Kallman are generally in harmony with the observations. The Bowen fluorescent mechanism operates not only in planetary nebulae but also in certain galaxies called Seyfert-type galaxies whose central regions show strong bright lines, in X-ray binaries, and even in the solar atmosphere.

Thus, in summary, lines in planetary and other gaseous nebulae are produced by (a) collisional excitation, as for forbidden lines in the ordinary optical region and ultraviolet permitted lines such as C IV, and (b) recombination followed by cascade as for the hydrogen and helium lines and numerous weak lines of carbon, nitrogen, oxygen, etc. Note that the Bowen lines occur only in sources where copious amounts of He II 303.78 Å are produced. They are not seen, for example, in diffuse nebulae.

Chemical compositions of gaseous nebulae

After describing the adventures of atoms and ions in the hot plasma of a gaseous nebula, let us return to Fig. 10.2 – the Karmic cycle of birth and death of stars and the steady changes that must occur in any active galaxy. We would expect a gradual increase with time of elements that had been manufactured in stars, so a vigorous galaxy committed to a strong program of star building should have an interstellar medium rich in elements such as carbon, nitrogen and oxygen, as compared to one with a low star-building rate.

Thus chemical compositions of gaseous nebulae are of interest in the context of the evolution of galaxies. A study of chemical compositions of emission-line nebulae (H II regions) may cast light on star-building rates, which can differ from galaxy to galaxy or even within a given galaxy. Analyses of planetary nebulae tell us what elements are built in stars, conveyed to the surface, and ultimately dispersed into the interstellar medium. Analyses of nebulae as well as of stars supply information, but limitations in accuracy can be quite severe.

Consider first planetary nebulae. Our observational data will consist of fluxes in emission lines, and various measurements of infrared and radio-frequency radiation; the optical and ultraviolet data often must be corrected for the effects of interstellar extinction. The emission in the hydrogen lines, e.g. Hβ, is proportional to $N(H^+)N_e$, where $N(H^+)$ is the number of hydrogen ions per unit volume and N_e the corresponding number of electrons. Since the recombination rate depends on the number of captures of free electrons by ions, it will also depend on the electron temperature, T_e, increasing as T_e decreases. The emission in optical region collisionally excited lines is going to depend on N(ion), on N_e, and very sharply on T_e, since the higher the value of T_e, the greater the collisional excitation rate. We get T_e and N_e

Table 10.2 *Composition of planetary nebulae (PN) compared with the sun*

	He	C	N	O	F	Ne	Na	Si	P	S	Cl	Ar	K	Ca
Mean PN	110 000	780	182	435	0.03	107	1.5			10	0.17	2.7	0.09	0.11
NGC 7027	110 000	540	120	350	0.032	84	2.1	5	0.35	8	0.18	2.0	0.17	0.4
NGC 6537	180 000	40	890	170										
IC 2165	101 000	870	95	275		86	4.2			31	0.13	2.3	0.06	0.10
Sun	98 000	362	112	850	0.036	123	2.1	35	0.28	16	0.30	3.6	0.13	2.3

from ratios of certain forbidden-line intensities as described above. Then, from the intensity ratio, $I(\lambda)/I(H\beta)$; for an appropriate line, λ, of a given ion, we find $N(\text{ion})/N(H^+)$, but we are unable to get $N(\text{element})/N(H)$ unless we observe all ionization stages or can allow for the fraction in the unobserved ones. In a stellar spectrum we find $N(\text{ion})/N(H)$ from lines of a given ion by means of a curve of growth. Then, given the electron pressure and temperature of the atmosphere, we could compute $N(\text{element})/N(H)$ with the aid of the Saha theory and a knowledge of the source of the opacity in the continuous spectrum. Alas, a gaseous nebula is very far from thermal equilibrium and no Saha-like theory is at hand. A rarefied gas is illuminated by a distant source whose radiant flux distribution may resemble that of a black body near 50 000–100 000 K, while the local gas temperature may be 10 000 K.

How may we allow for unobserved ionization stages? One procedure which works better for planetary nebulae than for diffuse nebulae is to calculate a theoretical model, adjusting the radius and radiative flux of the illuminating star, the thickness, density, and chemical composition of the nebular shell until the predicted and observed intensities agree. Hopefully, the model can then be used to estimate the populations in unobservable ionization stages.

Table 10.2 compares the relative numbers of atoms on the scale $N(H) = 1\,000\,000$, for the sun, an average of a large number of planetary nebulae (PN), the bright NGC 7027, (with a rich line spectrum), a C-rich object (IC 2165), and an N-rich object (NGC 6537). Generally, among planetary nebulae, the oxygen abundance tends to be reduced, the carbon and nitrogen abundances are often enhanced, and flourine, neon, argon, and potassium (K) do not differ much from the solar values. Sulfur and chlorine show fluctuations that may be due to uncertainties in the analysis. The C-rich planetary IC 2165 comes from a star that has evidently burned helium to carbon, while in N-rich NGC 6537, oxygen and carbon have been heavily depleted in the CNO cycle.

One point to be emphasized is that we measure elemental abundances for the gaseous phase. Our census does not count atoms in solid grains. Some of the carbon and a substantial fraction of the refractory elements silicon and calcium are locked up in solid grains. Oxygen and nitrogen remain mostly in the gaseous state.

What about analyses of glowing clouds of gas like Orion, and other diffuse bright-line nebulae in our own and other galaxies? Recall that these are basically dense, irregular clouds of dust and gas in which stars are suddenly ignited. Hence, unlike the planetary nebulae, there is no symmetry. If the newly born star is massive, it will emerge as a hot O or B star that proceeds to ionize and excite the left-overs of the original cloud from which it had been formed.

The various processes of photoionization, recombination, and collisional excitation of low-lying levels are similar to those occurring in planetaries, but shock waves from supernovae may play important roles in exciting the interstellar material. Abundances of He, C, N, O, Ne, S, Cl, and Ar have been found not only for H II regions in our own Milky Way system, but also in other galaxies such as the Magellanic Clouds, M33, M31, and M101. Substantial differences are found not only between the interstellar media of different galaxies, but also from point to point in a single galaxy. As long ago as 1941 it was noted that in the Triangulum Spiral, M33, the character of the emission-line spectra of faint H II clouds changed as you moved outward from the central core. Apparently, the rate of element building decreases with distance from the center. See Table 9.1, p. 169.

The interstellar medium, the gas between the stars

Where the gas constituting the interstellar medium is dense and illuminated by bright hot stars it becomes ionized and fluoresces in a manner similar to that exhibited by planetary nebulae. The vast bulk of interstellar gas, however, is at a very low density and is often very cold as well. By mass, about 99 percent of the interstellar medium is gas, and most of this gas is hydrogen. Some of it is ionized, some is neutral, and some is in the molecular form.

Ionization of atoms proceeds slowly because the necessary radiation comes from stars at great distances; consequently it is highly attenuated. Nevertheless, occasionally a high-frequency quantum or a cosmic ray will come along and detach an electron from an atom. Once ionized, an atom stays that way for a long, long time because at very low densities its encounters with free electrons are very infrequent. Consequently, most of the atoms of calcium, sodium, and other metals are ionized.

Radiation capable of exciting atoms to higher energy levels is likewise very weak, and collisions capable of lifting atoms from their ground levels to excited levels are rare. Consequently, an atom spends most of its time in the very lowest energy level. Most of the atoms remain in this placid state, where they are quite indifferent to visible light. They absorb radiation only from the very-short-wavelength region of the spectrum. This radiation is not transmitted by the earth's atmosphere; in fact much of it is absorbed by neutral interstellar hydrogen.

An important element in the interstellar medium is hydrogen, whose interstellar Balmer lines have never been observed in absorption. Most of the hydrogen atoms near hot stars must be ionized, but occasionally they recapture electrons in the second, third, fourth, or higher energy levels. As these electrons cascade to lower levels, they emit radiation. Now only atoms in the second level can absorb the Balmer series. Even though an electron may momentarily land in the second level, it will remain there but a hundred-millionth of a second before dropping to the lowest

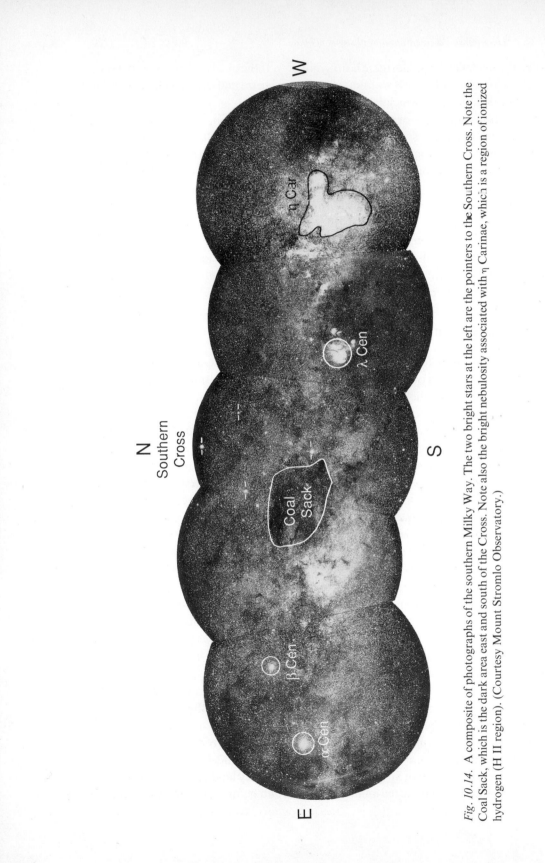

Fig. 10.14. A composite of photographs of the southern Milky Way. The two bright stars at the left are the pointers to the Southern Cross. Note the Coal Sack, which is the dark area east and south of the Cross. Note also the bright nebulosity associated with η Carinae, which is a region of ionized hydrogen (H II region). (Courtesy Mount Stromlo Observatory.)

level. In order for a Balmer line to be absorbed, a quantum of light of just the right frequency would have to come along during that time interval. The chance of that happening in interstellar space is negligibly small.

The downward cascade of captured electrons will produce a faint glow of light in the space between the stars. Struve and his co-workers detected this faint light as emission lines of hydrogen and oxygen ions in many regions of the Milky Way. These volumes, sometimes 80 to 250 parsecs in diameter, were often sharply bounded and associated with groups of hot O stars. The 3727-Å line of [O II] is nearly always present in hydrogen-emission regions and the green [O III] lines are sometimes observed.

These regions of rarefied, glowing gas can be observed directly by using red-sensitive plates and narrow band-pass filters whose transmission is a maximum at Hα, 6563 Å. In this way, Collin Gum found the largest galactic gaseous nebula known; it fills many square degrees of the southern sky. The main difference between these extended regions of faint emission and objects such as the Trifid or η Carinae nebulae would appear to be total mass and density. Conventional diffuse nebulae have densities of 100 to 10 000 atoms per cubic centimeter, whereas in objects such as Gum's Nebula the density may be 1 to 10 atoms/cm^3. Since, for a given temperature, the emission varies as the density squared, objects such as the Trifid or Lagoon (M8) nebulae will be enormously brighter.

B. Strömgren developed the theory of hydrogen ionization in the interstellar medium. He predicted that for a radius of about 30 parsecs near a Class O star, hydrogen would be completely ionized if $N(H) = 1/cm^3$. Then there would be an abrupt boundary within a fraction of a parsec, beyond which all hydrogen would be neutral. Within the volume of hydrogen ionization, called an H II region, emission lines would be produced as free electrons are slowly recaptured and cascade to lower levels. If the gas density was relatively uniform near a hot star, the latter would appear to be surrounded by a disk of faintly glowing hydrogen. These 'Strömgren spheres' have been observed on Hα photographs of the Milky Way and in other galaxies as well. Even objects such as the Orion Nebula are Strömgren spheres in a denser medium. The theory can also be applied to planetary nebulae, many of which consist of an inner ionized zone surrounded by an outer, cool shell that consists of neutral or molecular hydrogen. See Figs. 10.14 and 10.15.

Thus, quite generally, we might expect each Strömgren sphere to be surrounded by a neutral hydrogen zone. In fact, the normal state of much of the hydrogen gas in the Milky Way is neutral. When a hot star is nearby, the gas becomes ionized and we have an H II region in which the gas kinetic temperature is typically 6000–10 000 K. In the outer neutral hydrogen or H I regions, where electrons and protons are recombined to form normal hydrogen atoms (which may in turn form hydrogen molecules), temperatures of 60–100 K are typical.

Fortunately, nature has provided us with a powerful tool for studying the neutral hydrogen gas. In Chapter 3 we described how ordinary atomic energy levels were split by the magnetic effects produced by the spin of the electron and its orbital motion. The interaction energy is usually small compared with the energy difference

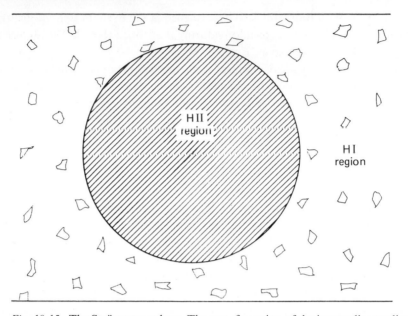

Fig. 10.15. The Strömgren sphere. The gas of a region of the interstellar medium is ionized by the hot, illuminating star within it. The temperature of the ionized gas (H II region) is about 8000 K, while the temperature may be less than 100 K in the unionized (H I) region. The edge of the H II region is rather abrupt. Actually, the interstellar medium is very inhomogeneous (see. Fig. 10.3), so the edge of the H II region may appear irregular. The density of the gas may vary from < 1 atom/cm³ in low-density regions to 10–100 atoms/cm³ in the clumps. A Strömgren sphere often is not a static structure; after the hot O or B star is 'turned on' the surrounding H II region grows until the number of its ionizing photons equals the total number of recombinations in the region.

between normal levels. In addition to the electron spin there is also a spin of the atomic nucleus, but the magnetic effects of this spin are much smaller. Nevertheless, interaction between the electron's magnetic effects and those of the nucleus produces a so-called hyperfine structure of spectral lines which is observed as a very small-scale splitting in lines of some atoms such as manganese.

In a hydrogen atom the magnetic interaction of the spinning electron with the spinning proton is such that in one direction of the electron spin the energy is slightly greater than in the other. The energy differences are small, about two million times smaller than the energy needed to detach the electron from the atom. Thus the neutral hydrogen atom in its ground level can exist in one or the other of two states whose energies differ by 0.000006 electron volt. When the atom flips from one state to the other it will absorb or emit a quantum of energy whose wavelength is 21 centimeters. This radio-frequency radiation was predicted by H.C. van de Hulst in Leiden and first observed by H. Ewen and E. Purcell at Harvard. The probability of a spontaneous flip is very low. If the atom is in the state of higher energy it remains there ten million years before flipping to the lower state by spontaneous emission.

The 21-centimeter radiation has been a powerful tool both for galactic structure research and for studying the properties of the interstellar medium. The displacement of the line tells the line of sight velocity of its source while the profile and central intensity of the line can give the number of atoms in the line of sight and the temperature of the cool hydrogen gas. Interpretation of the data can be complicated by temperature variations in the gas and superposition of contributions from various clouds moving with different velocities along the line of sight. Note that the line originates in regions of neutral atomic hydrogen; it is not emitted in H II regions nor where hydrogen exists only in purely molecular form as in clouds of heavy interstellar extinction. Normally the line is observed in emission, but if the hydrogen is found in front of an emitter of intense continuous radio-frequency emission, such as the Cassiopeia source (see Chapter 12), it may appear in absorption.

Although hydrogen is the most abundant gas in the interstellar medium it was not the first gas recognized there (except for bright emission nebulae). In fact, the vast extent of the interstellar gas cloud was first found as a by-product of stellar spectroscopic observations. Interstellar gas absorbs starlight in particular wavelengths in the same way that a stellar atmosphere absorbs light from a photosphere and thus imprints absorption lines of the familiar elements upon the continuous spectrum. So tenuous is an interstellar cloud, however, that an exceedingly long column of gas is needed to produce an absorption line of sufficient strength to be seen in optical region spectra.

The first element to be discovered in interstellar space was ionized calcium, found by J. Hartmann in 1904 in the spectrum of δ Orionis, a close double star. When the line of sight is nearly in the orbital plane of a binary star, the revolving components alternately approach and recede from the observer. The motion will be mirrored in the spectrum where, owing to the Doppler effect, the absorption lines will appear to oscillate in position with the same period as the revolution (Fig. 10.16). Absorption lines produced by interstellar gas, which does not share the orbital motion, will appear stationary, however. Since the K line of ionized calcium does not share the periodic motions of the other lines of δ Orionis, it must be produced in a gas cloud detached from the star. Furthermore, the interstellar lines are sharp, in marked contrast to the diffuse stellar spectral lines on which they are sometimes superposed.

Subsequently, at Lick Observatory in 1919, Miss M.L. Heger discovered the lines of interstellar sodium. Development by Theodore Dunham of the high-dispersion coude spectrograph made possible the detection of other elements and ions (Ca I, K, Ti, Fe) and fragmentary molecules (CH, CH^+, and CN) from their very weak absorption lines (T. Dunham, W.S. Adams, and A. McKellar). The Ca II interstellar lines have been the most intensively studied. They often show several components, corresponding to sheet-like clouds moving with different velocities along a line of sight (Fig. 10.17).

A great leap forward in studies of interstellar absorption lines came with spectrographs flown in satellites, particularly 'Copernicus', which measured brighter stars, and the International Ultraviolet Explorer(IUE). Lines arising from the ground levels of familiar atoms and ions of abundant elements appear in great numbers in the 1000–3000 Å region. One measures the equivalent widths (as with

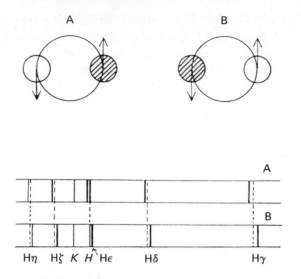

Fig. 10.16. How interstellar lines were discovered. In frame A, the brighter star is approaching the observer; in frame B it recedes. Consequently, stellar absorption lines are shifted towards the violet in the spectrum of A and towards the red in the spectrum of B, but the stationary H and K lines of Ca II, due to an interstellar cloud of gas, remain unchanged in position.

stellar spectra) and uses the curve of growth method. Light from an early-type star is extinguished by a cloud of absorbing atoms and molecules. Except for Lyman α, C II 1034 Å, and O I, 1307 Å, only Doppler broadening is important. The curve of growth gives the number of atoms in the corresponding stage of ionization. Allowing for other stages of ionization is a little tricky. We have to estimate the radiation flux to which the atoms are subjected from the types and colors of nearby stars so that we can calculate the rates of ionization. The electron density is obtained for example by comparing $n(\text{Fe II})/n(\text{Fe I})$. We use what amounts to a modified form of the Saha equation which enables us to calculate elemental abundances from ionic abundances in the interstellar medium.

In the detailed analyses carried out by L. Spitzer and E.B. Jenkins (1975) and by others, it is found that refractory elements such as Si, Fe, Ca, Ti, Mn, Al, and Ni are depleted by factors ranging from 10 to 1000; C, N, and O are depleted by factors of 2 or 3, but S, Ar, Cl, P, and Zn are not depleted at all! Most of the condensation must have occurred already in the ejecta of cool stars, but M. Jura among others concluded that some of it must take place in the interstellar medium itself.

To summarize the status of interstellar chemical composition:

(1)   From emission-line spectra we get the abundances of He, C, N, O, Ne, and a few other elements in the gaseous phase in our Galaxy and elsewhere.

(2)   Metallic absorption lines indicate a strong depletion of refractory elements; even C, N, and O are somewhat depleted.

(3)   In the cold molecular clouds, well shielded from starlight, there exist many molecular species, often highly reactive fragments that are not seen on the earth (see next section).

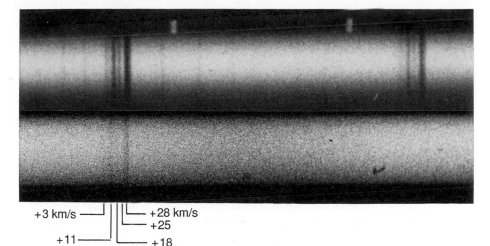

Fig. 10.17. The interstellar lines in the spectrum of ε Orionis show several components, corresponding to distinct clouds with radial velocities of +3, +11, +18, +25, and +28 km/s. (Courtesy of G. Munch, Mount Wilson and Palomar Observatories.)

(4)   Molecules offer opportunities to obtain isotope ratios, such as $^{12}C/^{13}C$. This particular ratio is hard to find because ^{12}CO lines are saturated, so that the equivalent width is insensitive to the number of molecules, while the corresponding ^{13}CO lines will be weak and their equivalent widths are proportional to the number of ^{13}CO molecules. Using bands of CH^{+} in the optical region, and a charge-coupled device as a detector, Isabel Hawkins and M. Jura found the $^{12}C/^{13}C$ ratio to be 45 in the interstellar medium, not 90 as in our solar system.

(5)   One of the most striking features of the interstellar medium is its chemical homogeneity; within a factor of two, the chemical composition of the gas and dust between the stars seems to be the same everywhere in our part of the Galaxy.

Molecules in the interstellar medium

Perhaps the greatest single advance in the study of the interstellar medium in the last 20 years has been made in the study of molecules and their relationship to cool, opaque, interstellar clouds. Why is the study of molecules so important and what can it reveal?

(1)   Molecules provide the tools to investigate the often totally unanticipated environment in which they are formed, and help us to understand the conditions under which stars can form from dense molecular clouds. Molecules found in carbon stars, OH–IR objects, and planetaries reveal conditions in envelopes of dying stars.

(2)   Many of the molecular structures found in these clouds are but fragments of types found on earth. In interstellar space, chemical reactions often do not run to completion, as they usually do on the earth (where they sometimes leave a sludge in the bottom of a test tube). In the interstellar medium, chemical reactions are interrupted by the effects of radiation, sudden shocking of the gas, lack of reaction partners, etc.

(3)   Hence there exists a time-dependent situation of extreme complexity; the chemistry has to be followed instant to instant by individual detailed reaction events. We cannot use the simpler approach where reactions run to completion.

(4)   In this world of 'dirty test-tube chemistry' we must consider not only reactions between unfamiliar fragments, but also various interactions with grains, which often act as catalysts, greatly enhancing the rate of a reaction.

(5)   Finally, in compact molecular blobs, some selected molecular transitions are enormously enhanced in intensity along certain directions. This is the phenomenon of masering which is explained in the next section.

We distinguish between molecules such as carbon monoxide (CO) found widely spread throughout the Milky Way in clouds of low density where the shielding from the interstellar radiation field is only moderate, and those molecules that can survive only in the dense molecular clouds. Here, all optical galactic radiation (starlight) and even some infrared radiation is cut out. Only cosmic rays can penetrate the dusty barriers. The grain temperatures drop to very low values, ~ 10 K, so they serve as cold traps on which a wandering molecule can freeze out. Presumably, collisions of blobs of material within clouds, cosmic rays, and shock waves provide energy that allows complex molecules to be 'shaken loose' from grains.

Molecular hydrogen dominates the clouds where molecules are formed. Usually we cannot measure it directly and have to estimate it from the concentration of CO, assuming a constant ratio of CO to molecular hydrogen of 8×10^{-5} for example. In cold clouds, such as Sagittarius B-2, many other molecules are formed. Some workers in the field have proposed structures call PAHs, polycyclic aromatic hydrocarbons. These do not show sharp millimeter lines but may be responsible for broad absorption features such as the 4430 Å 'dip' in the optical spectra of highly reddened stars and, quite possibly, sharp infrared emission features. A sample of the solid stuff of the interstellar medium might be relatively familiar to chemists studying various hydrocarbon pollutants, truly an interstellar smog!

Masers

Important clues to the understanding of star-forming regions and materials ejected by dying stars are provided by what are called masers, an abbreviation for 'microwave amplification of stimulated emission of radiation'. The physical phenomenon is similar to that occurring in lasers, which involve the normal optical region.

Let us consider a hypothetical atom with just two energy levels, a ground or lower level, L, and upper or excited level, U. If this atom is inclosed in a box whose walls are kept at a temperature, T, it may be excited from L to U by radiation or by collisions and we could write: Number of excitations = number of radiative excitations plus number of collisional excitations. Let us suppose that the density of the gas is so low that we can neglect collisional effects. Then the number of excitations per unit volume $F(L,U)$ will equal $N(L)B(L,U)I(L,U)$, where $N(L)$ is the number of atoms in level L per unit volume and $I(L,U)$ is the intensity of the radiation of frequency $v(L,U)$. The difference in energy between the two levels is $E(U) - E(L) = hv(L,U)$, by the quantum condition (chapter 3), and $B(L,U)$ is a constant for the particular line under question. It is related to the 'f-value' described in Chapter 5. Now when the atom is in the upper level, U, it can escape spontaneously to a lower level, L, by the emission of a quantum $v(U,L)$, and we might expect that the total number of downward jumps (with no collisions involved) would be: $F(U,L) = N(U)A(U,L)$, where $A(U,L)$ is the probability of spontaneous emission. Under conditions of strict equilibrium (where the walls of an enclosure are kept at constant temperature, T), $N(U)/N(L)$ will be given by a formula that depends only on T (Boltzmann's law, Appendix C), and $I(L,U)$ must be given by Planck's radiation law (Chapter 4 and Appendix C). When he imposed these strict requirements, A. Einstein found that he had to postulate a hitherto unknown physical process, induced emission or negative absorption. That is, an atom in level U, illuminated by radiation of exactly $v(L,U)$, is triggered to emit an additional photon which flies off in exactly the same direction as the incoming one.

Consider now an equilibrium situation with only radiative processes, i.e. there are no collisions. Then the number of absorptions from L to U will exactly equal the number of spontaneous emissions from U to L plus the number of induced emissions from U to L.

Suppose, however, that the number of atoms in level U would be maintained in excess of that given by this relationship – that is there would be an excess number of atoms in level U. We might imagine a stream of electrons moving with the right energy to excite level U. Alternatively, we might illuminate the atoms with radiation of exactly the right frequency, $v(L,U)$. Then a quantum v striking the atom in level U will trigger an emission of a new quantum of exactly the same frequency which will go off in the same direction as the triggering quantum, thus increasing the beam intensity. This phenomenon is the principle of the laser in optical regions and the maser in the microwave region. In order for a maser to work, we must have an excited energy level that readily emits a photon upon stimulation, but for which the probability of spontaneous emission is low. This condition is much more easily met in the far infrared and microwave region than in the optical region. Secondly, the atoms or molecules must be *selectively* excited to the higher level. The most promising mechanism is one whereby molecules are excited by infrared radiation or by collisions to a higher energy level from which they can cascade in such a way as to selectively populate the upper level of the masering transition, $v(L,U)$. The process bears some analogy to the Bowen fluorescent mechanism described earlier. As molecules accumulate in the upper level and their number exceeds that predicted by the Boltzmann law for an equilibrium situation, masering can occur. Mechanisms of excitation still present challenging problems for the theorist. Masers can enhance intensities a thousand to more than many million-fold. If the temperature of the background exceeds that of the maser, the latter can amplify the background. It can even amplify its own spontaneous emission.

The best-studied astrophysical masers are hydroxyl (OH), water (H_2O), and silicon oxide (SiO). Masering phenomena are found in dying stars, star-forming regions of the interstellar medium, and even in comets where molecules can be selectively excited by 'windows' between deep absorption lines in the solar spectrum. Masers are also found in external galaxies where they are sometimes stronger than in our own galaxy.

Maser lines can be measured with very high spectral resolution, and with the aid of the VLBI (very long baseline interferometer) we can construct angular images that are very much smaller than anything attainable optically. The spectral lines are relatively sharp and from their profiles the gas kinetic temperatures can be obtained (see Chapter 5). Measurements of circular polarization permit us to determine magnetic field strengths in molecular clouds from the Zeeman effect.

Fig. 10.18, adapted from one by Ray Norris, shows how an OH maser provides direct information on the expanding shell around a dying star such as a Mira or OH–IR object (Chapter 9). Expanding shells or blobs can also be found around stars in the process of formation. H_2O masers are observed in such regions but not in the same locales as OH masers. They occur in dense molecular clouds such as the Kleinman–Low object in Orion. A number of masers are observed to be moving rapidly away from a compact infrared source, IRC 2, which consists of a torus or ring, with material spewing outwards in the direction of the poles, thus forming a bipolar nebula, somewhat similar in appearance to some planetary nebulae, although here bipolarity is associated with a star-forming region. Compare Fig. 10.23.

A most striking feature of interstellar medium masers is their compactness; some have dimensions of the order of 100 astronomical units, i.e. about the size of the solar system. Their masses may be of the order of ten Jupiters, their total luminosities are of the order of 10 000 suns, and they are often associated with young, highly obscured stars. They are variable; a typical H_2O maser lasts about 3 months. In Orion, one water maser flared to make it the brightest object in the radio sky (a million janskys, 1 jansky $= 10^{-26} Wm^{-2}Hz^{-1}$); it would correspond to a 10^{20} megawatt radio station!

Radial velocities of masers are measurable by the Doppler shift very accurately, while very long baseline (radio-frequency) inteferometry (denoted as VLBI) can produce precise relative positions. Hence relative proper motions can be found. If a system of masers seems to be expanding from a fixed source, as is found in Orion, we can combine radial velocities and proper motions to get the distance. In this way, OH masers have given the distance of Orion as 480 ± 80 parsecs (cf. 400–500 parsecs by spectroscopic parallax methods), 7200 pcs to the galactic center.

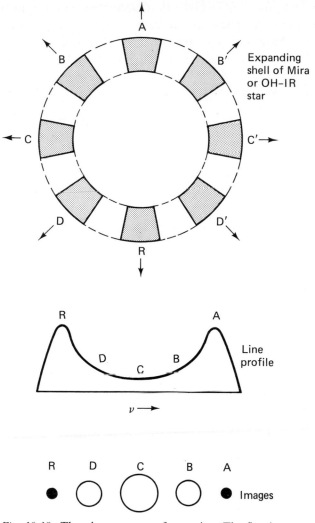

Fig. 10.18. The phenomenon of masering. The flux in a maser of any frequency depends on the number of molecules moving with the same velocity with respect to the observer. Larger numbers of molecules in approaching A or receding R shells produce peaks at A and R in the line profile. Smaller numbers of masers with similar frequencies are found at positions such as B, C, and D. When images are retrieved, A and R should yield small blobs centered on the star, B and D small, often broken, rings, while C gives the largest ring of all. By combining the velocities obtained from Doppler shifts of lines with the angular sizes of the rings, the distance of the star can be found. Time variations are also seen in the profiles. The effects are seen in A before they appear at R, indicating that the shell is expanding, not falling in. (Adapted from a diagram by Ray Norris, *Sky and Telescope*, March 1986.)

The actual physical processes involved in masering are not yet quantitatively understood, nor do we yet grasp why they seem to be so important in regions where stars form, nor why they so often appear when a star perishes.

The interstellar grains

A significant clue to the nature of the particles responsible for 'blacking out' distant stars comes from a study of the colors of stars that are only partially obscured. In Chapter 4 we saw how the color of a star is related to its temperature. The cool stars are red and yellow in color, whereas the hot stars are blue. The types of lines that appear in the stellar spectrum are a good indicator of the temperature, and therefore of the true color of the star. In many regions of the Milky Way one finds stars that show spectral lines characteristic of high temperatures, e.g. O stars, yet these stars appear red. We may, therefore, surmise that the light reaching us from these objects has not only been dimmed, but has been reddened as well. The phenomenon is not unlike the appearance of the sun as it sets, reddened, in a dusty or smoky atmosphere.

The fact that some interstellar particles possess the ability to redden starlight tells us that they must be smaller than about 0.03 millimeter in diameter, for larger objects, like meteorite fragments, will simply block starlight without affecting its color. At the other end of the size scale, we find that free electrons may also be ruled out as a factor in obscuring stars because they, too, are incapable of changing the color of light. Furthermore, although interstellar space contains many atoms and molecules, these cannot be blamed for the dimming of distant stars. Atoms and molecules in sufficient numbers to dim starlight are too efficient as reddeners to be the cause of the observed extinction. On the other hand, a mixture of particles with radii in the range 0.01–0.3μm can account for the observed scattering properties, reddening, and extinction of starlight. Particles as tiny as 10 Å also must exist.

Questions about the grains for which we seek answers are:

(1)   What are their optical properties; in particular what effects do they have on the extinction and polarization of starlight? What are their reflectivities or albedos?

(2)   What are the sizes and shapes of the particles? Are they flat, spherical, or elongated?

(3)   What are they made of and how are they constructed? Are they essentially small crystals or are they amorphous, i.e. disordered structures?

(4)   Presumably they originate mostly in envelopes of dying stars and are ejected into the interstellar medium. What is their subsequent history? They appear to grow, but certainly not to large sizes; are they perhaps cut down to size by shock waves and evaporation?

(5)   What role do grains play in the physics of the interstellar medium (e.g. as catalysts or sites for molecule formation)?

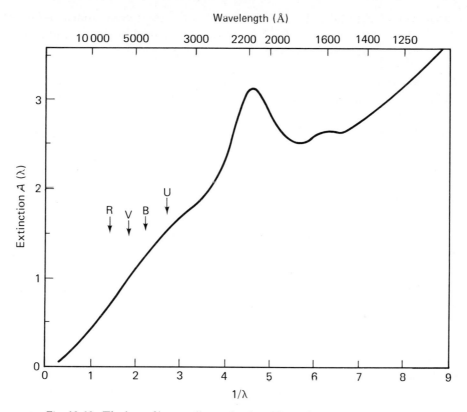

Fig. 10.19. The law of interstellar extinction. The extinction, $A(\lambda)$ is plotted against $1/\lambda$ (λ in μm) on the scale $A(5500 \text{ Å}) = 1.0$. One magnitude of visual extinction corresponds to an average distance of about 500 parsecs in the Milky Way, although the actual value may vary enormously from point to point. The upper scale gives the corresponding wavelength in ångströms. Note the strong extinction peak near 2200 Å. In the ultraviolet, the extinction curve shows considerable fluctuation with region in the Milky Way. U, B, V and R denote the wavelengths corresponding to different color systems. (This average curve is plotted from data by B. Savage and J. Mathis, 1979, *Annual Reviews of Astronomy and Astrophysics*, **17**, p. 84.)

The first clues to the nature of the interstellar particles came from studies of the extinction law, see Fig. 10.19. In principle, one compares the observed fluxes from two early-type stars of the same spectral class and energy distribution, one of which is hardly dimmed at all and the other which is heavily reddened. Early observations were confined to an optical range 0.3 to 2 μm over much of which the extinction, $A(\lambda)$, varies roughly as $1/\lambda$ and is the same throughout most of the galaxy. Extensions to the infrared and ultraviolet ranges gave additional clues. In particular, the ultraviolet observations showed $A(\lambda)$ to have a prominent maximum near 2200 Å (the shape of this feature resembles a broad absorption line, centered at 2170 Å, widened only by natural broadening (see Chapter 5) with a width of $\sim$480 Å.)

An important quantity is the ratio of reddening to total absorption. Suppose we measure the magnitudes, U, B, V, of a star in the three-color Johnson–Morgan system. The 'blue − visual' or $B − V$ color excess is defined by the equation:

$$E(B − V) = (B − V)_{\text{obs'd}} − (B − V)_{\text{true}}.$$

A method of determining the color excess for a star cluster is described in Appendix E. The ratio of total extinction to color excess $R = A(V)/E(B − V)$, where $A(V)$ is the observed extinction in the V system, is usually taken as about 3, although Harold Johnson suggested it changed from point to point in the galaxy. The average value of the extinction, $A(V)$, is about 1.8 magnitudes per kiloparsec, but the value fluctuates enormously and may amount to hundreds of magnitudes in cold clouds.

Of what are the grains composed? Candidates include ice (in cold clouds), amorphous silicates (a familiar example is quartz), carbon both in the amorphous form and as small ordinary crystals resembling graphite, and organic refractory compounds – possibly long-string polymers and polycyclic aromatic hydrocarbons. When a vapor condenses into solid particles it produces amorphous structures, although crystalline structures are in the long run more stable. Graphite is the most stable form of solid carbon. It consists of parallel thin slabs fitted together to form a crystal structure. Soot is the amorphous form of solid carbon; it is made up of randomly assembled thin platelets. When carbon abundance dominates, the particles are not likely to resemble the orderly structure of graphite; rather they may mimic something between graphite and soot.

Grains may consist of silicate or carbon cores surrounded by a mantle of refractory substances, possibly long polymers, or carbon chains. In cold clouds there may be a third component consisting of a shell of ices of water, CH, CO, etc. The grains evidently play important roles as catalysts to form molecules of methane, ammonia, water, and hydrogen.

Broad, diffuse absorption bands, of which the most striking is the 2200 Å feature, also include the 4430 Å band and several others than are correlated with color excess. In dusty clouds numerous absorption features usually attributed to grains and ices appear, although complex molecular structures may play a role. Most often observed are the 9.5 μm amorphous silicate line, the 3.08 μm band of water ice, and various bands at 3.3–3.6 μm, 4.6μm, and 6.0 μm, some of which are believed due to ices of CO, CN, water, ammonia, etc. Most fascinating of all are the polycyclic aromatic hydrocarbons (PAHs), to which several emission lines have been attributed. PAHs are well known as carcinogens, i.e. cancer-producing substances. They occur as waste products of our industrial society, e.g. smog. When samples of interplanetary dust and primitive carbonaceous chondritic meteorites are subjected to exciting radiation in the laboratory to produce what are called Raman spectra, they emit similar radiations, suggesting that the early solar system material may have contained contributions from dust clouds of dying stars.

Grains are not only non-spherical, they also appear to be elongated and aligned parallel to one another over large regions of space, presumably by a magnetic field. If light from a highly reddened star is observed through a piece of Polaroid, its brightness will appear to fluctuate by a small amount as the Polaroid is rotated. This means that the light is polarized and hence the obscuring particles scatter the light in a manner that depends on their orientation. To produce the observed effect, which was discovered by W.A. Hiltner and J. Hall, not only must the particles be elongated in form, but they must also be lined up.

L. Davis and J.L. Greenstein explained the mechanism of orientations. In the interstellar gas, collisions of atoms with particles set them spinning rapidly. The most likely mode of spin is about the shorter diameter, like a pencil turning end over end. Unless the long diameter of the grain is perpendicular to the magnetic field the grain experiences a change in magnetization as it rotates. This changing magnetization acts to dissipate the kinetic energy of the grain and causes it to line up with its short axis along the field. Then, if the magnetic field is perpendicular to the line of sight, we will see many of the grains broadside. Under these conditions, they will produce polarization. If we look along the direction of the field, we see the spinning particles tilted at every angle; there is no preferential direction and hence no polarization.

The existence of polarization tells us several things: (a) that there must exist magnetic fields in space; (b) that the grains of interstellar dust must be cooler than the gas, for if they were at the same temperature or hotter, they could not be aligned by the field; and (c) that since different types of grains would respond differently to magnetization and would have different polarization properties, we can narrow the range of prospective candidates. The actual amount of polarization is small and it shows a slight dependence on color.

The similarity of the extinction law from one part of the Galaxy to another suggests that grain size distribution must be rather similar over wide reaches of the Milky Way. If left undisturbed in the interstellar medium, grains would grow to sizes much larger than are observed, so some factors must limit their radii. Grains can be destroyed by blast waves from supernovae, by sputtering by fast-moving ions, by collisions between clouds, and by shock waves. Those in large molecular clouds might be protected from supernovae but would be lost in the process of star formation. A typical grain may last ten to a hundred million years.

Violence in the interstellar medium: the role of shock waves

If a compressed shell or sheet of gas travels into a rarefied gas with a speed exceeding that of sound in that medium, it produces what is called a shock wave. The atoms in the advancing shock front are all moving parallel to one another but the collision causes kinetic energy to be converted to heat. The effect is similar to that exhibited when a lead bullet hits a target and its kinetic energy is suddenly converted into heat sufficient to melt the bullet.

Fig. 10.20 depicts the situation. We can imagine ourselves traveling along with the shock wave so that the low-density gas flows into the shock front. A relatively cool gas is suddenly struck by a fast-moving blast wave. We define the ratio of the velocity of the shock wave to that of sound in the gas as the Mach number. Not surprisingly, the bigger the Mach number, the more dramatic the effects. As the low-

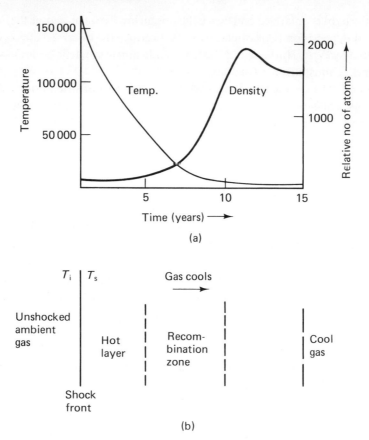

Fig. 10.20. Shock phenomena in a gas. We assume that the pre-shocked gas is at a temperature of 8000 K with a density of 10 hydrogen atoms/cm³, corresponding to a warm cloud. It is suddenly hit by a shock wave traveling with a velocity of 100 km/s. Part (a) depicts the sudden rise in temperature (so rapid that the temperature line rises almost vertically along the left axis), which is followed by a steady fall, while the density peaks and then declines slightly. In part (b) we imagine that we travel with the shock front and gas travels through it. The gas is quickly heated and becomes compressed. It recombines and gradually cools. T_i is the initial temperature of the gas before it is shocked, T_s is the temperature of the shocked gas.

density gas passes through the shock front, it becomes highly compressed and its temperature rises abruptly. The atoms become excited and many become ionized. Then, as the hot gas recombines, it radiates energy and cools. The denser the gas, the faster it will radiate and return to something resembling its initial state, but at a higher pressure. If the density is very low, the gas can remain hot for a long time.

In a strong shock wave in the interstellar medium we may see X-ray lines of elements such as neon, oxygen, and sulfur just behind the front. In the ultraviolet we observe lines of highly ionized atoms, e.g. C III, C IV, Ne III, Si III, N III, N IV, N V, and Lyα. In the optical region we will later see [O I], [O II], [O III] and [S II], and in the infrared, [S III]. As the gas cools still further the vibration-rotation lines of CO appear, as do the molecular lines of hydrogen. A given parcel of gas will experience this sequence of events as time goes on. Likewise, if we look at a shock running through the interstellar medium, we will see the successive stages of heating, compression, and subsequent cooling and recombination spread out in space behind the shock front. The spectrum of a shock-heated gas differs from that of a gas excited to glow by a hot star in that a much greater range of excitation is displayed, from X-rays to molecular bands.

Shock waves may be produced by stellar winds; for example molecular hydrogen emission may be produced in the cool, outer, neutral envelope of a planetary nebula by shocks. Collisions between high-velocity clouds can also produce shocks, but the most dramatic effects are caused by supernovae. A gas of only 0.001 atoms per cubic centimeter will not have time to cool between shocks from successive supernovae. A denser gas may be able to cool and produce several moving layers behind successive shock fronts, manifest as sheets, filaments, and shells often seen in neutral hydrogen.

Many diffuse clouds are actually sheets or thin slabs of gas. From curve of growth measurements of absorption lines we get the amount of material in the line of sight. From the ionization equilibria of C, Mg, S, and Ca, and by assuming that the gas is ionized by the diffuse light of Milky Way stars, we find the densities of electrons and hydrogen atoms. Since $N(H)L$, the number of atoms in the line of sight, is known from the curve of growth, we can solve for the thickness, L, of the cloud. For example, $L = 1.0$ parsec for the ζ Ophiuchi cloud. Since the same absorption lines are seen in the spectra of many stars scattered over a wide angle in the sky, we conclude that the width of the cloud may be several parsecs, indicating that we are seeing a sheet of material more or less perpendicular to the line of sight.

Not all clouds are formed by shocks and compression. Some are sufficiently massive for gravitational effects to be important. Great arcs, seen especially in the light of [S II] in our own and other galaxies have been identified as supernova remnants. They consist of material ejected from the supernova and material of the interstellar medium that has been compressed by the snow-plow effect of the advancing shock (see Chapter 11).

Shock-wave effects may occur in spectra where the principal source of the excitation is starlight, as in Mira stars, in the envelopes of other stars in advanced evolutionary stages, or in planetary nebulae. The rich emission-line spectrum of the bright, visible ring of NGC 7027 is primarily excited by radiation from the central star, but in the cool, optically invisible outer shell molecular hydrogen, H_2, is excited by shock waves. In fact it appears that shock waves are responsible for the excitation of narrow sheets and filaments in other planetaries, and for the excitation of the extended haloes of planetaries such as NGC 6543 and NGC 6826. Shock waves appear to be produced at the edges of ionized zones in H II regions, such as the complex Orion Nebula (Fig. 10.21), and probably play roles in the excitation of molecules in extensive cool clouds.

Magnetic fields in the Galaxy

Steady-state, large-scale electric fields probably do not exist in the Galaxy since no separation of electric charges can be maintained in an ionized gas. Magnetic fields

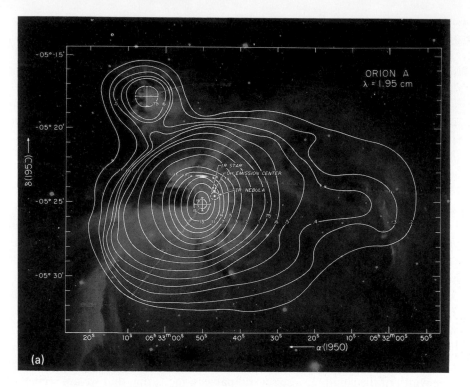

Fig. 10.21. Comparison of direct and radio isophotes of the Orion Nebula. (Isophotes are contours of constant radio flux or surface brightness.) Note that radio contours do not fit the optical pattern because of dust obscuration. The radio fluxes come from sudden accelerations of free electrons by protons (so called free–free emission or bremsstrahlung); hence they define the regions of ionized gas and tell us nothing about how the cold gas is distributed.
(a) Measurements at 1.95 cm by J. Schraml and P.G. Mezger with the National Radio Astronomical Observatory 42-m telescope are superposed on a direct photograph. Positions of some OH and IR sources are indicated. (Courtesy P.G. Mezger.)

almost surely do exist, however. We have already seen one piece of evidence – the polarization of starlight, the Hiltner–Hall effect. G.A. Shajn noted that in the regions where polarization was observed, filaments of ionized gas tended to be lined up along the corresponding inferred magnetic field direction. Magnetic fields tend to constrain electrically charged particles to motions parallel to the lines of force; hence a gaseous nebula would tend to deform along the direction of the magnetic field, rather than perpendicular to it.

Cosmic-ray physicists have postulated magnetic fields within the Galaxy in order to explain how high-energy particles are retained within it. Additional evidence is suggested by the presence of powerful emitters of radio-frequency radiation at very long wavelengths, the so-called non-thermal radio-frequency sources. This radiation is usually attributed to emission from high-energy electrons moving in magnetic

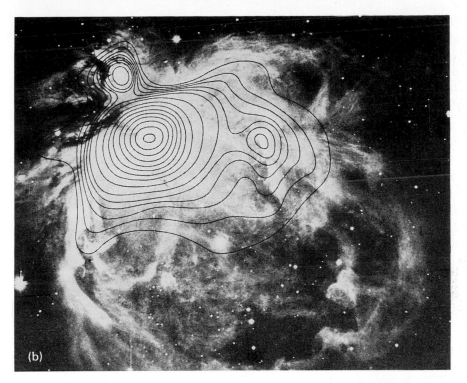

(b)

Caption for fig. 10.21 (*cont.*)
(b) Measurements at 74 cm (408 MHz) secured with the Mills Cross at Molonglo
Radio Observatory of Sydney University are compared with a Lick Observatory
photograph. Notice the significant differences in detail between (a) and (b). At 408
MHz the radio telescope sees through the nebula, but at 2 cm the nebula is opaque to
the emission and we can see only surface features in these regions. (Courtesy B.Y.
Mills and P.A. Shaver.)

fields. We would like to know the properties of the galactic magnetic field: how
strong is it, what is its large-scale structure, and how did it originate. The galactic
magnetic field is important for several reasons:

(1) It affects the motions of charged particles, so even an incompletely ionized
cloud could be significantly affected.
(2) It acts to constrain cosmic rays in the Galaxy, and perhaps also to accelerate
them – a very important effect (see Chapter 12).
(3) It may play a role in maintaining an extended, attenuated, halo of gas
around the Galaxy.
(4) Energy stored in the galactic magnetic field is comparable with that stored
in kinetic energy of particles in the interstellar medium.
(5) Near the galactic center fields of the order of a hundred times stronger than
the general galactic field are found. These will influence local gas motions
and may affect star formation.

All galaxies with interstellar media appear to have magnetic fields which are revealed by synchrotron radiation from high-energy electrons moving with a velocity approaching that of light (see Chapter 12).

How can we measure the magnetic field? One thinks of using the Zeeman splitting (see Chapter 2) but the field involved is of the order of 3 or 4 microgauss, that is a thousand million times weaker than in a sunspot or a hundred thousand times weaker than the earth's field. The problem would appear hopeless, except that the neutral gas is cool and the Doppler effect and other sources of line broadening are unimportant. If the magnetic field is along the line of sight the two components are circularly polarized in opposite directions. The Zeeman splitting of the 21-centimeter line gives the strength of the magnetic field in the cold neutral gas, not in the ionized gas.

Another method of evaluating the interstellar magnetic field is to measure the rotation of the electric vector of plane-polarized radiation of some distant source such as a quasar or pulsar. More than a century ago, Michael Faraday noted that if plane-polarized light was passed through a transparent substance in a magnetic field, the plane of polarization was rotated. The effect was a maximum if the light ray was parallel to the magnetic field; it depended on the square of the wavelength of light, the substance involved, and the strength of the field. The ionized gas of the interstellar medium produces a distinct Faraday effect on plane-polarized radio-frequency emission from distant, non-thermal sources, galactic or extragalactic, or relatively nearby pulsars (see Chapter 12). If one can estimate the electron density in the intervening interstellar medium, and know the path length through this gas, the magnetic field strength can be estimated. Observations at several wavelengths are needed. What we obtain is the field strength in the ionized component of the gas; the neutral portion gives a negligible Faraday effect. Observations are difficult because some Faraday rotation may occur in the source itself, or in the earth's ionosphere. Since the direction of rotation depends on the magnetic field direction, if the radiation passes through a twisted tangle of magnetic fields, considerable cancellation may occur.

The polarization of starlight and the presence of synchrotron radiation from non-thermal sources give the field topology relative to the line of sight, but say little about its strength. Stellar polarization data suggest that the field runs roughly along the spiral arms but is subject to many twists and kinks. The magnitude of the field is variable. Near the galactic center, radio-frequency observations by Mark Morris and his associates show a series of parallel aligned filaments emitting synchrotron radiation which is generated by electrons moving at nearly the velocity of light (see Fig. 10.22). Here the field is of the order of 0.001 gauss, as compared with 3–4 microgauss in our part of the galaxy! Such a powerful field strongly influences the local interstellar medium and may also act to inhibit star formation.

The problem of the origin of the interstellar magnetic field is far from being resolved. Presumably it is not a remnant of an original or primordial field, because during the lifetime of the Galaxy such a field would long since have become so

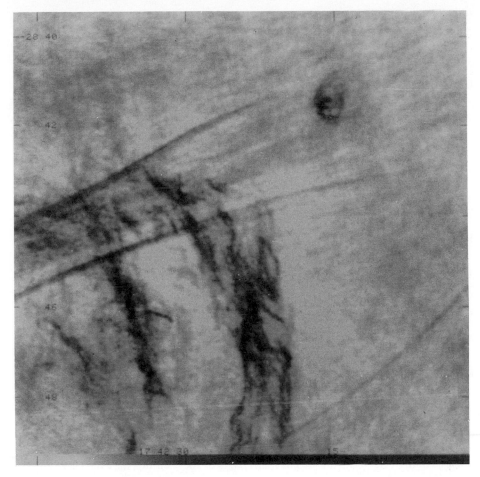

Fig. 10.22. Radio frequency continuum radiation near the galactic center. These filaments, some of which extend for up to 40 parsecs along a line perpendicular to the galactic plane, show polarization which indicates that they arise from motions of very fast electrons in a magnetic field which is much stronger than that found in the outer reaches of our Galaxy. This radiation differs fundamentally from that emitted by a hot ionized gas (cf. Chapter 12), but there is a large cloud of such 'normal' gas in the neighborhood. This region has been extensively studied by Yusef-Zadeh, M. Morris, O.B. Slee, G. Nelson, B.Y. Mills, and others. (Courtesy Mark Morris.)

tangled and twisted as to be undetectable. We suspect that some kind of a dynamo action operates as with the earth's field, but the question remains – How?

Molecular clouds

Certainly the most engaging objects in the interstellar medium are the molecular clouds which are found with a wide range in density, size, mass, and chemical complexity. Most interesting of all are the giant clouds where star formation takes place. Most of the clouds and all of the denser ones are found near the plane of the

Galaxy, but the IRAS satellite detected a wispy complex of clouds far from the galactic plane; these are the so-called high-latitude cirrus. They have relatively low densities, 30–500 atoms/per cubic centimeter, sizes of the order of 1–3 parsecs, and masses 1–100 times that of the sun. They may simply be diffuse clouds not involved in star formation.

Projected against bright nebulae such as the Lagoon Nebula (M8) one observes a number of small, round, dark clouds, commonly called globules. B.J. Bok tried to estimate their sizes and masses; typically they appear to be only a fraction of a parsec in size and contain less than a solar mass of material. Probably they are mostly too small to be gravitationally stable. Although some may be sufficiently massive and dense to evolve into stars, most stars must be formed from condensations in dense clouds where gravity plays a big role.

The extended clouds that fill so much of the constellations of Taurus, Auriga, and Perseus are complicated structures made up of individual components with masses between 1000 and 10 000 times that of the sun, and densities typically of the order of 100 to 1000 atoms per cubic centimeter. Individual clouds in the mixture may be 5 to 20 parsecs in size with small dense cores; possibly they are held together by gravitation. In these regions of extensive cool gas and dusty obscuring material are found small fan-like nebulae associated with variable stars of the T-Tauri or R-Monocerotis type. The bright-line component of their spectra has low excitation, often resembling those of planetary nebulae. The shapes of spectral lines originating in the star-like cores suggest that material may be falling into these objects, and also may be ejected in many instances, indicating complicated scenarios. These T-Tauri stars are objects in the early stages of their lives.

Giant molecular clouds are entities that are so large and dense that their gravity may not only hold them together, but may also strongly influence motions of stars and gas in their neighborhood. The main parts of these clouds may have sizes of 20–200 parsecs and involve cores with densities of roughly 10 000–100 000 atoms or molecules per cubic centimeter. The total masses of these complexes may amount to a hundred thousand to a million suns, or even more. Perhaps the nearest giant molecular cloud and region of star formation is the highly obscured ρ Ophiuchi structure. Its proximity makes it possible to secure measurements of high angular resolution and thus to observe the 'fine structure' of a star-forming region. Optically it has a radius between 3.5 and 6 parsecs, contains five compact H II regions, and a cluster of about 50 stars. Three of these are of high luminosity; one is hidden by 100 magnitudes of extinction in the optical region. Most of the imbedded stars, including some T-Tauri objects, are evolving towards the main sequence. Velocity studies give evidence for shock phenomena and suggest that the onset of star formation was triggered by shock waves. Here is a rich treasure-trove of molecules: OH, CO, SO, H_2CO, CN, CS, CH, NH, CH^+, HCO, HCO^+, HCN, N_2H^+, NH_3, and H_2O (observed as ice in the 3.1-μm feature). The magnetic field reaches a value of 0.0005 gauss in some regions of the cloud which are associated with X-ray sources and a 100 mega electron volt gamma-ray source. No neutron stars or black hole candidates have been found.

For some giant molecular clouds, detailed contour maps for different molecular species can be made, showing how different molecules may accumulate in different locations. Sometimes, as in the Kleinman–Low nebula in the Orion complex, striking chemical inhomogeneities can be found. The Orion complex itself, some 500 parsecs distant, is the most massive and engaging nearby region of star formation. The complex covers about 30 square degrees of the sky, the Orion Nebula involving only a small fraction of the total. This H II region is gradually eating its way into the huge dark bulk of material behind it. It is our good fortune that we are on the right side of this vast cloud. Within such a structure, very luminous stars can be formed, live their brief spectacular lives, and expire, completely unseen!

The giant molecular cloud Sagittarius B-2 (Sgr B-2) lies some 200 parsecs from the galactic center. It can be studied only by infrared and radio techniques since not only is the central region of the Milky Way hidden by some 25 magnitudes of extinction, but the extinction produced within the cloud itself amounts to ~ 3500 magnitudes or an attenuation of 14 000 decibels! This heavy shielding from radiation permits Sgr B-2 to be the richest source of interstellar molecules (see Appendix F). At the center the grain temperature is about 35 K, low enough to freeze out most gases, but the cloud temperature and density vary and different molecules tend to be concentrated in different areas. This innermost region is subject to a strong magnetic field and powerful tidal action from the very massive central core of the galaxy, which may actually be a black hole. Sgr B-2 is so massive (three to ten million suns) and so compact (diameter about 40 parsecs), that it is stable. Infrared measurements indicate a total luminosity of seven million suns.

Star formation: seen through the dust but darkly

As the demise of a star was hidden by the dust it had ejected in the last moments of its life as a red giant, so also is a star's birth enshrouded in dust, this time the dust of the interstellar medium. As long as our view was restricted to the optical spectral region there was little hope of getting any kind of a satisfactory answer to the question: how are stars formed? With the advent of infrared and high-spatial-resolution radio-frequency techniques, the veil was gradually pulled aside, but questions remain. Our life span is short compared to that of a star, but by observing many objects at slightly different epochs in the star-formation process and using the full gamut of observational techniques, we eventually may be able to give a satisfactory answer to the question.

Stars form from molecular clouds, but how do these clouds break up into clusters and associations of stars? For massive stars, those of 10–30 times the mass of the sun, we suspect that some triggering event such as shock wave from a supernova may be involved, but for stars of one solar mass a different mechanism may occur. Massive stars seem to form in certain giant molecular clouds, like Orion or in the 30 Doradus complex in the Large Magellanic Cloud (see frontispiece). In the Taurus cloud only stars of relatively low mass seem to form. The general predominance of stars of less than a solar mass suggests that the prevailing, overall, galactic star-building processes favor these objects, perhaps even in Orion.

A typical extended molecular cloud destined to spawn stars may have a mass

1000–10 000 times that of the sun, a density of 100 to 1000 atoms or molecules per cubic centimeter (mostly molecular hydrogen), a size of 2 to 5 parsecs, and an internal temperature of about 10 K. The cloud is likely to be clumpy; it may contain a magnetic field of 15 to 130 microgauss (revealed by the Zeeman splitting in lines of OH).

Extensive theoretical studies by Peter Bodenheimer, Frank Shu, D.J. Hollenbach, and their associates, as well as by other groups of workers, give some idea of the sequence of events leading to the formation of low-mass stars. A solar mass contained in a cloudlet of 0.1 parsec (20 000 AU), with a magnetic field of less than 15 microgauss can eventually collapse into a proto-star. The imbedded magnetic field can be moved out of the cloud in about ten million years, by the action of what are called magneto-hydrodynamical or Alfvén waves. These also carry away angular momentum, so the spinning mass can contract further. It is essential for the magnetic field to diffuse out of the mass before the core can collapse. The gas cloud still spins and carries considerable angular momentum. Thus it does not contract into a sphere, but into a disk structure that surrounds a denser core. A proto-star consists of an infrared source at the core and a thick dusty disk that forms in about a hundred thousand to a million years. As the core further contracts and heats up, a wind develops. In the equatorial plane of the thick disk the wind is quickly stopped, but in the polar regions it breaks out. Infall of gas and dust stops because of the wind and exhaustion of available material. The stellar core now becomes visible as a T-Tauri-type variable, far to the right of the main sequence in the HR diagram. The subsequent evolution of the dusty disk is of considerable interest. Energy is dissipated, mass is transfered inward and angular momentum outwards. A planetary system may be formed as solid particles in the disk grow into larger bodies by accretion. Eventually all the material is swept up into planets or smaller bodies, falls into the star, or is ejected from the system. Strong observational evidence supports this scenario.

The IRAS satellite has detected rings around more than 24 main-sequence stars including Vega and β Pictoris. Some giant (luminosity class III) stars have rings. Vega's ring has a radius of about 85 AU. If the particle size is about 1 millimeter, their total mass is less than 0.01 solar masses. Grains smaller than 1 mm would gradually be dragged into the star by what is called the Poynting–Roberston effect in less than a thousand million years. We may speculate that the material resembles that found in comets. Since the temperature of the Vega dust cloud is 60 to 130 K, it may consist largely of dirty ice grains or snowballs. We cannot rule out the possibility of larger bodies in the systems of Vega and β Pictoris. Massive star formation, which may also occur in regions where less massive stars are assembled (e.g. the Trapezium in Orion) is a more dramatic phenomenon. Fig. 10.23 shows a possible scenario.

The formation of binary stars remains a puzzle; the fission of a rapidly spinning star will not produce a binary. Probably the fragmentation occurs before the disk is formed. The frequent occurrence of binary systems suggests that angular momentum is shared between two bodies of comparable mass at a relatively early stage before the formation of the thick disk and infrared core.

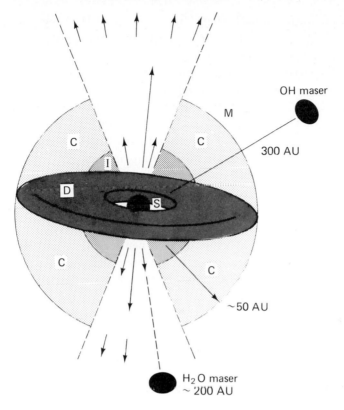

OH maser

M

300 AU

C

C

I

D

S

C

C

~50 AU

H$_2$O maser
~ 200 AU

Fig. 10.23. The emergence of a massive star. This schematic diagram depicts a 'Becklin-Neugebauer' infrared object which is believed to be a massive O or B star (S), nearing the main sequence. It is still surrounded by a thick circumstellar disk or torus (D), and immersed in a vast molecular cloud (M) which includes both OH and H$_2$O masers. An inner region, I, of radius R_0, is ionized. A zone, C, between the dense molecular cloud and the ionized region, still contains refractory material, but all ices have been melted. The sketch is not drawn to scale. Typical distances are given in astronomical units; they may differ considerably in a single object with time, or from object to object. (Adapted from T. Henning and J. Gurtler, 1986, *Astrophysics and Space Science*, **128**, p. 199.)

Massive stars originate in giant molecular clouds like Orion. They evolve quickly through the proto-stellar stage and, when the nuclear fuel is ignited, they become bright and hot, producing an H II region such as the Orion Nebula, which grows as the cloud is ionized. The glowing nebula grows until the total ionization rate by the photons from the hot star(s) equals the rate of recombination. At the interface between the H II region and the cool, molecular region, a steep jump in gas pressure appears since the H II region has a temperature of 10 000 K and the H I region is at about 80 K. Then shock waves may appear and travel both inward toward the star and outward into the cool, molecular cloud. From time to time other hot stars may form producing new contributions to the H II region or setting up entirely new ones. Very massive stars may die as supernovae, causing huge shock waves that may further compress the cloud and form new stars or may even destroy what remains of the giant cloud. Evidence of such effects appear in giant H II regions of other galaxies such as 30 Doradus or NGC 604 in the Triangulum Spiral, M33.

Research on the interstellar medium is now tremendously active as a consequence of new observational techniques, especially in infrared and radio-frequency ranges, and the development of new theoretical insights. It appears that we are beginning to understand the great complexity of the interstellar medium and even how single stars are formed, although binaries still elude us. The original interstellar medium consisted of the primordial hydrogen and helium from the Big Bang; it was later enriched by the ejecta of early generation stars, perhaps including many supernovae, modest stars that evolved into white dwarfs, etc. The solar system was formed some 4.5 thousand million years ago from an interstellar medium that was already 5–10 thousand million years old, and had been well 'salted' with the output from many long-defunct nuclear furnaces. What of the future of the interstellar medium? Remember that the formation of white dwarfs, neutron stars, and black holes removes material from the cycling forever, so the interstellar medium suffers from irreversible attrition. Unless great quantities of fresh gas are flowing in from the space between the galaxies, our Milky Way may be nearing the end of its star-forming capacity. M. Jura and also A. Sandage have estimated that at its present rate of attrition the interstellar medium can last only about another 1.5 thousand million years; 90 percent of its life is over. Our Galaxy might then become an S0-type spiral, virtually without dust or gas. We came along, perhaps, at just the right time, so we can still figure out how stars are formed – before the clues vanish.

11

Uncommon stars and their sometimes violent behavior

The nova phenomenon

On the evening of 8 June 1918, the noted American astronomer, E.E. Barnard, was driving through the countryside in a dejected mood. The spectacle of a clear, starlit sky only added to his dejection, for on that same afternoon cloudiness had thwarted his attempt to observe a total eclipse of the sun. As his eyes wandered aimlessly over the familiar star patterns of the Milky Way, they came to the constellation Aquila. Suddenly aroused, Barnard pointed upwards excitedly and exclaimed: 'That star should not be there!' Barnard's startled companions, fearing that his disappointment of the afternoon had unbalanced him, were nonplussed until they realized that he had discovered a nova. Indeed, this new first-magnitude star in Aquila was discovered independently by dozens of people on the same night. Older photographic records at some of the observatories showed that on 5 June the star was of the 11th magnitude, as it had been for the previous 30 years. Forty-eight hours later it was of the sixth magnitude – a hundred-fold increase in brightness. Still increasing on 9 June, its magnitude was -0.5, thus outshining all the stars in the sky but Sirius and Canopus. After attaining the peak of its splendor, the star began to fade, rapidly at first and then more slowly. In 18 days it had declined to the third magnitude, and it faded from naked-eye view about 200 days later. Its present magnitude is close to the original value before the outburst, but it is still in reality five or six times as luminous as the sun.

Before the advent of large telescopes, apparitions such as that described above were regarded as actual new stars, and even now the designation 'nova' remains in common usage, although the photographic records show that bright novae like Nova Aquilae 1918 existed as faint stars before their outbursts. All available evidence indicates that a nova is the result of a stupendous stellar explosion whereby the outermost layers of a star are wrenched away.

The conclusion that novae are the results of explosions in the surface layers of stars was deduced from the remarkable spectral changes that accompany such 'outbursts'. In a typical nova, such as Nova Cygni 1920, the changes may be described as follows. In the early stages, the ejected surface layer swells up like a balloon. As it grows larger, it becomes brighter in total light, although the apparent photospheric temperature actually drops. At maximum light, the spectrum is continuous and crossed by dark lines, as it was before and during the rise in brightness. Evidently the photosphere, although enormously distended, is still intact. The dark lines of hydrogen and ionized metals are produced in that portion of the atmosphere that is between the surface of the main expanding cloud and the observer. While the expanding shell grows rapidly, the atoms in the line of sight are in rapid motion towards the observer (Fig. 9.4, p. 179). Consequently, the absorption lines are shifted violetward from their normal positions by the Doppler effect. The magnitude of the shift gives the rate of expansion. Expansion speeds of 2000 kilometers per second are not uncommon.

When the ejected layer has expanded yet further, its gases become so rarefied as to be virtually transparent. Then we see through to the far side of the envelope and obtain a view of the whole expanding shell. The continuous spectrum fades and bright lines emerge, much widened by the Doppler effect. The star itself shines on, emitting a continuous spectrum. But the light from all parts of the expanding shell that do not lie between the star and the observer will produce a bright-line spectrum. Summing-up all of the contributions (see Fig. 9.4), we observe what is called a P-Cygni-type profile. See also Fig. 11.3.

After the original shell has become fairly rarefied, the absorption lines disappear. Then we observe diffuse, high-excitation lines of permanent gases that originate from a continuously ejected cloud close to the star, while the principal shell is now greatly expanded. As the gases become more transparent, the continuum weakens; the depth to which we can see becomes deeper as the amount of ejected material diminishes.

As the gas expands further, the bright metallic lines steadily weaken, while those of hydrogen (H), nitrogen (N), and oxygen (O) remain prominent. The behavior is easily explained. Metal atoms have low ionization potentials; their outer electrons are easily torn away. Multiply-ionized atoms absorb and emit light almost exclusively in the ultraviolet spectral region and can be observed only with space telescopes. 'Permanent gases' like N, O, and neon (Ne) are more difficult to ionize. Their lines remain long after those of the metals have disappeared.

Yet later, the most prominent lines are the 'forbidden' ones, characteristic of gaseous nebulae (see Chapter 10). At this stage the material is thinned out to a density of about a million million times less than that of the air we breathe. Meanwhile, the continuous spectrum of the star has fallen almost to the same feeble intensity it had before the outburst, while the bright lines of the nebular shell fade more slowly. The core of the nova may also show strong emission lines. Eventually, after the uproar has subsided, the gaseous shell has melted away into the interstellar medium, and the nebular spectrum disappears. The star returns to its normal, pre-outburst state, perhaps to erupt again after thousands or even millions of years. Some recurrent novae, though, repeat after 10 to 40 years.

Not all novae show the same type of light curves or spectroscopic behavior (see Fig. 11.1). Intrinsically more luminous novae decline more rapidly, whereas less luminous objects, such as DQ Herculis, decline more slowly. Just before Christmas

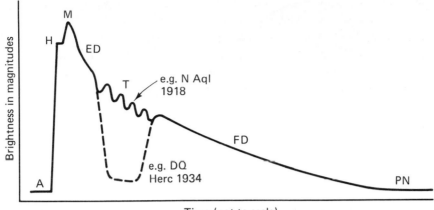

Fig. 11.1. Idealized optical-region light curve of a nova. The time scale is greatly expanded in the early stages; the decline to the final post-nova stage may require years. The letter A denotes the brightness of the pre-outburst nova. The initial rise, typically about nine magnitudes, takes place in 2–3 days. Then there is a 'stand still' or pre-maximum halt, H, just before the star reaches its maximum at M. Then follows an early decline (ED) of 3–3.5 magnitudes, which may occur in about 2 or 3 weeks for a fast nova or 2 or 3 months for a slow one. Subsequently, the star enters the transition period (T) during which it may fluctuate in brightness in a quasi-periodic fashion, as did Nova Aquilae 1918, or plummet perhaps a hundred-fold in brightness, as did DQ Herculis 1934, to partially recover later. At the end of the transition period, the star has fallen about six magnitudes below maximum. Then follows the final decline (FD) to the post-nova (PN) stage. The time scale differs enormously from slow to fast novae; the latter are intrinsically much brighter. Even though the visual brightness may decline precipitously, the bolometric luminosity may remain constant for a long period of time. (After Dean B. McLaughlin.)

1934, young observers working with an objective prism camera at the University of California's Students' Observatory in Berkeley were flabbergasted to see strong cyanogen (CN) absorption bands in the immediate post-maximum spectrum of Nova DQ Herculis. Up to this point the spectrum had looked normal. As the photospheric spectrum weakened, the bands disappeared as mysteriously as they had come. It was not realized then that novae can be copious producers of carbon.

The spectroscopic evidence that novae do indeed cast off shells of gas was confirmed by direct observations of Nova GK Persei (1901), Nova V603 Aquilae (1918), and Nova DQ Herculis (1934). Six months after the Nova Aquilae outburst, a faint greenish envelope or shell could be seen in the telescope; it has been expanding at about the rate of 2 arcsecs per year. Although the shell ejected by Nova Aquilae appeared roughly symmetrical, that of Nova Persei is strongly asymmetrical, as though most of the material came from a single hemisphere.

Expanding shells help us determine distances of novae. From the spectral line shifts, Doppler's principle gives us the actual rate of expansion in kilometers per second. If we can measure from direct photographs taken at different epochs the angular rate of expansion of the shell, we may compute the distance. The values

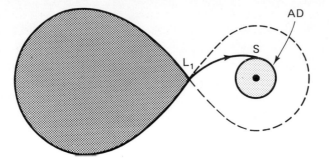

Fig. 11.2. Schematic diagram of a cataclysmic variable. The originally more massive star has evolved into a white dwarf which is surrounded by an accretion disk, AD. The companion, which normally would have evolved into a conventional red giant, is constrained by the presence of its companion, the white dwarf, to extend no further than its Roche lobe. Material stripped off the donor star through the Lagrangian point, L, in stream, S, feeds the accretion disk. Development of instabilities in the disk surrounding the white dwarf leads to various types of outbursts (see text).

obtained in this way for novae Herculis, Aquilae, and Persei are about 1300–1500, 1200, and 2000 light-years, respectively.

Novae and cataclysmic variables

From Chapter 8 we recall that the evolution of stars in close binary systems differed spectacularly from what happened with isolated objects. At a late stage in the history of the system, the principal (originally more massive) component will have evolved into a white dwarf, while its companion has swollen up to fill its own Roche lobe. Material from it begins to flow through the Lagrangian 'pinch' point, L_1, towards the white dwarf (see Fig. 11.2). The gas which now enters the gravitational field of the white dwarf does not score a 'bull's eye' by falling directly on the small target of the white dwarf, but circles around the latter in a thin, flat, 'accretion' disk. This gas carries angular momentum as well as mass. Slowly, viscosity in the disk causes angular momentum and a small amount of mass to be transferred outward as most of the material spirals inward.

The gas lost from the envelope of the secondary (giantish) star is hydrogen rich. This material is fed through the accretion disk, onto the white dwarf surface, where temperature and density increase until the point is reached where hydrogen may be ignited by very rapidly occurring (runaway) nuclear reactions. Much energy is liberated; a shell is ejected cataclysmically. Variable stars that follow this type of scenario are often called 'cataclysmic variables'. The detailed sequence of events depends on the type of star involved, separation of the components, etc.

Suppose, though, that the material did fall directly from the companion star onto the surface of a white dwarf of one solar mass, with a radius 1.6 times that of the earth. It would hit the white dwarf photosphere with a velocity of 3600 kilometers per second, liberating thirteen million kilojoules per gram, which would heat it to a temperature high enough to produce X-rays. Even with the accretion disk acting as a buffer, the infall is supersonic and shock waves are produced.

Our present model of a nova outburst is based largely on the work of S. Starrfield and his associates. We envisage the material as gradually being accreted on the white dwarf until it accumulates to a point where the temperature of the compressed gas exceeds a hundred million degrees. There then occurs what is called a TNR or thermonuclear runaway reaction, which differs fundamentally from the type found in normal hydrogen burning in stars (see Chapter 8). In a main-sequence star which burns hydrogen tranquilly by the CNO cycle, the rate of the reaction is fixed by the speed at which nuclei of C, N, or O can capture protons. The bottleneck of the process is usually the rate at which ^{14}N captures protons. At temperatures greater than two hundred million degrees proton capture is virtually instantaneous and the bottleneck of the process is the rate of the beta decays, 863 seconds for ^{13}N, 176 seconds for ^{15}O. Much heat is liberated at the base of the zone of accreted material, rapid convection occurs, and fresh C, N, O, and H nuclei are poured into the shell source. At the peak of the outburst the most abundant of the C, N, and O nuclei are those unstable to beta decay (e.g. ^{13}N which decays to ^{13}C). Lavish nuclear power output (10^{12} to 10^{13} erg/g/s) furnishes enough energy for the ejection of the overlying strata.

The light curve, maximum light, and velocity of ejection will depend on the amount of C, N, and O present in the envelope; the greater the number of these catalysts, the greater the amount of hydrogen that can be cycled. The nova is luminous in the ultraviolet, the extreme ultraviolet or EUV, and in soft X-rays. The light maximum is reached when the radius is between ten and a hundred million kilometers and the envelope temperature is 7000 to 9000 K. Then the photosphere becomes transparent, the visual luminosity declines, but the bolometric luminosity remains high as peak energy moves to the ultraviolet, permitting excitation of high-lying levels of atoms and ions of permanent gases. The spectra near maxima often show great complexity and sometimes resemble those of high-luminosity stars such as P Cygni (see Fig. 11.3).

Starrfield suggested that only 10 to 50 percent of the material actually accreted is lost in the initial explosion. The rest may escape by a vigorous wind and/or by the tidal action of the companion star. Eventually, the white dwarf succeeds in ridding itself of the excess material; it settles down to the pre-outburst phase as the accretion disk is restored. The system may evolve for another ten thousand or hundred thousand years before another outburst.

Novae show a great range in luminosity, ejection velocity, and in the chemical composition of the ejecta. The brightest recent nova appears to have been Nova V1500 Cygni 1975, whose maximum radius was about two hundred million kilometers, and whose luminosity exceeded a hundred thousand times that of the sun. The type of outburst depends not only on CNO enhancement but also on the white dwarf mass. The white dwarf actor in the nova drama has a typical mass greater than that of the sun (m(sun)), i.e. much larger than that of a planetary nebula nucleus. Starrfield found the mass of the slow nova HR Delphini 1967 to be 1.25

Nova Vulpeculae
(1968)

Nova Delphini
(1967)

P Cygni

29 Vulpeculae

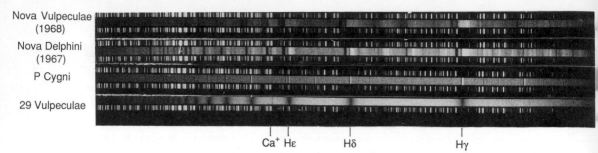

Ca^+ Hε Hδ Hγ

Fig. 11.3. The spectra of novae following maximum light. The spectra of two novae, Nova Vulpeculae (1968) and Nova Delphini (1967) are compared with the spectra of the very luminous emission-line star P Cygni and the normal star 29 Vulpeculae (spectral class A0V). Note the broad emission lines with absorption on their violet edges in the spectra of the two novae. Nova Delphini showed extremely complex changes in its spectrum during the many months it lingered near maximum light. This nova was unusual for the fact it remained near maximum brightness for months, in striking contrast to the behavior of conventional novae.

Bright hydrogen lines, displaying sharp absorption on their violet edges, are the most prominent features of P Cygni, a star which rose from obscurity in 1600 in nova-like fashion, faded a bit, brightened again and then finally settled down as a fifth-magnitude star; its emission lines are narrower and less prominent than in the spectrum of the novae. (Courtesy Ojai Observing Station, University of California, Los Angeles.)

m(sun), while a rapidly evolving object which plummeted eight magnitudes in one month, ejecting its shell with a velocity of 10 000 km/s had a mass of 1.38 m(sun), close to the Chandrasekhar limit (see later in this chapter). A low-mass white dwarf, such as DQ Herculis, with a mass 0.9 that of the sun, can produce only slow novae, no matter how much mass is added.

The chemical compositions of novae ejecta always differ from that of the sun. For example, the hydrogen-to-helium ratio is always different. Compared to solar values, C, N, and O are grossly overabundant in novae. G.J. Ferland and G. Shields found the amounts of C, N, O, and Ne were raised by factors of 25, 100, 20, and 20, respectively, in V1500 Cygni (Nova Cygni 1975). C. Sneden and D.L. Lambert found ^{13}C was greatly enhanced in DQ Herculis, which showed the strong CN bands temporarily in December 1934. The star faded slowly until the end of March 1935 and then plummeted as a cloud of soot was formed.

Other types of cataclysmic variables

Classical novae are the most dramatic examples of a class of objects generally called cataclysmic variables. All involve a white dwarf component and a late-type star which fills its Roche lobe. The classical novae have amplitudes of 9 to 14 magnitudes and are believed to repeat in intervals of a hundred thousand to a million years. Recurrent novae, such as T Coronae Borealis, have outbursts separated by 10 to 100 years and magnitude changes of 7 to 9. Nuclear energy release is involved in both types of phenomena. The temperatures attained by classical novae sometimes exceed those of the nuclei of planetary nebulae, as shown by the appearance in 1988 of [Fe X] in Nova Muscae 1983 with an intensity greater than Hβ.

Irregular nova-like variables of the SS Cygni or U Geminorum type, whose amplitudes are about 2 to 6 magnitudes, with the intervals between outbursts ranging from 10 days to 30 years, involve gravitational rather than nuclear events. The AM Herculis stars show strong magnetic fields.

The 'symbiotic' variables, such as Z Andromedae, CI Cygni, BF Cygni, and AX Persei, involve a white dwarf and a giant which appear to be well separated from one another. Strong winds produce a gas cloud that envelopes the whole system. The hot component often shows abrupt nova-like outbursts; the gaseous envelopes shows lines of [O III], [Ne V], and [Fe VII], characteristic of a gaseous nebula.

An essential property of these systems is that a cool giant or even a cool main-sequence star is paired with a white dwarf. Infall of gas on the white dwarf produces outbursts either by gravitational effects or by the release of nuclear energy. The mass of the white dwarf component may gradually increase until it approaches the Chandrasekhar limit, when a catastrophic collapse may occur.

X-ray binaries

In binaries with compact objects, white dwarfs, neutron stars, or black holes, X-rays are sometimes produced when infalling mass strikes the accretion disk, usually within a small area. AM Herculis, AN Ursae Majoris, Cygnus X-2 and Scorpio X-I are well-known examples of X-ray sources, while HZ Herculis is an X-ray-emitting neutron star that is periodically occulted by the accretion disk that surrounds it.

One of the most dramatic of X-ray binaries is Cygnus X-1. In the optical range one observes a primary O star, HDE 226868, with a period of 5.6 days, and a mass between 15 and 30 solar masses. It circles an invisible companion with a mass between 9 and 15 m(sun). This is not a normal star; it seems almost certain that it is a black hole! There may be a third body in the system as both optical and X-ray emission show a 294-day period. The X-ray luminosity is about 4×10^{30} watts; it flickers in less than 0.001 second, which implies an X-ray emitting region less than 300 kilometers across. Presumably the invisible black hole shreds material from the O star which falls with great speed on the accretion disk that surrounds the black hole.

Perhaps the most powerful X-ray source known is the object called A0538-66 in the Large Magellanic Cloud. It is one or two orders of magnitude more powerful than other X-ray sources. In its quiet phase the optical and X-ray luminosities are 12 500 and 2500 times the solar values, but when one of its periodic active phases occurs the X-ray luminosity is 250 000 times that of the sun! One model proposes a neutron star in an eccentric orbit about a mass-losing giant; the activity becomes enhanced when the neutron star sweeps by the outer envelope of the giant.

Some of the most engaging and remarkable objects in astronomy could be included in the category 'wonders of the invisible world'. One thinks of W-50, a giant

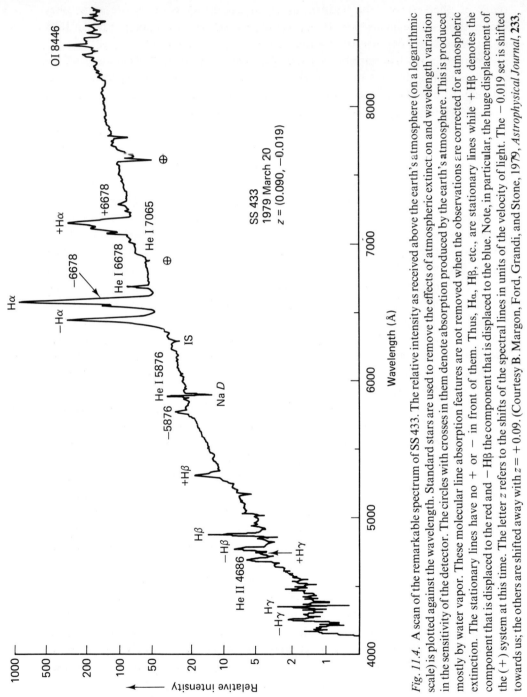

Fig. 11.4. A scan of the remarkable spectrum of SS 433. The relative intensity as received above the earth's atmosphere (on a logarithmic scale) is plotted against the wavelength. Standard stars are used to remove the effects of atmospheric extinct on and wavelength variation in the sensitivity of the detector. The circles with crosses in them denote absorption produced by the earth's atmosphere. This is produced mostly by water vapor. These molecular line absorption features are not removed when the observations are corrected for atmospheric extinction. The stationary lines have no $+$ or $-$ in front of them. Thus, Hα, Hβ, etc., are stationary lines while $+$Hβ denotes the component that is displaced to the red and $-$Hβ the component that is displaced to the blue. Note, in particular, the huge displacement of the $(+)$ system at this time. The letter z refers to the shifts of the spectral lines in units of the velocity of light. The -0.019 set is shifted towards us; the others are shifted away with $z = +0.09$. (Courtesy B. Margon, Ford, Grandi, and Stone, 1979, *Astrophysical Journal*, **233**, L63.)

H II region more splendid than Orion, of the center of our Galaxy, and of the tantalizing source Cygnus X-3. All of these sources share one property in common – extinction by dust grains (25 magnitudes for W-50, a comparable amount for the galactic centre, and a large but unknown value for Cygnus X-3). The last-mentioned seems to be a binary system, comprising a presumably normal star, paired with a compact object, probably a neutron star. The X-ray emission varies with a period of 4.79 hours. Cygnus X-3 is also a variable infrared and radio-frequency source, with the infrared luminosity varying in phase with a period of 4.79 hours. In September 1972, the radio-frequency emission increased a thousand-fold in a sudden outburst. But the most intriguing feature of Cygnus X-3 is that it is a powerful source of gamma rays, whose energies range up to 10^{16} electron volts. We defer to Chapter 12 more discussion of this tantalizing object that lies about 40 000 light-years from us in the remotest reaches of our galactic system.

An incredible stellar system, SS 433

My candidate for the most fantastic object in the sky is SS 433, many of whose properties were unravelled by my colleague, Bruce Margon, then at University of California, Los Angeles. The mystery of this star is told with all the drama of a well-written detective story by David H. Clark in his *The Quest for SS 433*. N. Sanduleak and C.B. Stephenson had listed it in their catalogue of objects with a bright Hα, while J.I. Caswell and D.B. Clark had detected it as one of several point-like radio sources seen in or close to the edge of supernova remnants.

Spectroscopy of the star revealed three sets of strong emission lines of H and He (see Fig. 11.4). One set is nearly stationary; the other two sets show large velocity amplitudes, indicating material moving with velocities approximately a quarter of the velocity of light! Spectral data, largely obtained by B. Margon *et al.*, and supplemented by Italian observers at Asiago, and by others, were interpreted as arising from what was presumably a neutron star source with an accretion disk and two jets of material ejected with a velocity a quarter that of light (A. C. Fabian and M. Rees, and M. Milgrom). As the neutron star and its accretion disk precess with a period of 164 days, the jets sweep around and the radial velocity varies, in a manner recalling a spectroscopic binary, but here it is line of sight velocity of the jets and not that of the stars that varies. These jets are also observed in the infrared and the radio-frequency ranges at great distances from the source. The so-called stationary lines show a small shift with a period of 13.1 days, which is actually that of the binary. The more massive star has a mass approximately 15 times that of the sun. There is a strong wind, much of which is captured by the accretion disk whose luminosity is comparable with that of the hot star. The material in the wind produces relatively sharp 'stationary' lines. As the stars move about, the lines and continuum vary as partial eclipses of the large star and the accretion disk occur. Fig. 11.5 depicts a possible model of the system. The precession of the jets with a period of 164 days causes a garden-hose effect.

X-ray pictures with the Einstein satellite and M.S. Hjellming's radio images recall

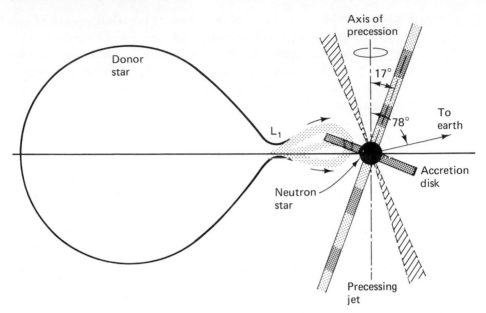

Fig. 11.5. A possible model for SS 433. A plausible representation of the SS 433 system, as compiled from the work of many investigators including B. Margon, G. Abell, D.H. Clark, P.G. Murdin, D. Crampton, M. Rees, M. Milgrom, and A.C. Fabian is depicted. A large, hot, donor star of about 15 solar masses feeds material to the accretion disk of the neutron star. This material is ultimately ejected in jets along the neutron star axis with a velocity of about 80 000 km/s. The neutron star, with its accretion disk and jets, precesses like a gyroscope with a period of 164 days about an axis making an angle of 78 degrees with respect to the direction of the earth; the cone of the jet makes an angle of 17 degrees.

the jets and lobes that are characteristic of certain radio galaxies. The optical jets extend over some fifty thousand million kilometers, but extensions seen in X-rays stretch out over parsecs.

Although the spewing of material from a donor star to an accretion disk surrounding a white dwarf or neutron star is not unusual, what is unique about SS 433 is the remarkable high-velocity jet system. The material flowing through the jets comes from winds from the hydrogen-rich envelope of the hot star. The fast-moving gas leaving the neutron star is very hot and scoops up the relatively cool hydrogen-rich gas. Collisions with atoms of hydrogen and helium give the bright-line spectrum. Further out in the stream, the hydrogen becomes ionized and is no longer visible optically, but the X-ray emission from the hot jet remains detectable to a great distance. The gas does not flow out as a uniform stream, but discrete clouds are often found.

There remain a number of mysteries; three are mentioned below. How is the jet collimated? One might expect the gases to spread out in all directions, but they seem to be well focussed – possibly by a magnetic field that may be fairly strong. How can the material be accelerated to a velocity of 80 000 km/s? A power output equivalent to a hundred suns would be needed for this task alone. The prevailing belief is that radiation pressure does the job, but how? What specific atomic mechanisms are involved? Finally, what causes the relatively rapid (164-day) precession of the neutron star and the accretion disk? The precession is not uniform. Margon found a nodding or nutation with a period of about 63 days. Could it be caused by a variation in the strength of the force causing the precession? The precession of the earth's axis is well known to be caused by the gravitational attraction of the sun and moon on a non-spherical earth, but in SS 433 the origin is not identified.

Margon, Clark and others have suggested that SS 433 may provide clues to our understanding of quasars and radio galaxies, but the scale of these phenomena is very different. Radio galaxies show jets analogous to those in SS 433, but on a scale of a hundred thousand or a million parsecs. The tremendous power in these distant objects is believed to be derived from a black hole in the neighborhood of which some particles are accelerated to high velocities and escape, while the bulk of the mass is swallowed in the bottomless pit. Surely, here are challenges to match the efforts of theoreticians for many years to come!

Super luminous stars

The various novae and catastrophic variables that we have described on previous pages are largely examples of dramatic late-evolutionary stages of initially rather prosaic binaries which would have attracted little attention during most of their lives. Late stages of evolution of the most massive stars, however, can sometimes also be even very spectacular, but for different reasons to those described above.

First, let us consider this question: how massive can a star be? Intuitively, it is easy to understand that instability must increase with mass. The luminosity of a star is not linearly proportional to mass but goes rather as the third or fourth power thereof. Following A. Eddington, we can suppose that an upper mass limit will be reached when the force exerted by radiation on the material in the atmosphere becomes comparable with the attractive force of gravity. The maximum allowable mass is estimated to be from 60 to 120 solar masses. As the limit is approached, the atmosphere becomes increasingly unstable. Instability can be manifested in several ways: orderly winds, turbulence, and large-scale mass motions which detach great amounts of stellar material at truly catastrophic rates. In our discussions of stars of moderate mass, we saw that as hydrogen in the core became exhausted, the star developed a deep convective envelope and evolved away from the main sequence as it became a giant or supergiant. In stars more massive than about 30 suns, vigorous turbulent convection may occur. The chaotic motions of large blobs of material produce a turbulent pressure, somewhat like the gas pressure produced by the motions of gas molecules, and this turbulent pressure is added to the gas pressure and radiation pressure already active. It must be compensated by the weight of overlying layers for the star to remain stable. If the star is evolving to the domain of lower temperature (i.e. towards a red giant phase) the radiation pressure becomes unimportant, but the chaotic turbulent pressure may now dominate the situation and cause the envelope to be detached. The more massive the star, the more pronounced the turbulent pressure is likely to be, and the more violent the mass loss as the star tries to evolve towards the red giant or supergiant stage.

Evolutionary tracks calculated by A. Maeder showed that if the mass is less than 15 times that of the sun, the star can reach the red supergiant stage without significant mass loss. If the stellar mass exceeds about 60 times that of the sun, substantial mass loss occurs via winds and the instability engendered by turbulence. Eventually, the hydrogen-rich envelope is completely lost in the wind and the star may be stripped down to the region where the hydrogen has been burned to helium by the CNO cycle. Such an object might be expected to show a bright-line spectrum due to the surrounding shell of rapidly expanding gas. If the star is stripped down to layers where the CNO cycle has operated, we might expect strong lines of nitrogen and helium but no hydrogen. Carbon and oxygen lines would be relatively weak since most of their atoms would have been converted to nitrogen, the bottleneck of the CNO cycle. As the star continues to tear itself to pieces, the CNO cycling region would have been ejected. The strata now exposed would reveal the products of helium burning into carbon by the triple-alpha reaction and the subsequent building up of oxygen. Eventually, the core may settle down as a neutron star or become a supernova. Fig. 11.6 illustrates some possible evolutionary tracks.

Is there observational support for these conjectures? Indeed there is. Some rare and fascinating objects are found in the upper stretches of the HR diagram. These include the unstable, bright, Hubble–Sandage variables (HSVs), identified as among the most luminous stars in nearby galaxies, the prototype of a mass-losing star, P Cygni, the Wolf–Rayet stars, and the bizarre star known as η Carinae.

Wolf–Rayet stars have long been known but their evolutionary status has not been clarified until quite recently. They show broad-emission-line profiles indicating huge mass-loss rates of the order of 10^{-5} to 10^{-4} solar masses per year. The lines are often of the P-Cygni type; the profiles are broad with relatively sharp absorption features on the short-wavelength edge. These dips are produced by gases moving rapidly towards the observer. Wolf–Rayet stars have large masses, indicating that they have lately evolved from even more grandiose stars that have shed much of their outer envelopes in brisk winds. We never see to the photospheres of these objects, if indeed they have photospheres in the ordinary sense.

The Wolf–Rayet stars fall into two categories; the C–O sequence and the N sequence. The C–O group shows strong lines of carbon, oxygen, and helium, with weaker lines of silicon, and embrace a range of temperatures. At the top of the temperature scale one finds the high-excitation group, showing emission lines of O VI. Curiously, this type of Wolf–Rayet star was first recognized among nuclei of planetary nebulae. The N stars show nitrogen and helium with a trace of carbon; they also cover a range in excitation. The two sequences can be understood in terms of the evolutionary scenarios described by Maeder and others.

Recall that planetary nebulae nuclei often exhibit spectra of the Wolf–Rayet type. The ejected envelope of the highly evolved star forms the planetary nebula as the core settles down on its way to becoming a white dwarf. The spectra of C-type planetary nebulae, such as those of NGC 40 or BD $+ 30°$ 3639, are very similar to those of classical, massive, Wolf–Rayet stars, but while the planetary nebula is

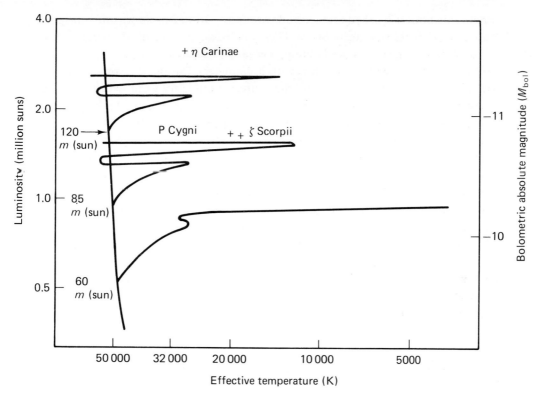

Fig. 11.6. Evolutionary tracks of very massive stars. Luminosity (in units of one million suns) and bolometric absolute magnitude are plotted against effective temperature. (Notice that the scales are logarithmic). Evolutionary tracks are calculated assuming rather large mass-loss rates (in the neighborhood of 0.0003 to 0.0004 m(sun)/year.) Note that the more massive stars never reach the red supergiant domain but turn back towards the main sequence as they lose their outer envelopes. The Wolf–Rayet stars fall in this region, as do the extremely luminous objects such as P Cygni, ζ Scorpii, and η Carinae. (After A. Maeder.)

doomed to die as a white dwarf, the classical Wolf–Rayet star may depart in a burst of glory as a supernova; we do not yet know.

It must be emphasized that for very massive stars, mass loss will severely modify the evolutionary picture. The star never becomes a red supergiant but gets 'turned back on its tracks' to become possibly a Wolf–Rayet star or eventually even a supernova. Maeder concluded that turbulent mixing combined with mass loss keeps the internal chemical composition quasi-uniform, so the surface layers reflect the effects of internal nuclear reactions. The most remarkable of all these massive, highly evolved stars is probably η Carinae, to which we now turn our attention.

η Carinae

One of the most remarkable variables in the sky, η Carinae was first described by J. Bayer (1603) and E. Halley (1677). It gradually brightened from about the third magnitude until it reached a magnitude around -1 in 1843. It was then brighter than any star in the sky except Sirius. Then it faded quickly, reached a magnitude of 7.7 by 1900, and has been brightening gradually since then.

Visually, one sees a fuzzy 'star' with a diameter of about 1.5 arcsecs. This is actually a dusty cloud surrounding the source of a mighty wind with the highest known rate of mass loss, 0.001 to 0.01 solar masses per year! Surrounding this is a shell with a diameter of about 10 arcsecs, the so-called homunculus, which expands with a rate of about 500 kilometers per second. It contains a number of nebulous blobs which E. Gaviola found to be moving outwards across the sky at a rate of about 5 arcsecs per century. (R. Gehrz and E.P. Ney (1972) found that the expansion probably began at the time of the 1843 outburst, but other less spectacular ejections may have occurred.) Then, quite outside this little 10-arcsec disk, occur fainter condensations whose spectra show lines of He II and nitrogen in various stages of ionization but with no trace of carbon or oxygen. Finally, η Carinae is associated with an extended beautiful nebula near which it must lie, since this great η Carina Nebula shows changes that are obviously caused by the light variations in η Carinae itself. Distances of 1500 to 2800 parsecs have been quoted; the latter value is often preferred.

The spectrum is presently dominated by numerous emission lines of iron, particularly Fe II and [Fe II], with lines of ions of He, N, Mg, Si, S, Ca, Sc, Zr, V, Cr, Co, Ni, and Cu, but *no* oxygen! The spectrum of the nucleus is extremely complex (see Fig. 11.7). The stronger lines show broad emission profiles, often with sharp, narrow emission spikes. There is often violet-displaced absorption, as in 1961 when the Hγ profile showed dips corresponding to shells moving outward with velocities of 48, 120, and 480 kilometers per second. Complexities in the profiles may well correspond to different brightening events and complex ejection patterns.

Although the visible region continuous spectrum shows a flux distribution of a type appropriate to a hot gas capable of producing the observed bright-line spectrum, the infrared emission must arise from a quite different source: hot dust. J.A. Westphal and G. Neugebauer (1965–68) found that when observed at 20 micrometers, η Carinae was the brightest object in the sky outside the solar system. Furthermore, the diameter of the infrared source depends on wavelength; the greater the wavelength, the greater the diameter of the source. We are observing emission from a dust cloud whose temperature varies with distance from the central star; the cooler dust which emits preferentially at the longer wavelengths is found further from the star. Now 90 percent of the energy is emitted longward of 1.65 μm. If you assume that the distance is 1500 parsecs and that the visual extinction is 3.6 magnitudes, you find that from 0.4 to 20 μm the luminosity of η Carinae is about two million suns! More likely the total energy output is closer to four million suns, the absolute bolometric magnitude is then -12 (1969), which is close to L. Gratton's estimate of -12.5 for 1843.

The interpretation is that the star has been shining at about the same rate all along, but that in 1843 it temporarily got rid of its dusty cocoon, which has since been re-established. The most plausible model is that a hot star with a temperature

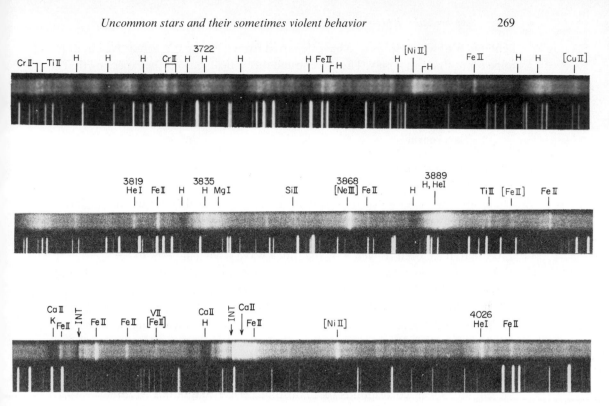

Fig. 11.7. A portion of the spectrum of η Carinae. Note the extremely complex character of the spectrum which arises from the central condensation and the rapidly expanding surrounding envelope. Strong absorptions occur on many of the lines, particularly those of H and Ca II. The dark lines marked 'INT' indicate interstellar absorption. (From a spectrogam obtained in May 1961 with the coude spectrograph designed and built by T. Dunham at Mount Stromlo Observatory.)

of about 30 000 K shines on a thick dust cloud composed of solid particles, and what we observe is mostly the heat radiation from these grains which are presumed to be silicates.

What is the nature of η Carinae? *Ad hoc* hypotheses to the effect that it was some kind of non-thermal object or left-over from a peculiar type of supernova can be disregarded. Gratton's hypothesis that it might be a young, massive star approaching the main sequence would require it to have a chemical composition appropriate to that of the interstellar medium. Earlier work by A.D. Thackeray and others had noted a deficiency of oxygen, but the critical test was an analysis of the outer condensations by K. Davidson, N.R. Walborn and T.R. Gull (1982), who used data from the Cerro Tololo Observatory in Chile and the International Ultraviolet Explorer. They found five ionization stages of nitrogen, none of carbon or oxygen. All the C and O got processed to N by the CNO cycle in an advanced evolutionary stage. The grains would appear to be silicate grains, but rather unusual silicates, since they contain no oxygen.

A century ago, Camille Flammarion asked: 'Will it awake, this sun of Argo, will it revive completely and project anew around its brightening sphere the radiation of light and heat which seemed to have departed from it forever? We may, we ought to hope for it.' Indeed, it may. There are sound theoretical reasons for expecting that η Carinae, which is intrinsically one of the brightest most massive stars in our Galaxy, will explode as a supernova.

Supernovae

One of the most spectacular events in the cosmos occurs when a star blows up, sometimes becoming temporarily brighter than the whole galaxy that contains it! These are the supernovae, furnaces in which are forged most of the chemical elements heavier than nitrogen and carbon. These explosions produce vast upheavals in the interstellar medium and may induce star formation. Our solar system appears to have been triggered by the shock wave from a supernova.

The distinction between novae and supernova was not clearly understood until well into the twentieth century. In classical novae the superficial layers of a compact star in a binary are torn off in an explosion. In a supernova the whole star is blown up. These events are rare. None have been seen in our own Galaxy since 1572 (Tycho) and 1604 (Kepler). There have been two during recent historical times in galaxies of the local group. In August 1885, a nova-like object appeared in the Andromeda Galaxy, M31. It declined rapidly from its maximum around the sixth magnitude, which corresponds to an absolute visual magnitude of about -19.2. The very brightness of this star seemed to argue against M31 being a galaxy like our own, if we assumed that the object was an ordinary nova. A generation later, ordinary novae were found in M31 and the 1885 object was recognized as truly abnormal.

Investigations of properties of supernovae had to come from studies of remote galaxies. We are indebted here to the extensive surveys by Fritz Zwicky and studies by W. Baade and R. Minkowski. Although Zwicky deduced that there were several distinct types of supernovae, most studies until recently found that a majority of them could be put into two types, type I and type II, on the basis of their spectra and light curves (see Fig. 11.8 and Table 11.1).

Other types have since been recognized, such as a group that is helium rich but with no hydrogen, thus resembling type I. These objects have been studied particularly by Craig Wheeler and his associates at the University of Texas.

Type I supernovae appear to originate from binary stars whose detailed evolution preceding catastrophe poses some engaging challenges. Type II supernovae come from the collapse of a single massive star. Let us now describe the probable late, spectacular stages of a type II supernova's history.

The type II supernova event

In the preceding sections we saw how massive stars evolve to Wolf–Rayet stars, Hubble–Sandage variables, or even exotic objects like η Carinae. Yet others become red supergiants, losing their outer envelopes into space. If the massive star becomes a supernova, this residual material might form a buffer with which the rapidly moving

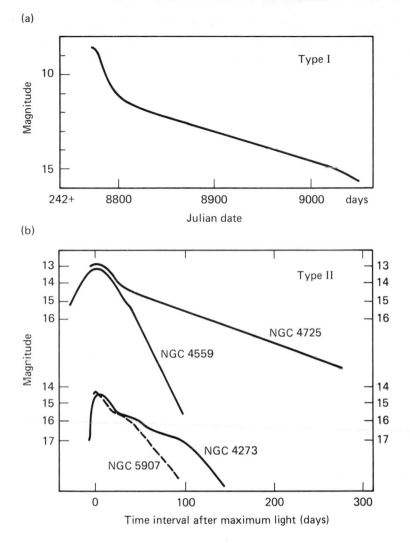

Fig. 11.8. Light curves of supernovae.

(a) Light curve of the type I supernova in IC 4182. Magnitude is plotted against Julian date. (Julian dates are days numbered consecutively starting long before any recorded astonomical observations. They facilitate the calculation of time intervals between observations.) The light curves of type I supernovae are remarkably similar. (After D. Hoffleit.)

(b) Light curves of type II supernovae. Here magnitude is plotted against the time interval after maximum light. Note that light curves of supernovae of type II show considerable variations in shape. (After Fritz Zwicky.)

ejecta of the dying star might collide. The pre-supernova may be a red supergiant, or in the case of supernova 1987A it might have started back toward the main sequence before it blew up.

The exciting events are those that are taking place in the ever compressing, ever shrinking, and ever hotter core. During a star's life as a luminous main-sequence star, it shines by turning hydrogen to helium in its core. With the core H exhausted, it

Table 11.1 *Characteristics of supernova type I and type II*

	Type I	Type II
Light curves	Regular with smooth decay	Variable decay rates
Ejection velocity	10 000 km/s	Time dependent[a]
Absolute magnitude (at maximum)	−19	−17.5
Spectra	No H nor He, but metals are present	Metals and H, 'normal' composition
Occurrence	Elliptical, irregular, and spiral galaxies	Spiral arms, associated with young, massive stars
Ejected mass	0.1 to 1 solar masses	>3 solar masses
Frequency of occurrence	$\gtrsim 100$ years	$\gtrsim 10$ years[b]

Notes:
[a] Typically velocities of 2000 km/s are observed, but velocities of 30 000 km/s can be observed in the outer thin shell.
[b] A 10 year estimate, by D.N. Schramm, is from the galactic element building rate.

converts H to He in a shell, and the temperature and density of the He core gradually rise until the star can convert He to C by the triple-alpha reaction. The evolutionary process was described in Chapter 9. In a very massive star, the central temperature can climb to a hundred million and thence to a thousand million degrees. Energy continues to be liberated as C is converted to O, and ultimately to Ne, Mg, Si, and S.

Evolution now occurs at an ever quickening pace as fusion reactions build Si, S, Ca, etc., up to Fe (see Fig. 11.9). By the time we have a Si-burning shell surrounding an Fe core, the stage is set for some interesting developments. Energy cannot be extracted by building elements beyond iron. On the contrary, as the temperature is raised, the Fe nuclei are shattered to He nuclei as large amounts of energy are soaked up from the surroundings. The local temperature falls and gas pressure cannot support the weight of overlying strata.

Disintegration of nuclei by the absorption of gamma rays (photo-disintegration) occurs on a time scale of 10^{-10} seconds. Nuclear reactions keep the chemical composition close to equilibrium values appropriate to the local temperature and density. A catastrophic collapse now occurs on the time scale of milliseconds (10^{-3} second). Although the progenitor of the supernova had evolved over an interval of millions of years, the star collapses under its own weight in less than one second. The mechanical, radiative, and nuclear effects are traumatic.

The sudden, high compression has other consequences. Electrons are jammed into protons and other nuclei, especially iron, with the liberation of neutrinos, which carry away most of the energy of the supernova event. Most of the neutrinos are produced, however, in the intensely hot, dense medium 1 to 10 seconds after the collapse. Once the density exceeds a few times 10^{11} g/cm^3, the neutrinos cannot easily escape and slowly diffuse through the material. G. Gamow referred to the copious production of neutrinos at this stage as the URCA process, in honor of a Brazilian gambling casino where the players were relieved of their cash with the same efficiency as a star is relieved of its energy by neutrinos in the supernova process.

Collapse continues until the density exceeds that of matter in a large atomic nucleus, 2.7×10^{14} g/cm^3. Now the core approaches incompressibility, and material falling inward upon it generates mechanical waves. The core can be compressed to a density about 200–300 percent greater than that of matter in an atomic nucleus before it bounces back to produce a sound wave that evolves into a shock wave. These shock waves play a crucial role in supernova pyrotechnics. In the simplest models they originate about half-way out from the center to the edge of the core and then run upward with velocities between 30 000 and 50 000 km/s, quickly passing from the iron core, and thence through the main structure of the star to tear off the stellar surface in a few days.

There are some worries. Detailed calculations by J.R. Wilson, W.D. Arnett, H. Bethe, and others indicate that the energy of the shock wave could be sapped away by such processes as the breaking up of iron nuclei to form helium nuclei, neutrons, and protons, so that the wave then might never escape and the star may fall in upon itself into the bottomless pit of a black hole. Under plausible conditions, however, Wilson finds that neutrinos may interact with matter in the iron–alpha particle conversion zone and set free enough energy to 'save the wave'. Competition between processes tending to dampen out the wave and thus swallow up the star in a black hole and those permitting the wave to reach the surface may depend delicately on initial conditions at the start of the supernova countdown.

The mass of the progenitor plays a critical role. Very massive stars would produce very large iron cores that then may collapse directly to black holes. In other scenarios, the material within 50 kilometers may condense as a neutron star, while all the rest is expelled, carrying with it the signature of nuclear events. Calculations by T. Weaver and S.E. Woosley, by K. Nomoto, and by others, indicate that for masses between 8 and 11 times that of the sun the core does not form Fe but only builds up to O and Si. As the core becomes compressed, shock waves can be initiated. These suffice to rip off the envelope since energy has not been dissipated *en route*. It is suggested that the progenitor of the Crab Nebula may have had a mass of about 9 suns.

Many years ago, S. Chandrasekhar showed that the upper limit to the mass of a degenerate gas star or white dwarf would be about 1.35 suns, beyond which point the pressure exerted by degenerate electrons would not suffice to support the weight of the overlying material crushing down upon it. Nuclear forces could hold out against gravity up to a mass around 1.8 times the sun, beyond which point the material would be crushed into a black hole. The core of a star evolving to a supernova loses most of its energy via escaping neutrinos. All theories agree in demanding a huge neutrino flux at the time of stellar collapse, a prediction obligingly confirmed by the behavior of SN 1987A (see later in this chapter).

Building of heavy elements in supernovae

Most atomic and nuclear species are produced in just a few moments in the life of a star. What elements can be made depends on the mass of the star. We have seen that nuclei up through nitrogen can be built in stars of rather modest mass (up to ~ 2

m(sun)), but formation of oxygen, neon, and most heavier elements requires more massive stars in whose deep interiors huge temperatures and densities are developed.

Many nuclei beyond the iron (Fe) peak can be built up by successive neutron capture in a star of relatively modest mass, provided we have a good supply of neutrons and Fe. It seems that Fe can be made only in supernovae. The fact that events transpire so quickly in the supernova explosion means that many isotopes are built that would not otherwise occur in nature. Fast neutron capture means that a nucleus does not have a chance to undergo beta decay before it captures yet another neutron. The process goes on until the neutron-happy nucleus can accept no more and must undergo beta decays until a stable nuclide is formed. The rapid neutron capture (r-process) nuclides are well known; their presence in the solar system attests to the importance of supernova events.

The very high temperatures attained in cores of massive stars imply intense radiation fields, rich in gamma rays, that can photo-dissociate nuclei, eject neutrons, and produce relatively rare, proton-rich nuclei. Supernovae are also the main sources of elements like oxygen and neon. Analyses of meteorites have revealed the complexities of some of the processes whereby elements of the solar system were formed. For example, meteorities enriched in pure ^{16}O reveal a chapter in element building somewhat different from the events that produced the bulk of our world.

Note that the material ejected in a supernova contains both unstable nuclides, such as ^{22}Na (half-life = 2.6 years), and stable ones, such as ^{28}Si. Material in solid grains of some meteorites contains the gas ^{22}Ne, which is the decay product of ^{22}Na. How did it get there? Evidently dust grains formed quickly in the ejecta of supernovae and these contained metallic ^{22}Na, which decayed to ^{22}Ne that remained entrapped in dust that later formed meteorites. Note that the time interval between the formation of radioactive sodium and its incorporation in solid material must have been short. The solar system contains products of slow neutron capture that must have been formed in S-type stars as well as supernovae debris. Extinct radioactivity provides a powerful tool for studying the pre-history of the solar system. It can be shown that the interval between the time when material destined to form the solar system was last subjected to supernova radiation and the actual formation of meteorites was of the order of a hundred thousand to a million years, a short time indeed compared to the system's 4.5-thousand-million-year age.

The ejecta of supernovae contain contributions from both the exploded star and the neighboring interstellar medium. If you catch the phenomenon early enough, you can identify individual blobs corresponding to different depths in the pre-supernova. Some blobs may be oxygen-rich, others correspond to strata where oxygen was burned to silicon; such studies give checks on the theory of supernova evolution.

Type I supernovae

Type I supernovae radiate more energy in a few months than the sun does in a thousand million years. Their characteristics set well-defined constraints on their progenitors. One clue to their origins is given by the fact that they occur in elliptical galaxies where new star formation is no longer occuring. In these ancient populations, all single, high-mass stars have long since vanished. A second important clue is the similarity of the light curves and the spectra of type I supernovae. They show no evidence of hydrogen; thus they must originate from highly evolved stars that have lost their outer envelopes and now consist primarily of carbon, oxygen, and heavier elements.

The most likely progenitor of a supernova of type I is a massive binary that evolved quickly and then settled down as a close pair of degenerate stars. All supernova type I scenarios require a binary system; we describe an evolutionary sequence proposed by I. Iben and A.V. Tutukov. Their sequence starts with a massive binary with components between 5 and 9 solar masses, a separation between 70 and 1500 solar radii, and a period between 0.1 and 6 years. The stars evolve through two stages with common envelopes as first one of them and then the other fills the Roche lobe. All the outer envelopes with all the hydrogen are lost and there emerges a pair of C-O white dwarfs with masses between 0.7 and 1 times that of the sun, a separation of 0.2 to 3.5 solar radii, and an orbital period between 12 minutes and 14 hours. General relativity predicts that in such a system where the stars move so rapidly, angular momentum is carried away by gravitational waves. The stars are gradually forced closer together over an interval between a hundred thousand and ten thousand million years.

Thus a pair of massive stars formed early in the history of elliptical galaxies may only now be spiralling in to their mutual doom in an apocalyptic explosion. As the two stars are gradually dragged together, a point is reached when the less massive white dwarf is shredded by the tidal action of the more massive one. The debris forms a thick disk, composed mostly of carbon and oxygen, about the survivor. The stage is set for the grand finale!

Accretion from the thick disk raises the mass of the white dwarf to exceed 1.35 m(sun), the Chandrasekhar limit. Consequently, the material of the white dwarf becomes even more compressed than the limit allowed for degenerate matter. The temperature rises giddily, and nuclei of carbon and oxygen collide with each other with sufficient vigor to release energy via thermonuclear reactions. The carbon burning does not occur with the ferocity of events transpiring in an iron-core collapse in a type II supernova. The theory by Nomoto and Woosley, and by Craig Wheeler, would suggest that the burning is not actually explosive. The wave of fusion reactions propagates through the star in a manner more nearly reminiscent of the burning of a fuse than the explosion of the dynamite. It is called a deflagration, not a detonation, but the white dwarf is nevertheless thoroughly destroyed with the material hurled upwards with speeds of about 11 000 km/s. Detailed theoretical studies predict expansion velocities, light curves, luminosities, and spectra (the strongest constraint) in accord with observations.

Such an explosion of a white dwarf star would produce an evanescently brilliant object, but detailed calculations do not indicate that it would be as bright as observed – unless we suppose that the build-up of elements does not stop with the creation of sulfur and silicon but proceeds onward to nuclides of the iron group. It is the radioactive decay of these nuclides that supplies the bulk of the luminosity of the type I supernova. Detailed theoretical studies indicate that these nuclear events can create as much as 1 solar mass of ^{56}Ni that decays to ^{56}Co and finally ^{56}Fe over a period of several months.

The great similarities between light curves and spectra of type I supernovae suggest a uniform pattern for these events. Some have proposed that the progenitors have the same mass when they explode and that they attain the same maximum luminosities. If, indeed, type I supernovae are identical we could use them as distance indicators for remote galaxies, a possibility first suggested by F. Zwicky. A. Sandage and G.A. Tammann found a value of -19.74 for absolute magnitude at maximum light – in good agreement with the Craig Wheeler theoretical value of -19.7. A critical evaluation of type I supernovae as distance indicators of galaxies merits further careful study.

Helium-rich supernovae

A necessary and sufficient condition that a supernova be of type I is that its spectrum shows no evidence of hydrogen. The classical type I supernova (or type Ia if it closely follows the above description) also shows no helium. In recent years a type Ib has been recognized. It shows helium lines and tends to occur in spirals like type II. Light curves and spectra of supernova Ib stars have been successfully interpreted by Craig Wheeler and his associates. Analyses of the chemical composition of these objects from their spectra are very difficult. One must calculate the flow of radiation through a model atmosphere which is a rapidly expanding envelope. Results suggest a variety of chemical compositions; samples analyzed range from 90 percent He, 1 percent C, and, 9 percent O to 10 percent He, 10 percent C, and 80 percent O. As the envelope expands and the density falls, we observe forbidden lines of heavier elements. Supernova Ibs may produce copious amounts of oxygen.

Progenitors of supernova Ibs may have masses in the range of 8 to 25 m(sun) which would suggest putting them in the type II group, even though they had no hydrogen. A binary system may be invoked but, in this case, the companion strips the star down to the helium zone before the supernova event. Thus Ibs differ from Ias by having a companion that assists in removing the hydrogen envelope. But could we not get a supernova Ib by turning a Wolf–Rayet star into a supernova?

The great bonanza; the Magellanic Cloud Supernova (SN) 1987A

The brightest supernova in 400 years happened in 1987 at just the right time when it could be observed throughout the electromagnetic spectrum from the radio-frequency range to gamma rays. It was even studied with newly commissioned neutrino telescopes, and gravity wave detectors. For the first time astrophysicists had their cameras ready for one of the most dramatic events in nature. In particular, they could identify uniquely the progenitor star, follow the light curve in detail over a broad spectral range, and detect the neutrino flux. The supernova substantiated beautifully some theoretical predictions, but it added a few puzzles of its own.

The B3-type progenitor star, Sanduleak $-69°$ 202, with an effective temperature of 18 000 K, a luminosity 100 000 times that of the sun, and a probable mass roughly 20 times that of the sun, seemed at first to present a puzzle, as conventional thinking had dubbed M supergiants as the immediate progenitors of supernovae. Theoreticians favor an explanation based on the fact that in the Magellanic Clouds the metal-to-hydrogen ratio is lower than in our Galaxy.

The outer layers of Sanduleak $-69°$ 202 had no inkling of impending doom. The star had evolved to the point where it finally converted silicon and sulfur into iron and then took the next, fatal step. Its probable internal structure just before the explosion is depicted in Fig. 11.9, due to S. Woosley *et al.* As the mass of the iron core hit the critical 1.4 m(sun) value, the iron disintegrated into helium. Thus a dense ball of degenerate iron about the size of planet Mars crumbled in a few seconds to a 70-kilometer diameter sphere at nuclear density. Material falling on this core produced shock waves that went back up through the star, blasting away the surface, and tearing the star asunder.

The gravitational potential energy (GPE) given up by the collapsing core amounted to about 3×10^{53} ergs – or the equivalent of converting a tenth of the solar mass into energy. This GPE amounted to about a hundred times the energy in the shock wave, and about ten thousand times the visible light. Most of the energy escaped by neutrinos, whose flux was actually observed by a detector near Cleveland in the USA and the Kamiokande II detector in Japan. The total number of counts; 8 in the USA and 11 in Japan, was small but showed that the neutrino flux in the range between 6.3 and 40 million electron volts corresponded with the predicted temperature of the core and total flux of 3×10^{53} ergs. The pulse duration was consistent with the collapse of the core to neutrons and indicated that, if the elusive neutrino had a mass, it could not exceed about 20 electron volts.

Although clearly a type II supernova, SN 1987A was atypical. It was much too faint for either a regulation type II or Ib. Possibly we have not seen such objects in distant galaxies because they would be too dim. Following the explosion, the star then brightened as the expanding, ejected envelope radiated powerfully. As the envelope expanded, we were able to see the deeper and hotter layers. Were this the whole story, the star would have faded rapidly, but the ejecta contained radioactive isotopes whose decay fed energy into the shell. After June 1987, the light curve followed what is called an exponential decay (see Fig. 11.10). The star faded with a half-life time constant of 77 days which corresponds to the decay of ^{56}Co to ^{56}Fe. Thus do the details of the light curve confirm supernovae as suppliers of ^{56}Fe to the universe.

How are these radioactive nuclei produced? The 'hard core' of the supernova becomes a neutron star, but the shock wave that runs out after bouncing on the core passes through sufficiently dense layers for the temperature and density to be raised enough temporarily to build ^{28}Si and eventually up to ^{56}Ni, burning about 0.07 m(sun) to completion according to the calculations of S.E. Woosley. The process

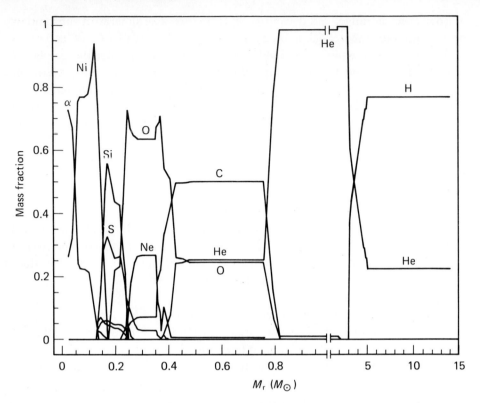

Fig. 11.9. Probable final composition of supernova 1987A just before the explosion. As is customary in stellar model calculations, the mass within a radius r, m_r rather than r itself is used as the abscissa. Note the change in scale at $m_r = 1$. Also note the change in composition from the outer envelope (which is the original composition of the star and therefore mostly hydrogen and helium) as we go inward. The mass is measured exterior to the neutron star remnant of 1.27 m(sun). Density rises rapidly with decreasing m_r. At the time it exploded the star had a radius of about 4.2 solar radii and it released about 2.9×10^{51} ergs. The edge of the helium zone probably lies at a radius of 0.7 that of the sun, while the outer edge of the silicon/sulfur zone lies at a radius of about 0.8 that of the earth. The temperature reaches ten thousand million degrees as the star verges on collapse. (After Woosley, Pinto and Ensman, 1988, *Astrophysical Journal*, **324**, p. 466.)

creates ^{56}Ni, which decays to ^{56}Co with a half-life of 6.1 days. Then ^{56}Co decays to ^{56}Fe and as it does so it produces an 847 000-eV gamma ray that is degraded to X-rays and powers the light curve. X-rays have been observed by satellite, and gamma rays were observed in August 1988. Eventually a central, spinning, neutron star or black hole may be unveiled. It may supply power to the supernova remnant.

Supernovae ejecta and remnants

Supernova remnants are among the aesthetically most beautiful nebulae in the sky. They include the Network (or Veil) Nebula in Cygnus, IC 443, and many delicate features registered on Hα photographs of the Milky Way. These nebulosities pump

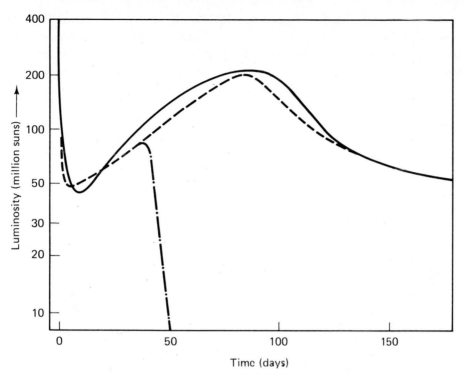

Fig. 11.10. Light curve of Supernova 1987A. The luminosity, in terms of a million times that of the sun, is plotted against time in days following core collapse. The solid curve depicts the observations, the dashed curve was calculated by Woosley *et al.* on the assumption that 0.07 *m*(sun) was converted by the star to ^{56}Co which ultimately decayed to ^{56}Fe. This decay supplied most of the luminosity beyond 50 days after core collapse. If no radioactive decay of cobalt had occurred, the luminosity of the star would have followed the dot–dash curve.

into the interstellar medium fresh supplies of elements from oxygen to iron and many other nuclides as well, and as they push against and compress the interstellar medium they can lead to the formation of stars and solar systems. One of the most exciting possibilities is that cosmic rays are produced in the ejecta of supernovae, a topic to which we shall turn in the next chapter.

The Crab Nebula (M1) represents one type of supernova remnant, called a plerion, which is brighter in the center, shows no shell structure, emits synchrotron-type radiation (see Chapter 12), and seems to be relatively short-lived. It is presumed that such remnants originate from type II supernovae. The Crab Nebula is powered by a central pulsar or neutron star.

More common are the hollow shell structures seen in projection against the sky, such as the Network Nebula in Cygnus. Most of the historical supernovae, such as 1006, 1572 (Tycho), 1604 (Kepler), and Cassiopeia A, are of this type. The Network Nebula, the Vela Nebula, and IC 443 correspond to supernovae that occurred before historical records. Similar filamentary supernova remnants are seen in the Magellanic Clouds and in the Triangulum Galaxy, M33.

These structures are formed when the rapidly moving material from the supernova blast impinges on the interstellar medium. Both the supernova shell and the interstellar medium are inhomogeneous so filamentary rope-like structures are formed. Behind the shock front the temperature can rise to very high values, of the order of 1 000 000 to 2 000 000 K, but the gas is very rarefied and glows faintly in soft X-rays. It also emits 'coronal'-type lines such as [Fe XIV] very feebly. The spectra of the filaments themselves show a very great range in excitation because shock effects are involved (see Chapter 10). Often [S II] is very strong; monochromatic photographs taken in this line are often used to identify supernovae remnants.

Initially, the chemical composition of the supernova remnant corresponds to that of the star that had been blown to bits, but as time goes on the stellar material is mixed more and more with the interstellar medium. As a result, the chemical composition of aged supernovae remnants, such as are observed in the Magellanic Clouds and in M33, corresponds mostly to that of the local interstellar medium, a fact confirmed by comparing compositions of the two.

Lifetimes of supernovae remnants, such as the Network, can be as much as 50 000 years, and in this interval other supernovae may have occurred in the same region. With each supernova producing its own cavity or 'bubble' in the interstellar medium, the different events may overlap and produce a 'superbubble', a volume of space much larger than any produced by a single supernova. Several examples of these superbubbles are found in our Galaxy.

The structure of the world in which we live and our presence here was possible only because supernovae existed. We find Iron-group elements in the most ancient of stars. True, some of these metal-deficient stars have low ratios of iron-to-hydrogen, perhaps only one-hundredth of the solar value or even less. It seems unlikely that they picked up iron, etc. from the interstellar medium at a comparatively recent date. Evidently massive stars must have been formed at the very dawn of the star-formation process. Many of them died as supernovae and as they did so they poured forth quantities of iron and other elements that were to be incorporated in the bodies of all stars since then.

Supernovae explosions do more than seed the interstellar medium with radioactive debris. Their explosions have a profound effect on the state of the interstellar medium and the formation of stars and even solar systems. Stars of modest mass supply carbon and nitrogen to the interstellar medium and, because of the vast number of these common folk of the heavens, we might expect huge quantities to be manufactured. Yet carbon is less abundant than oxygen which is not made in such mundane stars. A single supernova supplies much processed material: O, Ne, Fe, etc., to the intersteller medium. Analyses of meteorites show that both supernova ejecta and debris of gently dying stars are important in the ongoing drama of cosmochemistry.

It is in the context of cosmic rays, gamma rays, and X-rays that supernovae and the flywheels called pulsars that power some of their remnants offer some of the most enticing puzzles of astrophysics.

12

High-energy astronomy

Insofar as its gross properties are concerned, a star is one of the simplest objects in nature. A high internal temperature corresponds to large velocities of the particles that provide the pressure that supports the weight of the overlying layers (see Chapter 7 and Appendix A). At these enormous temperatures $\sim 10^7$ to $> 10^8$ K, protons move with such high speeds that they may penetrate the nuclei of heavier atoms, permitting the release of large amounts of energy. The nuclear furnace thus maintains the high temperature, sustaining a flow of energy to the glowing surface of the star. The astrophysicist refers to the radiation from the bright photospheres of the sun and similar stars as thermal radiation, that is, radiation emitted by ordinary matter simply because it is hot. Likewise, radiation from a normal gaseous nebula such as that in Orion can be understood. In simple terms, ultraviolet energy received from nearby, hot, illuminating stars is degraded by well-understood atomic processes.

While this simple picture of stars and gaseous nebulae was being explored and refined, other observations were being made with different tools that pointed to a greatly different and awe-inspiring vision of the observable universe. Early in this century it was noted that an isolated charged electroscope would gradually lose its charge. This was attributed at first to ionization produced by natural radioactivity in the earth; the charged particles of appropriate sign would be attracted to the leaves of the electroscope and gradually neutralize the charge that had been placed there. A balloon experiment by V.F. Hess in 1912 showed that the radiation came from above the earth and could truly be called cosmic radiation. High-energy particles impinge on the earth's atmosphere or on solid structures on its surface, shatter atomic nuclei and produce hosts of secondary or shower particles. It is these fragments or secondaries that are customarily observed. To measure the primary particles one must fly detectors in balloons, rockets, or satellites at heights above about 25 kilometers. These energetic atomic nuclei, mostly protons, move at speeds near the velocity of light. Unlike electromagnetic radiation, their paths are strongly bent by magnetic fields; the amount of the bending depends on the energy of the particle and the strength of the magnetic field. The paths of many low-energy particles are so severely deflected that they never reach the earth's surface at all.

Radio astronomy supplied the next significant clues. As mentioned in Chapter

10, radio observations revealed the expected quality and quantity of thermal emission from hot, ionized gas clouds like the Orion Nebula. But the radio observations also revealed something else, something quite unexpected. Early observations of the Milky Way at long wavelengths showed its brightness temperature (i.e. the black-body temperature that corresponded to the observed flux at these wavelengths) to exceed 100 000 K! This result was quite impossible to interpret in terms of thermal emission from heated dust. The mystery deepened when the first individual sources were detected by Australian and by British radio astronomers; these sources corresponded to no obvious optically observable galactic objects. The first optical object definitely identified as a radio source was the Crab Nebula, which turned out to be the Rosetta Stone of the subject. The Russian astrophysicist, J.S. Shklovsky, suggested that synchrotron radiation produced by electrons moving very nearly with the velocity of light could produce both optical and radio continuum emission from this supernova remnant. His prediction that the optical radiation would be polarized was confirmed by Walter Baade and by V.A. Dombrovsky.

Observations made originally with rockets in 1960 by R. Giacconi, H. Gursky, and B.R. Rossi, and more recently with satellites such as Uhuru and Einstein, have revealed numerous X-ray sources which embrace a wide range of object types – from binary stars and pulsars (i.e. neutron stars) to clusters of galaxies. Many, but not all, have been identified with optical sources. With a gamma-ray telescope flown in the Orbiting Solar Observatory III satellite, G.W. Clark, G.P. Garmire, and W.L. Kraushaar (1968) observed gamma rays with energies exceeding fifty million electron volts from the galactic disk. This discovery was quickly confirmed by other workers. Thus has high-energy astrophysics emerged as a most important field.

At the opposite extreme from X-rays and gamma rays is the 2.7 K microwave background (i.e. in the radio and far infrared regions) which was discovered by Arno Penzias and Robert Wilson. The energy distribution corresponds very nearly to that of a black body and has been found by measurement to be very nearly uniform in all directions. It has been studied by detectors flown in balloons and Cobe satellite and by its excitation of low lying energy levels of interstellar molecules such as cyanogen (CN). Presumably, this radiation is a residue of the 'primeval fireball'; that is, it is a relic from the earliest stages of the universe. Thus the 2.7 K microwave background is the Doppler shifted emission from the surface of the great expanding fireball itself, radiating at a local temperature of about two million degrees, but red-shifted into the radio-frequency spectral region.

It is of interest to note that the average temperature of the universe is not far from 3 K, except where matter is held in strong gravitational fields (e.g. stellar interiors), where it is considerably hotter, or in the neighborhood of stars, as on the cozy earth, or in gaseous nebulae. We do not see many stable surfaces with temperatures exceeding a million degrees, but with stars we see an accommodation with the cold universe at their surfaces, which are usually at between 1000 and 100 000 K. The surface temperature is determined by the combined effects of the energy generation rate in the star, the radius and the atmospheric opacity.

High-energy astronomy pertains to situations where matter is not in quasi-equilibrium or situations which involve effects of very large gravitational fields. We might divide high-energy astrophysics into three general areas.

(1)   Very hot objects. These usually involve some form of mechanical energy input as opposed to radiative sources. Examples are coronas of later type stars, supernova remnants, surfaces of very compact objects (see Chapter 8), and gas falling downwards under the influence of strong gravitational fields as in clusters of galaxies. Most of the emission appears in the far ultraviolet, X-rays, and gamma rays.

(2)   Non-equilibrium processes which, in the conversion of ordered mechanical motions into random motions (heat), produce large numbers of ultra-high-energy particles (i.e. particles with energies much greater than those corresponding to local temperatures). In many cases these processes involve the interaction of magnetic fields (which may have been amplified by the mechanical flows). The transformation process may involve direct conversion of mechanical to high energy (e.g. in shocks) or the combined effects of magnetic fields and fluid flows in what is called magneto-hydrodynamics. Although these processes are not considered 'normal', they are ubiquitous in the universe. We see the results in cosmic rays, X-rays, gamma rays, and (via the synchrotron processes) at radio wavelengths from all parts of the universe.

(3)   Objects that seem to involve extremely large gravitational fields, and as a consequence produce very large amounts of energy compared to a normal star. Examples include radio galaxies, galaxies with active nuclei (AGN), related processes in centers of normal galaxies (like the Milky Way), quasistellar objects, and even the universe itself at the time of the Big Bang. These objects emit energy all over the electromagnetic spectrum, and much of the interest in them revolves around the very high total energies suggested by the observational data.

Properties of matter and radiation at high energies

In examining the behavior of matter which is moving near the velocity of light, we have to apply the principles of special relativity which show that no material particle can travel with a speed equal to that of light, c, nor can any signal travel with what is called a group velocity exceeding c. The group velocity is that with which information is carried. A given signal may involve the superposition of many individual waves, each moving with what is called the phase velocity. For most purposes, what is important is the superposition (composite effect) or group velocity with which information is transmitted.

 As the velocity of a particle approaches that of light, many phenomena occur that appear bizarre by the standards of mass and motion to which we are accustomed. Within any material, light travels with a phase velocity, $v = c/n$. The phase velocity,

v, is generally less than c, since n, the index of refraction, exceeds unity at optical wavelengths. In some cases the phase velocity may exceed c, as for a radio wave traveling through an ionized plasma, but the group velocity is always less than c.

A speedy particle impinging on the earth's atmosphere may be moving with a velocity approaching c, and greater than that of light in air. The particle generates a bow shock wave, producing what is called Cerenkov radiation. The phenomenon is seen in the wake of a boat that is traveling faster than the speed at which the most prominent water waves move over the surface of the sea. Another well-known manifestation is involved in a sonic boom.

The observational data pertaining to high-energy particles and photons associated with them consist of measurements of cosmic rays, polarized, non-thermal, radio-frequency emission, X-rays, and gamma rays. High-energy electrons and cosmic-ray protons are associated, as also are gamma rays and cosmic rays. The direction whence the cosmic rays come gives us no information about the source that generates them, since the paths of these charged particles are twisted and turned by magnetic fields throughout the Galaxy. Only for solar cosmic rays, where we can associate the sudden burst of energetic particles with particular events on the sun, such as flares, can we be reasonably sure of their origins.

Let us consider first the high-energy particles. They are detected experimentally in three ways: (1) by ionization of atoms, (2) by destruction of molecular and crystal structures, and (3) by destruction of atomic nuclei themselves.

The total energy of a cosmic ray can be measured by how far it will travel in a solid before it is brought to rest. The 'stopping power' of various materials for high-energy particles can be found experimentally with the high-energy machines of modern physics. As a particle is slowed down, it may shatter the nuclei of many atoms in its path.

When a proton of, say, 1 giga electron volt (GeV) of energy, scores a direct hit on a nucleus, it brings in an energy supply far in excess of the binding energies of the individual protons and neutrons of the target nucleus. We can think of the incident proton as being repeatedly scattered by collisions with the protons in the nucleus and leaving the nucleus in an excited, unstable state from which it subsequently decays. Fig. 12.1 depicts some of the events often involved. For example, an oxygen nucleus battered by a speedy proton may emit a lithium or beryllium nucleus, an alpha particle, neutrons, and other fragments. Notice in particular that cosmic-ray collisions produce copious amounts of gamma rays. The positrons and electrons annihilate to give gamma rays of 0.511 mega electron volts (MeV) which are seen in solar flares, in gamma-ray bursters, neutron stars, and in the galactic center.

X-ray absorption

We recall that beyond the Lyman limit the interstellar medium is opaque because of the absorption by atomic hydrogen, but the absorptivity, α_ν, falls off quickly with increasing frequency, ν. In fact, $\alpha_\nu \sim \nu^{-3}$, so that at a frequency twice the Lyman limit the absorptivity has fallen to only about 12.5 percent of its original value. Because of their great abundance, hydrogen and helium play a big role, but in the region of soft

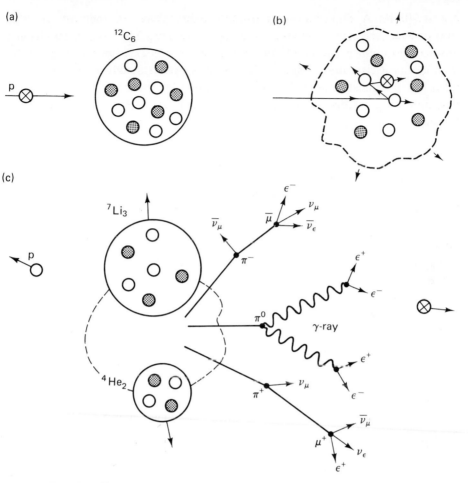

Fig. 12.1. Shattering of a carbon nucleus by a cosmic-ray proton.

(a) A proton of 1 GeV energy hits a target carbon nucleus. It has an energy exceeding the binding energy of all the neutrons and protons in the latter.

(b) As the impinging cosmic-ray proton (open circle with cross) ricochets from one neutron (filled circle) or proton (open circle) to another, it releases enough energy to finally shatter the carbon nucleus.

(c) In the scenario depicted here, the carbon nucleus is split into a nucleus of ^{7}Li, an alpha particle ^{4}He, a proton and a number of light evanescent particles (leptons) and ultimately gamma (γ) rays. The secondary evanescent particles include neutrinos, pions (π^0, π^-, π^+), and mesons (μ) that are associated with nuclear forces. The neutral pion (π^0) decays on a time scale of 1.8^{-16} seconds to gamma rays. The charged pions (π^+ or π^-) quickly decay to muons, muon neutrinos (v_μ), and muon anti-neutrinos ($\bar{v}_\mu$), with a mean lifetime of 2.5^{-8} seconds. The muons can lose energy by collisions with other particles. Eventually the μ^+ decays to a positron, an 'electron neutrino' and a muon anti-neutrino, viz. $\mu^+ \rightarrow \epsilon^+ + v_\epsilon + \bar{v}_\mu$, while correspondingly $\mu^- \rightarrow \epsilon^- + \bar{v}_\epsilon + v_\mu$. These events were studied in cosmic-ray showers in the earth's atmosphere before they could be produced in high-energy machines. In high-energy cosmic rays, muons, which have a mean lifetime of 2×10^{-6} seconds can reach the earth's surface. Note that if a gamma ray has sufficient energy, $> 2m_\epsilon c^2$, it can create an electron–positron pair! Strange particles can also be produced.

X-rays, 10–100 Å, photoionization from the inner 'K' and 'L' shells of metals like iron can be very significant in spite of their relatively low abundance. That is, absorption of optical radiation can remove the most weakly bound electrons from the outermost electronic orbits or shells, but X-rays tear tightly bound electrons from the innermost ones. Soft X-rays are soon absorbed, but hard X-rays may penetrate to great distances because the absorptivity falls off so steeply with frequency.

Bremsstrahlung (free–free radiation)

In our discussion of processes occuring in H II regions and in planetary nebulae (see Chapter 10) we described what is called 'free–free' absorption and emission. In the neighborhood of a positively charged atomic nucleus, an electron becomes accelerated and radiates energy; the rate of radiation of energy is proportional to the square of the acceleration, so that an accelerated charge always loses energy and slows down. But a free electron in the neighborhood of a positive charge may also absorb energy from radiation that falls upon it; this is the inverse of free–free emission. A distant encounter of an electron with a positive charge is more likely than a close encounter and less energy is gained or lost. The smaller the amount of energy emitted or absorbed, the lower the frequency of the radiation. Free–free emissivity or absorptivity depends only weakly on frequency (v), until $hv > kT$ (h = Planck's constant, k = Boltzmann's constant, T = temperature); thereafter it depends strongly on v.

Now free–free emission can also occur in a very hot gas, manifest as, for example, X-ray emission in binaries, and as diffuse X-ray emission from intergalactic gas in clusters of galaxies, radiating at a temperature of 100 000 000 K! X-ray workers have traditionally called this radiation, *Bremsstrahlung*, but this formidable German word need not frighten us; it is simply free–free emission coming from a very hot gas. When the speeds of the electrons approach that of light we must use the theory of special relativity in making calculations. This 'relativistic bremsstrahlung' is important for cosmic-ray radiation.

There are other radiation processes that occur in the high-energy domain for which there are no low-energy analogues. These include pair production, Compton scattering, inverse Compton scattering, and synchrotron radiation. They are described briefly below.

Pair production

The Einstein relation, energy = (mass) × (velocity of light)^2, expresses the equivalence of mass and energy. We have seen examples in previous chapters where mass was converted to energy, as in the fusion of our hydrogen atoms to make one helium atom with 0.7 percent of the mass disappearing as energy. But the process can go both ways with particles and anti-particles being produced in pairs. In particular, if the energy of a gamma ray exceeds $2m_e c^2$, where m_e is the mass of the electron and c

the velocity of light, then sufficient energy is available for the photon to be converted to an electron–positron pair. This process usually can take place only in the neighborhood of a charged nucleus, but actually it also could occur in an incredibly strong gravitational field. Showers composed only of electrons, positrons, and gamma rays, but no muons, pions, or other particles, can be attributed to gamma rays, as cosmic rays would yield not only these but a variety of debris, sometimes including strange particles (see Fig. 12.1).

How hot can a gas get if energy is poured into it without any efficient mechanism for energy loss so the temperature just continues to rise? Eventually, if the energy of an average electron exceeds 1 MeV (corresponding to a temperature of about 1.1×10^{12} K), the interactions are so severe as to lead to rapid pair production; the temperature cannot rise further.

Compton scattering

An incoming photon may collide with a stationary or slow-moving electron and transfer some of its energy and momentum to the electron; hence the frequency of the photon is decreased and its wavelength increased. In this process, high-energy quanta are degraded to produce energetic electrons (see Fig. 12.2a).

Inverse Compton effect

A high-energy electron may collide with a low-energy photon. The electron departs with greatly reduced energy and a high-frequency photon is created (see Fig. 12.2b). For example, an electron with a thousand million electron volts of energy may collide with a photon of energy 0.001 eV (corresponding to 3 K) to produce an X-ray photon! A similar electron colliding with a photon of starlight corresponding to a temperature of, say, 4000 K would create a gamma ray.

In summary, Comptonization (the Compton and inverse Compton effects) can become important at energies exceeding about 10 kilo electron volts (keV). High-energy photons encountering slow-moving electrons speed the latter up. Conversely, encounters between fast electrons and low-energy photons transfer energy to the latter. From the standpoint of the theory of special relativity, to an observer riding with the electron, the Compton and inverse Compton effects are equivalent.

Synchrotron radiation

The continuous spectrum stretching from the radio-frequency to the X-ray region in the Crab Nebula and the radio-frequency emission from old supernova remnants arises from what is called synchrotron radiation. A synchrotron is a device for producing high-speed electrons for investigations in nuclear physics. The emission pattern of an accelerated electron whose velocity approaches that of light is quite different to that of a more slowly moving one. In a synchrotron the electrons are constrained to move in a circular path by the action of a magnetic field perpendicu-

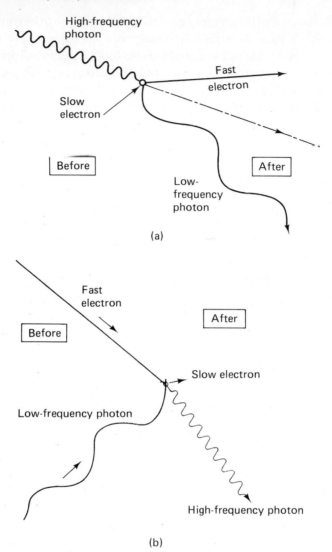

(a)

(b)

Fig. 12.2. The Compton and inverse Compton effects.

(a) In the Compton effect a high-frequency or X-ray photon collides with a slow electron. There results a fast-moving electron and a low frequency or infrared photon.

(b) In the inverse Compton effect a collision between a fast electron and a low frequency (infrared) photon results in the production of a slow electron and an X-ray or high frequency photon.

Comment for those familiar with special relativity:
In the frame of reference of the electron, the two processes are actually identical! The relativistic electron 'sees' the (to a rest-frame observer) infrared photon as a highly blue-shifted X-ray photon. The electron undergoes a Compton scattering and suffers the appropriate change in momentum. The scattered photon has less energy (as seen in the reference frame of the electron). An external observer sees a low-energy photon change into a high-energy photon.

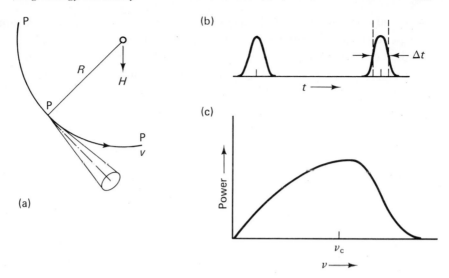

Fig. 12.3. Some properties of synchrotron radiation

(a) An electron which is moving at 'relativistic' speeds such that $E > m_e c^2$, radiates in a small cone of vertex angle, $\theta = mc^2/E$. The electron moves in a circular path (indicated by P's) of radius R. The magnetic field, H, is perpendicular to the plane defined by v and R. In the part of its orbit where it is approaching the observer, the electron nearly catches up with its own radiation.

(b) An observer watching a single electron would see radiation pulses each of duration: $\Delta t = (R\theta/c)(mc^2/E)^2$. Actually, there will be a whole host of electrons radiating non-coherently, so there is a superposition of flickers.

(c) This is the frequency distribution of the radiation emitted by a single electron. Note that it differs considerably from a black-body curve. Most of the energy is radiated in the neighborhood of a critical frequency:

$$\nu_c \text{ (Mc/s)} = 16.08 \ H \text{ (gauss)} \ E^2 \text{ (MeV)}.$$

In actual practice, the electrons are continuously losing energy so that any synchrotron-radiation-emitting ionized gas or plasma will show a distribution of energies. If the electron distribution follows what is called a power law, the emitted spectrum will also be a power law. Such a distribution is shown by the spectrum of the Crab Nebula (see Fig. 12.9).

lar to the plane of their motion. Because the electrons move in a curved path, they are constantly being accelerated; thus they must continuously radiate energy. The light is emitted, not primarily in a direction perpendicular to the electron's motion, as would be true for slow-moving charges, but rather in a narrow cone in the direction of motion, due to special relativity effects. Furthermore, the light is polarized. See Fig. 12.3.

We may visualize the electron as a toy engine running on a circular track. The beam from its headlight (across which a strip of Polaroid is placed) represents the emitted light. A given electron will appear to emit a momentary pulse in the direction of the observer once each revolution, but there are hordes of electrons and, since we observe the combined efforts of this vast multitude, the flicker will be wiped out. The radiation is emitted at the expense of the electron's kinetic energy, which decays away continuously. The rate of energy loss from a single electron is proportional to the square of the energy multiplied by the square of the magnetic field. Therefore, the speediest electrons will lose energy most rapidly. The time required for the decay of the energy of an electron emitting in the optical region is about ten years. Thus we easily see that in a source such as the Crab Nebula, where the energy distribution in the spectrum has remained nearly constant, a stream of high-energy electrons must be continuously supplied. For more than ten years (until the discovery of pulsars) the energy source of the Crab Nebula was one of the embarrassing unsolved mysteries. The energy distribution of the emitted radiation differs very markedly from that of a black body; there is a sharp high-frequency cut-off that depends on the orbital period of the fastest electrons, and thus on the magnetic field.

In particular, suppose there is what is called a power law in the number distribution of electrons with energy, E, that is, $N_\epsilon = \text{const } E^{-\beta}$. Intuitively, the justification for such a law is not hard to grasp. Suppose very-high-energy electrons are injected into a magnetized, largely ionized gas. They will radiate at a rate proportional to the square of the magnetic field multiplied by the square of their energy. They lose energy quickly and will tend to pile up in the low-energy domain radiating at low frequencies. If we plot the number of particles against their energy, we will get a curve that rises rapidly towards low energies. The spectral energy curve of the emitted radiation therefore differs markedly from that of a black body. This characteristic, together with the polarization of the emitted radiation, enables us to distinguish synchrotron from thermal radiation. In any real situation an exact calculation of the emissivity may be difficult. For example, the finite density of the ionized gas must be taken into account. Collisions of relativistic electrons with slow-moving electrons that may be in their way may degrade the energy of the fast-moving particles.

In discussing the origin of the characteristic spectrum of synchrotron sources, we see that:

(a)   The frequency of emission of the maximum radiation goes as HE^2 (where $H = $ the strength of the magnetic field and $E = $ energy).

(b)   Both cosmic-ray nuclei and electrons typically have a power law spectrum, $N \sim E^{-\gamma}$. (E is energy and γ is a constant over a large range of energy). This is observed directly in the cosmic rays at the earth, and we infer it from radio (synchrotron) emission from elsewhere in the universe.

(c)   The underlying reason for the characteristic spectrum of synchrotron emission is the electron power spectrum; there are fewer high-energy electrons which are able to radiate at the higher frequencies than low-energy electrons that emit at lower frequencies.

Although radiation losses might account for some of the decrease in numbers of high-energy electrons, cosmic-ray ions also have a similar power law energy spectrum, and radiation losses are relatively unimportant for the ions. It appears that the acceleration mechanism (shocks?) must tend to produce this type of energy spectrum in a natural way, since many different types of objects exhibit this type of spectrum, e.g. the Jovian magnetosphere, our galactic disk, supernovae remnants, and radio galaxies. The problem is that no one has figured out how the acceleration process produces the observed spectra! This is another example where nature's subtleties still elude us.

Anyway, the observed energy spectrum of any synchrotron source involves the interplay of three factors: (1) energy loss by radiation, (2) energy loss by collisions between synchrotron electrons and passing, slow-moving, thermal electrons, and (3) the input of high-energy electrons through acceleration processes. In many sources, such as the radio-frequency bursts in the solar corona and certain supernova remnants, we are probably observing the decay of an initial input of high-energy particles, the decay being fast in the sun and slow in the interstellar medium. There is evidence that particle acceleration is still going on in at least some supernova remnants. In the case of the Crab Nebula, the pulsar continues to be active. In Cassiopeia A shocks apparently form between fast-moving filaments, and a film record of knots and filaments that brighten and disappear has been made with the Very Large Array (VLA) in New Mexico.

To summarize: emission of radiation from a heated source (i.e. thermal emission – be it from a tungsten light filament, star, or gaseous nebula) is a relatively simple process governed basically by Planck's law (see Chapter 4 and Appendix B). Randomness of energy exchanges prevails and can be handled by simple, well known laws of chance and probability.

On the other hand, synchrotron radiation is not random. Particles are accelerated to high energies by a spinning neutron star or by some other means. Heavier particles, such as protons, may escape as cosmic rays but they, like the electrons, must lose energy. In a magnetic field electrons are constantly being accelerated and therefore they must radiate energy. This radiation is not emitted uniformly in all directions, with the consequence that the radiation is polarized. The rate of energy loss depends on the strength of the magnetic field and the square of the electron's energy. The faster moving electrons lose energy more copiously. No matter how efficient the generator of fast electrons may be, there will always be an accumulation of electrons towards the lower energies. Since the frequency of the emitted radiation depends on the particle energies, the energy flux will rise steadily as the wavelength increases. The resultant energy curve will not look at all like that of a black body (compare Fig. 12.9 with Fig. 4.1). These two properties of synchrotron radiation, polarization and a strikingly non-black-body energy distribution enable us to distinguish clearly between thermal and non-thermal radiation.

Synchrotron radiation is observed in distant galaxies and quasi-stellar objects. If the thickness of the radiating gas is large, synchrotron self-absorption can occur. We must take this effect into account in our interpretations of radiation patterns of active galaxies, with their vast volumes of high-energy gas.

X-ray astronomy

In Chapter 10 we mentioned how gas behind the shock front of a supernova remnant could be heated to temperatures as high as a million degrees and that such cavities from successive supernova explosions could overlap to produce vast regions of very high temperatures that would radiate X-rays. In Chapter 11 we described how material shredded from a companion star could be accreted by a white dwarf, a neutron star, or even a black hole. The actual site of the emission may be accretion disks, especially for small objects such as neutron stars or black holes. The virtually free-falling gases liberated their gravitational energy upon impact. Local regions could be heated to enormous temperatures and radiate copiously in the X-ray region. X-ray luminosities up to 10 000 or 100 000 solar luminosities could be attained in some objects.

Actually, X-ray emission from the solar corona had been predicted by B. Edlén. It was observed in 1949 by R. Tousey and his associates at the US Naval Research Laboratory who fired rockets above the earth's atmosphere. The sun is a relatively feeble source of X-rays. In 1960, however, R. Giaconni, H. Gursky, F. Paoline, and B.R. Rossi discovered a powerful X-ray source in the constellation of Scorpius. They and other groups found many other sources with detectors flown in rockets, but rapid expansion of our knowledge came only with the HEAO (High Energy Astrophysical Observatory) satellites. The second of these, the Einstein Observatory, was a particularly sophisticated instrument that acquired images with the resolution of a few seconds of arc, provided a detection sensitivity a thousand times better than previously accomplished, and permitted meaningful spectroscopy.

X-ray astronomy presents wondrously difficult observational challenges. Conventional optics are useless; X-rays are usually not reflected by a glass or metallic mirror, they simply go right through it. Only low-energy X-rays can be reflected at grazing incidence. Hence the Einstein telescope, designed with this type of optics, could secure data only in the soft X-ray range (0.1 to 4 keV). Flux measurements of high-energy X-rays require special combinations of filters and detectors; the spectral resolution is very poor.

Consider the whole gamut of celestial X-ray emission: the coronas of the sun and similar stars, intergalactic gas in clusters of galaxies, supernova remnants, accreting white dwarfs, neutron stars, black holes, active galactic nuclei, X-ray bursters (see later in this chapter), and a mysterious X-ray background. Some sources are rarefied gases of vast extent, others are dense regions. Hence a range of physical processes must be taken into account. The common factor that underlies all X-ray sources is that they are hot, typically a million to a hundred million degrees. The simplest systems are probably quiet stellar coronas. At the other extreme are dense gases in active galactic nuclei or in the vicinity of black holes where effects of radiative flow must be considered.

The quiet solar corona, with a temperature of approximately 2 000 000 K, with less than 10^8 electrons per cubic centimeter, and with lines of highly ionized atoms, e.g. [Ca XV], [Fe XIV], Si X, etc., was the first to be investigated. A steady state is set up wherein all the ionizations which are due to collisions are exactly balanced by recombinations. Unlike the situation in stellar atmospheres, photospheric radiation plays a negligible role. X-ray emission is found from stars all over the Hertzsprung–Russell (HR) diagram, except red giants. In the solar corona, X-ray emission is often prominent in highly excited loop structures associated with magnetic activity. We presume that X-ray emission from other solar-type stars may often originate from magnetized regions, and may serve as a measure of magnetic activity. The amount of emission seems to depend on the age of the star, being larger in younger stars.

An extreme example is provided by the intergalactic gas which is spread diffusely in clusters of galaxies. Recombination times are long, but the cluster lifetime is even longer, so we assume a steady state. Although the gas appears to be at a temperature of about ten million degrees, the presence of the X-ray '*K*' line of very highly ionized iron suggests that the material has been expelled from stars; it was not all left over from the Big Bang.

X-ray observations show most supernova remnants to have a shell-like structure that can be explained in terms of a spherically symmetrical blast wave. In type II supernovae, stellar material amounting to several solar masses is violently ejected into the local interstellar medium which may have a patchy structure. The chemical composition of young remnants corresponds to that of the supernova, but, after several centuries, the observed remnant consists mostly of interstellar medium material swept up by the blast wave.

The X-ray images of Tycho's 1572 supernova resemble radio maps which show a bright circular shell of radio emission, produced as the shock front blasts into the interstellar medium. The X-ray spectrum shows strong lines of Si XII, S XV, S XVI, and Ar XVIII, interpreted to indicate that these elements are six times as abundant with respect to hydrogen as compared with the sun; they were produced copiously in the supernova event.

Extracting the chemical composition of the gas from observed line intensities is far more difficult than for a normal gaseous nebula even though we can observe more stages of ionization than is usually possible for the latter. No equilibrium exists; there is a violent event in which the gas temperature (as measured by ionic speeds) is suddenly greatly increased behind the shock front. The electron temperature takes a time to catch up. One has to work out detailed scenarios of excitation, cooling, gas motion, etc. The problem is easier for older supernova remnants, but they contain much material from the local interstellar medium. Fig. 12.4 shows a spectrum of the supernova remnant Puppis A.

Much more complicated examples of X-ray emission are found in accretion disks of white dwarfs and neutron stars where stellar winds may also occur. One must now consider the flow of X-rays through optically thick slabs of material and effects of mass motion. All of the physical processes we have cited may occur: Bremsstrahlung, Comptonization, and even pair production.

A magnetic field produces yet further complications and roadblocks to our understanding of the phenomena. First of all, note that it will constrain the flow of ionized gas. A slow-moving electron will move in a spiral path around the line of force with a frequency $v_H = (eH)/4\pi\,mc$, H being the field strength in gauss, e and m the charge and mass of the electron (cgs. units), and c the speed of light. Because of the acceleration imposed on it by the magnetic field, the low-energy (non-relativistic) electron will emit radiation of this gyromagnetic or Larmor frequency, v_H. Note that harmonics can also occur. Unlike the synchrotron radiation emitted by

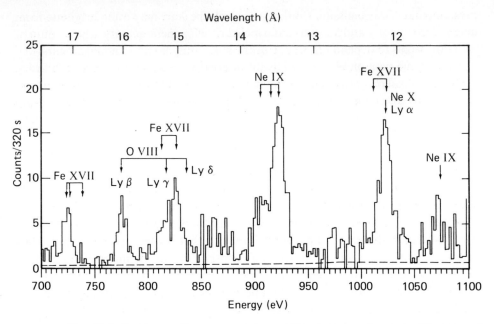

Fig. 12.4. Spectrum of supernova remnant Puppis A, as observed with a crystal spectrometer on the Einstein satellite. The dashed line indicates the background level. (From Winkler *et al.*, 1981 *Astrophysical Journal Letters*, **246**, L27.)

electrons moving with nearly a velocity c, gyromagnetic radiation is emitted nearly uniformly in all directions and is usually circularly polarized. For solar magnetic fields this radiation is in the radio-frequency range, but in certain white dwarfs, with fields of 10^6–10^8 gauss, it can fall in the optical region. It is believed to account for the strongly polarized optical continuum of the highly magnetized white dwarf, AM Herculis. In a neutron star, for which models predict a field strength of 10^{11}–10^{12} gauss, the Larmor frequency can appear in the X-ray range. There is some evidence that it has been actually seen in Hercules X-1.

Interesting examples of X-ray sources are included among novae, cataclysmic variables, combination variables, accreting neutron stars, and the black hole Cygnus (Cyg) X-1. Cataclysmic variables involve white dwarfs. Combination variables are binaries where a red cool star and a highly evolved hot blue star, such as a white dwarf, are paired. Optical and ultraviolet observations suggest that white dwarfs without magnetic fields have accretion disks, while those with strong fields, like AM Herculis, allow material from the companion to fall almost directly to the small star, sometimes producing, as in AM Herculis, an X-ray luminosity exceeding the visual solar luminosity.

More than a hundred compact X-ray sources are known with luminosities from 200 to 200 000 times that of the sun. Optical counterparts are known for about half. About 20 of them pulsate with periods between 0.71 and 835 seconds. Her X-1, with a luminosity 5000 times the sun (if the distance is 6 kiloparsecs), shows a cycle of 35 days which can be interpreted as oscillations of a pulsating source occulted by an accretion disk. The low-mass binary system Scorpio X-1, the brightest X-ray source in the sky, with an effective temperature of 10 000 000 K, emits soft X-rays in the 2–10 keV range. These are possibly caused by Compton scattering in an extended corona. Cygnus X-1 is a binary evidently involving a black hole of 10 solar masses. Its X-ray luminosity of 10 000 suns, has been attributed to Comptonization of low-energy (soft) photons as they diffuse through a cloud of electrons whose average energy is 30 keV.

Conventional galaxies like our own are not important sources of X-rays, but active galactic nuclei are powerful emitters at all wavelengths: radio, infrared, optical, ultraviolet, and X-rays. They are very conspicuous in the radio and X-ray regions. The energy is believed to come from infall of matter upon a central black hole whose mass exceeds a million suns. Although the radio emission arises from synchrotron radiation, the X-ray emission cannot come from this cause but seems to be produced by inverse Compton scattering of low-frequency photons by fast electrons, or direct emission from a very hot corona or accretion disk.

Among the more remarkable X-ray sources are the bursters, which are characterized by sudden rises in luminosity to as much as 250 000 suns with durations of 10–1000 seconds. The energies involved are 1–3 keV. X-ray bursters are thought to come from thermonuclear explosions on the surface of an accreting neutron star.

Finally, there is the mysterious diffuse X-ray background which covers the whole sky much as does the 2.7 K 'afterglow of the Big Bang'. Did the X-ray background originate in discrete sources at an early epoch or is it a smooth distribution arising from a mechanism yet unknown?

Gamma rays and cosmic rays

The most energetic processes in the universe, running the gamut from individual particle interactions to stellar explosions, active galactic nuclei and quasistellar objects are revealed via gamma rays, high-speed electrons involved in synchrotron radiation, and cosmic rays. Here we examine the evidence from the various ultra-hard, high-energy, electromagnetic radiation and high-energy particles known as cosmic rays. Gamma rays often act as tracers of cosmic rays, and relativistic synchrotron electrons are representative of the high-energy electrons associated with cosmic rays.

First consider some properties of gamma rays. Those above 100 MeV are emitted by mesons formed in nuclear interactions between cosmic rays and luckless atomic nuclei, mostly of hydrogen, that lie in their path. Just as galactic radio synchrotron radiation gives clues to the deployment of cosmic rays and magnetic fields within the Galaxy, so does the distribution of gamma rays serve to trace the cosmic-ray protons and the location and amount of interstellar gas. Lower energy gamma rays (< 100 MeV) may be produced by the inverse Compton effect or 'relativistic bremsstrahlung' – the free–free quantum emitted by an electron moving with nearly the speed of light and accelerated by a charged nucleus.

The gamma-ray–cosmic-ray relationship is not an even trade-off. For every

Table 12.1 *Some reactions which produce gamma rays*

Reaction	Half-life	Source or remarks
^{56}Ni$\rightarrow$^{56}Co	8.8 days	Supernovae (e.g. SN 1987a)
^{56}Co$\rightarrow$^{56}Fe	114 days	Supernovae (e.g. SN 1987a)
^{26}Al$\rightarrow$^{26}Mg	1 100 000 years	^{26}Al is produced in novae by ^{25}Mg$+$p$\rightarrow$^{26}Al

thousand cosmic-ray nuclei (92 percent H; 6 percent He) there is one potent gamma ray, perhaps ten relativistic electrons, a few positrons and, once in awhile, an anti-proton or exotic particle.

The great advantage of gamma rays is that over a range from 10^7 to 10^{11} eV they suffer very little extinction from the interstellar medium. They reach us from the galactic center and from heavily obscured objects such as Cygnus X-3. The earth's atmosphere stops them cold, although if they have high enough energy they can produce mesons that can reach detectors at the surface. If it has an energy greater than 10^{11} eV, the incoming gamma ray can induce Cerenkov radiation because the secondary particles produced move downward with a speed exceeding the velocity of light in air. Since Cerenkov radiation can also be produced by cosmic rays, we must be able to separate the two sources. However, gamma-ray studies are difficult. Most detectors must be flown above the earth's atmosphere; the photons are very few in number even though each one carries a terrific punch.

In addition to a continuous spectrum due to the decay of evanescent pion (π°) particles, inverse Compton effect, or rarely synchrotron radiation (as in the Crab Nebula), gamma rays also exhibit a line spectrum. The most common line corresponds to the annihilation of a positron and electron which gives two 0.511 MeV gamma rays. As X-ray lines arise in transitions between energy levels of tightly bound, inner atomic electrons, so do gamma rays originate from transitions between nuclear energy levels. Table 12.1 lists some examples for the production of gamma rays.

The galactic center is a well-known source of gamma rays. They are produced in binaries, in pulsars, and possibly in other galaxies. We often identify the source if it is periodic and the periodicity agrees with that established for X-ray or optical data.

In the early 1970s it was recognized that Cygnus X-1, a binary system about 2300 parsecs from the earth, was a good candidate for a black hole. One component was a bright supergiant of about 30 solar masses, but its invisible companion had a mass of about 10 solar masses. The maximum mass of a white dwarf is about 1.4 m(sun), that of a neutron star about 2 or 3 m(sun). Any cold condensed object of greater mass would have such a powerful self-gravity that even the strong nuclear repulsive forces could not prevail against it. Hence all material caught in such a predicament would be crushed in a 'bottomless pit' from which nothing could ever be rescued. For a 10 m(sun) black hole, the 'Schwarzschild radius' (the radius from within which no particle or signal could ever escape) would be about 30 kilometers.

An isolated black hole floating freely in space could reveal its presence only by its gravitational effects, but a black hole in a binary system or at the center of a galaxy can produce some spectacular effects as it gobbles up material or shreds whole stars. In Cygnus X-1, the supergiant companion is gradually losing mass to a thick accretion disk around the invisible black hole. Within this disk, temperature, pressure, and viscosity build up as the gas spirals inwards towards the Schwarzschild radius. Before the gas is lost in the bottomless pit, it may reach a temperature of a thousand million degrees and emit a copious supply of X-rays, gamma rays, and positron–electron pairs. For a period of two weeks in 1979, MeV gamma rays amounted to about half the energy output of Cygnus X-1, suggesting that the inflow of material is not constant. Studies of this object may help us understand the complex processes that can occur in active galactic nuclei with their presumed much more massive black holes.

Cygnus (Cyg) X-3, which was discovered by detectors flown in rockets in 1966, is seen in projection along a spiral arm, apparently behind the so-called Cyg X complex of H II regions. It is a very luminous source hidden behind dust clouds about 12 000 parsecs distant, in the outer fringes of our Galaxy. The X-ray emission varies with a period of 4.8 hours. Presumably Cyg X-3 is a binary, not visible optically because of the heavy interstellar extinction. There is also a bright infrared source which sometimes varies in phase, also with a period of 4.8 hours. This periodicity has also been observed in the radio region.

In September 1972, there occurred an intense radio outburst that made Cyg X-3 one of the brightest objects in the sky for a few days. Other, less intense, outbursts have subsequently occurred. Possibly they come from variable accretion rates of material on a condensed object, probably a neutron star. J. Grindlay has suggested that the radio flux arises from jets similar to those found in SS 433.

From very painstaking measurements of Cerenkov radiation caused by fast particles traveling through the atmosphere, it has been possible to make an unequivocal identification of Cyg X-3 as a powerful variable gamma-ray source. What clinched the matter was the close match of X-ray and gamma-ray periodicities. Whatever made the X-rays also made gamma-rays. Furthermore, the production of gamma rays up to an energy of 10^{16} eV implies a concurrent production of cosmic rays up to 10^{17} eV. On the basis of the observed gamma-ray intensities, investigators at the University of Leeds calculated that Cyg X-3 is such a powerful source that only one such object would be needed to produce all galactic cosmic rays in the 10^{17} eV range. The estimate may be optimistic, but certainly Cyg X-3 is a powerful cosmic-ray factory.

The origin of cosmic rays

Are all cosmic rays produced in sources such as Cyg X-3, presumably by a rapidly spinning neutron star? Probably not. Some low-energy cosmic rays are ejected in violent flare activity on the sun and similar stars. Could such particles later be accelerated to much higher energies?

The production rate and energy problem of cosmic rays may be considered as follows. We measure the overall flux of cosmic rays, and from chemical analyses (discussed in the following paragraphs) one can determine the amount of material through which the cosmic rays have traveled. Then, using the average density of the interstellar medium, one can determine the age of the average cosmic ray and ascertain the production rate required to account for the observed flux.

A mechanism suggested by E. Fermi starts with solar-flare-like cosmic rays and accelerates them by collisions with moving clouds of magnetized gas, or strong shocks in stellar winds or supernova remnants. In particular, the Fermi mechanism may be important in collisions of the charged particles with filaments in supernova remnants. Certainly, the cosmic rays have a long random walk through the Galaxy in the course of which they gradually pick up energy. Jean Paul Meyer noted that the heavy element composition of the solar corona, solar wind, solar cosmic rays and galactic cosmic rays are similar. With respect to the solar photospheric composition they show an underabundance of heavy elements with a first ionization potential greater than 9 eV relative to elements with a lower first ionization potential by a factor of four to six. He suggested that a neutral atom–ion separation occurs as the gases arise from the underlying chromosphere, so the elements of low ionization potential could escape more easily. The implication is that most galactic cosmic-ray nuclei were extracted from ordinary stars by stellar flares and later accelerated to GeV energies by a Fermi-type mechanism.

Ultra-high-energy cosmic rays with an energy greater than 10^{18} eV may come from outside the Galaxy, possibly from active galactic nuclei or extended lobes of radio galaxies, such as Cygnus A (see Fig. 12.5 and p. 301). Shocks in supernova remnants are believed to play a role, but the best bet for many cosmic rays would appear to be magnetized clouds of charged particles (magnetospheres) of rapidly spinning neutron stars. One important clue is that cosmic rays have been with us for a long time. A cosmic ray passing through a meteorite damages the crystal structure, thereby giving a fossil record of cosmic-ray activity. From meteorites of known age it is found that the density of cosmic rays has remained constant over some thousands of millions of years.

In summary, although some cosmic rays are produced in violent flare activity on the sun, most cosmic-ray particles – indeed all of the high-energy ones – come from outside our immediate neighborhood. Magnetic fields of the order of 10^{-4} to 10^{-5} gauss with irregularities of a size comparable with the earth–moon system tend to deflect away cosmic-ray particles with energies less than 10^8 eV. Hence we have accurate data only for cosmic rays of greater energy. The observed time-dependent variations are all due to the action of magnetic fields in the solar system.

The energy of a typical cosmic-ray particle is about 10^9 eV. Their density is about one per thousand cubic meters, but since they travel with nearly the velocity of light, every second about six of them would travel through an area of about one square centimeter in space. Cosmic rays of low energies appear to come from all directions of space; there is no preferred part of the sky. Their actual trajectories are modified, of course, by the earth's magnetic field. They diminish rapidly in numbers with increasing energy. That is to say, if we define N_E as the number of particles with energy greater than E, their distribution very closely follows the power law, $N_E \sim E^{-\alpha}$, where $1.5 < \alpha < 2.1$ for energies per particle between 10^{10} and 10^{19} eV.

Cosmic rays and the production of lithium, beryllium, and boron

The chemical composition of cosmic rays (at least for heavier particles – with possible exceptions as noted by J.P. Meyer) mimics that of the Galaxy, so they must have originated from objects of normal composition. An acceleration mechanism working in stellar flares may have speeded up samples of local material to high velocities. Galactic cosmic rays show an abnormally high ratio of the helium isotopes (^{3}He/^{4}He), and an enhancement of nuclides just lighter than iron, but the most striking feature is the substantial number of the cosmically rare nuclei of lithium (Li), beryllium (Be), and boron (B). These are produced by the shattering of heavier nuclei struck by cosmic rays and of heavier cosmic-ray particles by encounters with interstellar atoms.

The increased abundance of Li, Be, and B gives an idea of the amount of matter penetrated by high-energy particles. First we must measure experimentally the interaction cross-sectional areas for spallation (shattering) events. These turn out to be appropriate for nuclear sizes ($\sim 10^{-26}$ cm^2) rather than atomic sizes (10^{-14}–10^{-16} cm^2). We thus estimate the amount of material that must have been penetrated to produce the observed abundances. It turns out to be of the order of 4–6 g/cm^2 – equivalent to a sheet of steel approximately 1 cm thick. Since the density of normal atoms in the galactic plane is known, we can estimate the length of time these high-energy particles must have spent there, and get an average of about twenty million years. By combining the lifetimes and densities of cosmic rays we can estimate their total energy content. Certain nuclides, such as magnesium, silicon and iron, appear to be slightly more abundant than in the solar system; perhaps they were produced in supernovae. The cosmic-ray abundances seem to show some dependence on the energy involved; perhaps this fact can supply clues to the mechanism involved.

The mystery of extended radio sources

A discussion of galaxies lies outside the scope of this book, but some 'small scale' phenomena in our own Galaxy, such as SS 433 or perhaps Cygnus X-3 or Scorpio X-1, may be profitably studied on the grand scale in giant elliptical systems and quasistellar objects. An isolated, optically observed galaxy that may show no particularly strange features may be flanked (often symmetrically) by two huge blobs or lobes that emit intense radio-frequency radiation. These lobes have diameters that range from 10 000 to 100 000 parsecs; they are typically separated by 150 to 350 kiloparsecs, although separations greater than a megaparsec are known. See Fig. 12.5.

A powerful central source of less than a parsec diameter, quite possibly a very massive, spinning black hole in an active galactic nucleus, spews out streams of relativistic particles in both directions along the axis of spin. The radio structures tend to be aligned along the projected minor (smaller) axes of elliptical galaxies, just what we would expect if the rotational axis of the galaxy is parallel to that of the central source.

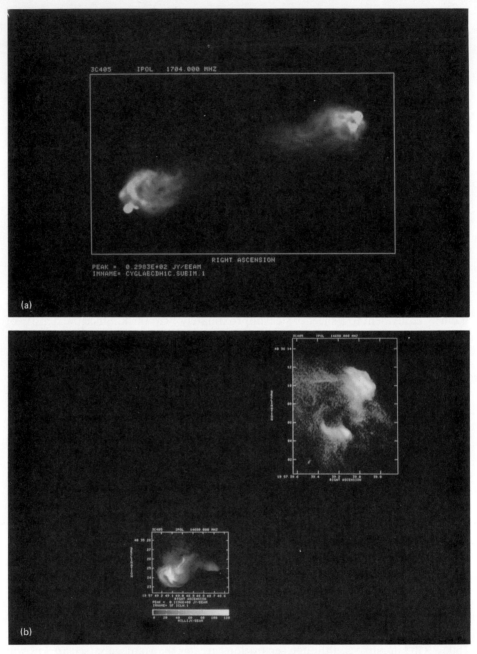

Fig. 12.5. The radio galaxy, Cygnus A, as observed with the Very Large Array (VLA).

(a) The double source is observed at 17 cm with a resolution of 1 arcsec. Note the filamentary structure of the lobes and 'hot spots' at the outer edges.

(b) Details of the images at 2 cm, with a resolution now of 0.1 arcsec, which is vastly superior to optical observations. Note the structure at the point of impact between the jet and the intergalactic medium. These observations illustrate the impressive capabilities of the VLA. (Courtesy C.L. Carilli, NRAO–VLA and Massachusetts Institute of Technology.)

In the region of greatest activity, near the central powerhouse, one often observes a strong continuous optical spectrum upon which are superposed hydrogen and helium emission lines broadened by Doppler effect due to rapid turbulent motions of clouds of hot gas. Also narrow, familiar, forbidden lines, such as [O III], often appear; these originate in extended, slowly moving, low-density clouds. The optical spectrum is produced by physical processes similar to those found in galactic planetary nebulae, except that the energy of the nebula here is derived from the non-stellar, non-black-body central source.

What is remarkable about these vast extended structures is that they often greatly exceed the size of a galaxy. Great jets extending from the core galaxy terminate in huge, diffuse lobes – all emitting in the radio range from 'relativistic electrons', i.e. electrons moving at nearly the velocity of light. The diffuse lobes are not uniform in brightness, but often contain 'hot spots' about 1 kiloparsec in size (see Fig. 12.5). Radio observations with high spatial resolution often reveal intricate structures in the filaments, suggesting that the emitting power source may not be emitting uniformly. The amazing alignments of these structures imply that the axis of the spinning source may remain fixed over the lifetime of the blobs, i.e. for millions of years. In Cygnus A the core is aligned to within a few degrees of the outer lobe, over a scale of about 5 000 000 to 1. Some radio sources show twisted filaments, suggesting that precession of the source can occur.

In addition to the radio emission, the giant elliptical galaxy M87 shows optical radiation not only from the central core, but also from lobes and hot spots; the patterns of optical and radio emission closely match one another. Polarization data indicate that both radio and optical radiation are of synchrotron origin. A somewhat similar match is found for the jet of the quasistellar object, 3C 273.

From these facts there emerges the picture of a powerful 'generator' that sends tightly focussed or collimated beams of high-energy particles that may exceed a million light-years in extent. Perhaps the intergalactic gas slows down these beams and causes them to fan out to produce huge lobes radiating strongly in the radio-frequency range.

The physical processes involved are most engaging. The radio flux is certainly synchrotron radiation rather than bremsstrahlung, which would be unpolarized. The occasional presence of optical emission in the lobes supplies another precious clue. Some of the optical emission may be inverse Compton scattering of the 2.7 K microwave background; this would be unpolarized. Most of the optical emission here must be of synchrotron origin, but since the decay times of an optical synchrotron electron due to radiative losses is about 10 years, these electrons must be accelerated to relativistic energies out in the filaments and lobes themselves! In other words, the central generator does not speed the particles up to their final energies and turn them loose. Localized particle acceleration must occur out in the boon-docks. The currently favored picture is that the intergalactic medium or the remnants of older ejecta get in the way of the fresh material (which is traveling faster). The result is a strong shock which converts the bulk motion of particles traveling at nearly the velocity of light into random motions corresponding to very high temperatures!

Radio-frequency intensity and polarization observations can be made over the extended surface of the source. From these data we can calculate the magnitude and direction of the magnetic field and ultimately the electron density (within the constraints imposed by a few necessary, plausible assumptions). First, we suppose that within the filaments and lobes there are relatively few low-energy or thermal electrons. Then measurements of the polarization give us the direction of the magnetic field at several frequencies so we can allow for the Faraday effect (rotation of the plane of polarization by electrons in magnetized gas along the line of sight). Combination of polarization and intensity measurements yield a relation between the magnetic field and the number of electrons along the line of sight in the lobes.

The filaments and lobes also emit X-rays. The same, speedy electrons that provide the synchrotron emission will scatter photons of the 2.7 K microwave background by the inverse Compton effect to produce X-rays. Since the energy density and number of photons of the microwave background is accurately known, we can calculate the X-ray emission provided we know the total number of electrons along the line of sight. Thus, from measurements of the surface brightness in both the radio and X-ray ranges, we can calculate both magnetic field and the number of electrons along the line of sight.

In making these calculations one usually assumes that the total energy is evenly divided between the accelerated charged particles and the magnetic field. This assumption leads to the lowest calculated total energy for the emitting regions. This equipartition appears to occur between the magnetic fields and cosmic rays in our own interstellar medium. In many active galaxies and quasistellar objects, however, various lines of evidence suggest that the energy in particles may be appreciably larger than in the magnetic fields, thus making the energy estimated by this hypothesis considerably lower than its true value. Also, some of the X-ray emission may come from bremsstrahlung (free–free) processes whose input must be estimated. One obtains a lower limit to the magnetic field. For Cygnus A the field, H, is 1.6×10^{-6} gauss, not a strong field by galactic standards, but when the enormous extent of the lobes and filaments is taken into account the total energy involved is staggering! The electron density is between ten and a thousand per cubic meter. The total mass of the ejected material, estimated as a hundred million million suns, is greater than that of some galaxies. In the most luminous sources, the energy stored in high-energy electrons and the magnetic fields needed to constrain them amount to about 10^{60} ergs. If all the mass of the sun could be converted into energy (by Einstein's formula (mass of sun) × (velocity of light)^2 = energy) the total amount of energy will be about 10^{54} ergs. Thus the energy stored in these radio-galaxy blobs must be equivalent to the total annihilation of a million stars like the sun – and much of this is invested in high-energy particles.

By combining the sizes of typical objects with estimated outflow speeds, one obtains ages of a hundred million years for these objects. Thus an energy equivalent to a total annihilation of the sun every hundred years would suffice. If nuclear processes were involved more than a solar mass per year would be required and all of this energy would have to be converted to high-speed particles.

Probably the energy is supplied by feeding stars and interstellar gas into a black hole. Here gravitational energy is somehow converted to the power needed to run the ejection machine, but the electro-magneto-hydrodynamical mechanism involved must be one of staggering complexity.

There remain a number of puzzling questions. What confines the gas to the filaments and lobes – a magnetic field (likely), gravity (unlikely), or pressure of gas in the intergalactic medium? If there was no confinement we would expect the gas to disperse at the speed of sound. The magnetic field is generally directed along the filaments (which is helpful). In some rich galaxy clusters, such as Virgo, where there are radio galaxies with diffuse lobes, X-ray observations indicate an intergalactic plasma at a temperature of $30\,000\,000$ K with a density of about 3×10^{-28} g/cm^3. In some instances the lobes are not arranged along a straight line with the central galaxy as was true for Cyg A. Instead, the whole configuration is distorted as though it was plowing through some intergalactic gas. The central elliptical galaxies barge ahead, but the lobes are dragged behind by friction with the intergalactic gas.

M87 is yet another type of radio source. The optical jet of length of about 2 kiloparsecs was discovered by H.D. Curtis in 1918. It was subsequently observed in the radio and X-ray regions. A pronounced radio and optical polarization indicates a non-thermal emission. Similar jets are observed in other systems; NGC 315 has one extending to 260 kiloparsecs.

Jets are not uniform but can show concentrations and gaps as though the powerhouse operated sporadically. The M87 jet shows an ordered magnetic field that changes direction from one blob to another. The optical and radio emissions are closely linked together and originate in the same spatial regions. The cores that power the jets in M87 and similar galaxies have sizes of the order of a few parsecs, thus differing from galaxies with compact, presumably black hole cores.

The Crab Nebula

The Crab Nebula, so named by Lord Rosse, who made a remarkable drawing of it over a century ago, is one of the most interesting of all gaseous nebulae. Direct photographs with appropriate filters show that it consists of a rather diffuse or amorphous mass upon which is superposed a network of intricate filaments that is slowly expanding much like the shell around a nova (Figs. 12.6 and 12.7). Measurements by J.C. Duncan suggested that the mass had expanded from a single origin, probably an exploding star, and that the outburst occurred approximately 900 years ago. This time interval is important because in 1054 Japanese and Chinese astronomers independently noticed a bright, temporary star in the same part of the sky. It became brighter than Jupiter, reaching a visual magnitude of -3.5, dimmed gradually, and disappeared from view after about 650 days.

Spectrographic observations have yielded valuable information on the expansion of the Crab Nebula. The filaments give a bright-line spectrum resembling that of a planetary nebula; the amorphous mass, which contributes about 80 percent of the light from the nebula, yields a continuous spectrum. An early spectrogram obtained

Fig. 12.6. The filamentary structure of the Crab Nebula, photographed by Guido Münch on a 103a-E emulsion at the prime focus of the 200-inch Mt Palomar telescope on 22 September 1966. He used an interference filter with a 50-Å band-pass and an exposure of 70 minutes to obtain the filaments in the light of Hα and forbidden radiation of ionized nitrogen. North is at the top; east is to the right.

by N.U. Mayall at the Lick Observatory is shown in Fig. 12.8. The slit of the spectrograph is placed along a diameter of the nebula. Notice that the strong 3727-Å line is bow shaped. This effect is just what we would expect in an expanding shell of gas. At the center of the nebular 'disk', the gas on the side toward us is rushing in our direction and the line is shifted toward the violet, while the material on the opposite side is moving away and the line is shifted toward the red. Hence the spectral line appears split in the central regions of the nebula. At the edges of the nebula, the material is moving across the line of sight, and the line-of-sight speed of the expanding shell is zero; hence the shift of the spectral line is zero here. Now the maximum separation of the two components of the line is proportional to twice the speed of expansion of the shell. On the other hand, measures on direct photographs

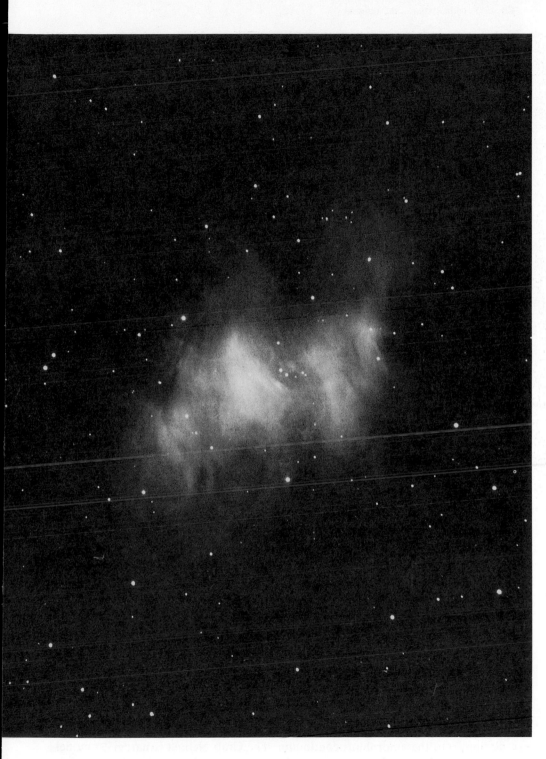

Fig. 12.7. The 'amorphous mass' of the Crab Nebula, photographed by Guido Münch on a yellow-sensitive (103a-D) emulsion at the prime focus of the 200-inch Mt Palomar telescope on 22 October 1963. He used a yellow (Corning 3484) filter and Polaroid filter so oriented as to give minimum transmission in the north–south direction. The photograph registers essentially the continuum radiation; only the strongest filaments can be seen faintly in the light of the forbidden radiation of neutral oxygen. North is at the top; east is to the right.

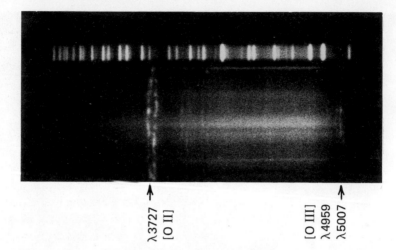

λ3727 [O II]   [O III] λ4959 λ5007

Fig. 12.8. An early spectrogram of the Crab Nebula, obtained in the 1930s by N.U. Mayall with the nebular spectrograph on the Crossley Reflector at Lick Observatory; notice the strong continuous spectrum and bright bow-shaped forbidden lines of ionized oxygen [O II] at 3727 Å. More recent work by Virginia Trimble shows that in addition to the expansion there are complex internal motions.

give the apparent (angular) rate of expansion. Clearly, if we know the actual speed of expansion and the angular speed of expansion, we can calculate the parallax. There is a slight complication in that the nebula is elliptical in shape; it appears to be an ellipsoid whose longer (major) axis expands at a different rate to the shorter (minor) axis. Allowing for these effects, the velocity of expansion along the major axis is 1720 kilometers per second, which corresponds to an observed angular rate of expansion of 0.2 arcsec per year. Now, at a distance of 1 parsec, a transverse (proper) motion of 1 arcsec/year would correspond to 1 astronomical unit/year $= (1.49 \times 10^8$ km/yr) $(3.156 \times 10^7$ arcsec/yr$) = 4.74$ km/s. Hence a linear velocity of V km/s corresponds to an angular transverse motion of μ arcsec/yr at a distance of d parsecs, in accordance with the relation $V/4.74\mu = d$ (parsecs). Putting in the numerical values for V and μ, we get $1720/(4.74 \times 0.20) = 1800$ parsecs for the distance. Then, if we allow for space absorption, the absolute visual magnitude of the star at maximum must have been about -18.2. In any event, the star must have been many times as bright as an ordinary nova, and hence it probably was a supernova.

The system of filaments of the Crab Nebula defines a thick shell of about 2 parsecs outer and 1.4 parsecs inner diameter. It probably has a mass about twice that of the sun, and may consist largely of matter swept up from material ejected from the star before it became a supernova. The emission-line spectra of the filaments have been studied intensively; the power output in the main lines exceeds that of the sun; the luminosities of e.g. [SII], [OII], and [OIII] are 40, 80, and 90 times those of the sun, respectively. Ionization and excitation in the filaments is probably produced by radiation from the amorphous continuum. The Crab Nebula is harder to model than a planetary nebula. It appears to be helium-rich, but good data for other elements are hard to get, thus it is difficult to identify the supernova type involved.

The diffuse or so-called amorphous mass originates mostly inside the ellipsoidal network of the filaments. It shows a distinctive structure and exhibits brightness variations near its center. These are manifest as 'ripples' which appear to travel with a speed of about 47 000 km/s. This flickering may be a symptom of energy transfer from the central source to power the nebula.

Early radio investigations showed the Crab Nebula to be an intense source of radio-frequency continuum; it was one of the very first of such sources to be identified with an optical object. Attempts to interpret the radio flux as thermal emission failed miserably; the nebula was much too intense for its optical brightness and the radio-frequency flux distribution did not fit the observations at all. A significant clue was provided by V.A. Dombrovsky's discovery (1954) that the light of the Crab Nebula was polarized. W. Baade secured a remarkable series of photographs through a Polaroid filter set successively at different position angles; these showed striking changes. Polarizations up to 60 percent are readily recognized; some regions may approach 100 percent. The radio data also show significant polarization effects.

I.S. Shklovsky suggested synchrotron radiation as an explanation for the emission from the amorphous mass, an hypothesis that has stood the test of time. The spectrum of the Crab Nebula has been observed from 30-meter radio waves to the gamma-ray region, with large gaps in the far infrared and ultraviolet, due to extinction in the earth's atmosphere and interstellar medium, respectively (see Fig. 12.9). The total emitted power is about 10^{38} ergs/s, or 25 000 times the luminosity of the sun! Nearly all of this energy is radiated as synchrotron radiation.

There are several points to be mentioned. First, the actual mass of the non-thermal particles that provide the synchrotron radiation is less than 0.00001 m(sun). In fact, more mass ($\sim 0.0006\,m$(sun)) may be involved in dust grains in the filaments. The total energy invested in relativistic particles ($\sim 10^{49}$ ergs) is comparable with the kinetic energy of expansion.

Since the rate of energy loss by synchrotron emission goes as the square of particle energy, the observed spectrum would show pronounced changes with time if new particles were not fed into the process. In particular, gamma-ray and X-ray emitting electrons would be degraded very quickly. The Crab Nebula appears to be nearly in a steady state although radio data suggest a decline in radio-frequency flux of less than 1 per cent a year. Hence there must exist a mechanism that supplies energy to accelerate electrons to very high energies at this rate; presumably it also accelerates heavy particles as well. The ultimate powerhouse appears to be a rapidly flickering source of radio-frequency (and in this instance also optical) radiation known as a 'pulsar', an incredibly dense, rapidly spinning neutron star. Basically, the energy of rotation of a dense ball about ten kilometers in diameter is gradually dissipated, like a flywheel that is running down. The actual mechanisms by which mechanical energy is transformed to accelerate charged particles to nearly the velocity of light are likely to be rather complex. The essential clue is that the spinning neutron star is strongly magnetized with a field of 10^{12}–10^{13} gauss. This magnetic field is anchored in the star and, as it sweeps around, the rapid motion of the lines of force generates

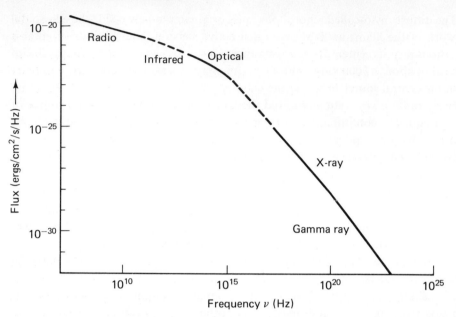

Fig. 12.9. The energy distribution in the Crab Nebula. To display data over a range of many octaves in energy, logarithmic scales are used. The flux declines with increasing frequency, slowly at first and then more steeply from 10^{-20} ergs/cm²/s/Hz at $\nu = 10^9$ cycles per second (Hertz) to 10^{-32} ergs/cm²/s/Hz at $\nu = 10^{23}$ Hz (gamma-ray region). Data come from several sources. Note large gaps (shown as dashed lines) between the radio and infrared–optical regions and between the latter and the X-ray region. The radio-frequency–infrared gap is due to extinction in the earth's atmosphere, the other break is due to interstellar extinction. Over large ranges the emission fits a power law.

an electric field of about 10^{12} volts/cm, sufficient to accelerate charged particles to cosmic-ray energies. Let us now examine more closely essential properties of neutron stars which have been mentioned previously in discussions of gamma rays and X-rays.

Pulsars or spinning neutron stars

The discovery of pulsars is a deservedly oft-quoted example of scientific serendipity. In the early 1960s, a sharp-eyed, attentive research student, Jocelyn Bell, working for Antony Hewish with a Cambridge radio telescope (which had been built for a very different job) noticed a series of rapid, regularly spaced oscillations on the analogue observing record. Many observers might have attributed it to outside interference or equipment malfunction, particularly since it vanished for a time. The signals reappeared, however, and their remarkable regularity inspired a careful study; soon other examples were found. Although the Crab and Vela pulsars can also be observed optically, for the others we are restricted to radio-frequency observations.

More than 95 percent of the time a pulsar emits no radiation towards the observer. Then it radiates abrupt pulses (Fig. 12.10) with durations of the order of 0.002 to 0.01 second (typically 0.01s). Pulses can be complex, showing oscillations of duration less than 0.001 s. In some instances the intensity can vary from zero to its full value in 0.0001. Although individual pulses show considerable variations, the average pulse profile remains constant for any given pulsar. The time interval between pulses is typically about 1 s but it can range from 0.0015 to 3.75 with a surprising constancy of period. For Cambridge pulsar CP 1919, the period is 1.3370113s, for CP 0328 it is 0.714518603 s; that is in the first approximation the pulsar behaves like a fantastically accurate clock.

Pulsar periods are not immutable, however. There is a tendency for them to increase by rates ranging up to 0.001 s/day. Discontinuities in the rotation rate, called glitches, also occur. The rotation rate speeds up very abruptly and then begins to slow down gradually again. First observed in the Vela pulsar, these glitches have been seen in many objects. They have been attributed to star quakes which change the pulsar's moment of inertia and to interactions between the pulsar's solid surface and a presumed superfluid interior. The radio-frequency radiation of pulsars shows linear polarization. In the Vela pulsar, for example, within each pulse the plane of polarization turns through nearly 90 degrees while the amplitude shows considerable variation.

One by-product of pulsar research is their use as a tool for probing the interstellar medium. Hewish and his associates noted that pulsars emitted over a wide range in radio frequency, but the lower the frequency the later the pulse is received. They concluded that this 'dispersion' or dependence of travel time on frequency was caused by interstellar electrons. The amount of delay depends on the total number of electrons in the line of sight. Thus, time delays were larger for pulsars near the galactic equator. Since a given receiver accepts radiation over a finite band width, this dispersion effect makes it difficult to observe distant pulsars; each pulse tends to be confused with the following one if the pulsar period is short and the dispersion is large.

If we know (or think we can guess) the electron density along the line of sight, these dispersion effects can be used to estimate the pulsar distance. For this purpose let us approximate the interstellar medium by cold, dense clouds intermixed with a hot, ionized, tenuous gas with a density of less than 0.1 electron per cubic centimeter. Observed dispersions show that typical pulsars lie at distances less than 1000 parsecs. Some pulsars are observed behind cold, neutral, hydrogen clouds that produce 21-centimeter absorption. Provided the distance of the cold hydrogen cloud can be found by other means, that of the pulsar can be estimated. The best method of getting distances is by optical identification of the sources.

The plane-polarized radio-frequency radiation of a pulsar will suffer rotation of the plane of polarization as it passes through an ionized gas in a magnetic field. The amount of this Faraday rotation (see Chapter 10) depends on the square of the wavelength, the total number of electrons in the line of sight, and the magnetic field strengths. The total number of electrons in the line of sight is known from the dispersion effect and, since we can measure the rotation of the plane of polarization at several wavelengths, we can determine the magnetic field. For example the Vela source gives a magnetic field of 2.5 microgauss, but one has to be wary of magnetic-field direction cancellations. That is, if in one region along the line of sight the magnetic field is directed, say, towards the observer, and in another region directly opposite, the Faraday rotation from the first region will be at least partially cancelled by the contribution from the second. Then the derived 'average' field-strength would turn out to be too small.

To get back to the pulsars themselves, what are they? No progress towards answering this question could be made until pulsars were identified with optically known objects. The first real breakthrough came when M.I. Large, A.E. Vaughan, and B.Y. Mills identified one pulsar in the middle of the Vela supernova remnant. It has a period of 0.089 second, the shortest known up to that time. The Vela pulsar was observed optically some years later. Then D.C. Staelin and R.C. Feifenstein found within the Crab Nebula a pulsar with a period of 0.033 s. W.J. Cooke, M.J. Disney, and A.J. Tayler at the University of Arizona found optical pulses that were similar to the radio pulses, while J.S. Miller and E. Wampler at Lick Observatory were able to actually identify the pulsar with the star originally suggested by Baade.

The large amplitude and short period of the pulse outbursts mean that the source must be small since otherwise radiation from different parts of the spherical globe would be blurred out by a factor of at least the ratio of the radius of the star to the velocity of light. If variations occur in 0.0001 s, the diameter of the source cannot much exceed 30 kilometers. Per unit area, the energy output rate exceeds that of the sun by many millions of times.

The Crab pulsar is slowing down. If we interpret it as a spinning body with a radius of 10–15 km, the energy loss is about 10^{38} erg/s i.e. about 20 000 times the energy output of the sun. This energy supply could account for the total energy emitted by the Crab Nebula if rotational energy is efficiently converted to particle energy, but how can this be done? A pulsar's presumably huge magnetic field ($\sim > 10^{12}$ gauss) plays the key role.

The blinking power source of the Crab Nebula like that of other pulsars is a neutron star. Neutron stars were proposed by F. Zwicky and Baade half a century ago as the 'hard-core' remains of supernovae. Long before the discovery of the first pulsar, J.R. Oppenheimer explored some of their theoretical properties.

In Chapter 9 we described white dwarf stars whose densities could be 100 000 times that of water, representing a situation where the pressure was so great that electrons were completely stripped from atomic nuclei. Then electrons and nuclei would move quite independently of one another and would obey different gas laws. The nuclei continue to follow the familiar perfect gas law, but the electrons obey what is called the degenerate gas law. In an object of solar mass, bereft of nuclear or gravitational energy sources, the electrons would be able to exert enough pressure to balance the heavy weight of the overlying layers. This situation would hold until the mass reached a limit of about 1.35 m(sun), the Chandrasekhar limit.

Eventually, when the density was pushed up to and beyond a million million times that of water, the very nuclei would be forced into contact with one another,

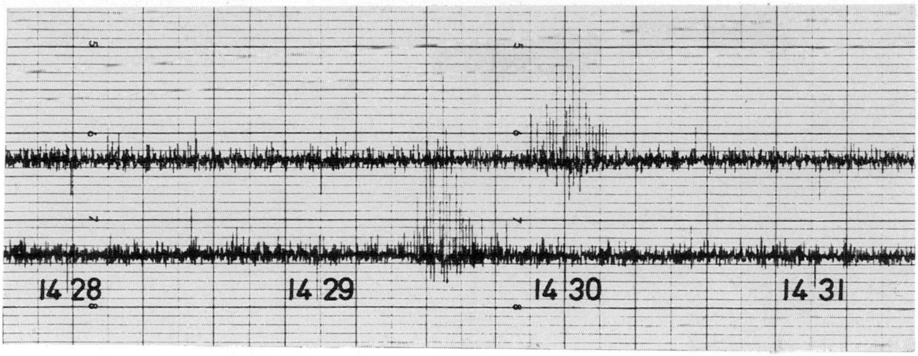

Fig 12.10. A photograph of a chart-recorder trace showing the discovery of the pulsar MP 1426 in October 1968. The observation was made with the east–west arm of the Molonglo radio telescope in New South Wales, Australia. The lower trace records the signal from a beam directed slightly east of the meridian, the upper trace the signal from a beam slightly west of the meridian. Strong pulses can be seen on both channels. The difference in transit times on the beams is proportional to their separation, thus proving that the signals originate outside the earth. This pulsar has been observed on many subsequent occasions, but not as well as on the original record shown here. It lies close to the Southern Cross in the sky but has not been identified with any optical object. (Courtesy Michael Large, University of Sydney.)

and the electrons would be pushed back into them. One by one, the positive charges in the nuclei would be canceled and neutrons would be created. At this stage we might anticipate a rapidly spinning object consisting of nothing but neutrons and presumably carrying a strong magnetic field, but the actual situation is rather more complicated (see Fig. 12.11). If the mass is pushed up beyond some not yet accurately specified critical value between 1.7 and 3.7 m(sun), the crushing force of gravity overwhelms the strongest of nuclear forces and the object collapses into a black hole.

The mass of the Crab pulsar has been estimated from its rotation, from glitches, from the total power output supplied to the nebula, and from the theory of neutron stars. The derived values lie between 1 m(sun) and 2 m(sun), with the odds on it lying in the range 1.4 to 1.5 m(sun). Direct estimates are possible for the masses of a few binaries, for example Hercules X-1 = 1.3 m(sun), Centaurus X-3 = 1.4 m(sun). Some neutron stars may have masses less than the Chandrasekhar limit. Why, then, did they not simply develop as white dwarfs? Presumably, they never had a chance; neutron stars originate from supernova events in which the core of the star is compressed by an implosion to very high densities. From this condition it can then never escape by expansion.

Other properties of neutron stars are fantastic by terrestrial standards. The surface gravities lie near 10^{11} times that of the earth and the gradient of the gravity is such that any object brought near such a compact mass would be immediately shredded by tidal action.

Two factors are of special importance; the influence of the high density on the state of matter within the object and the imbedded magnetic field. In an ionized gas the magnetic lines of force are anchored in the material. If a mass is compressed, the

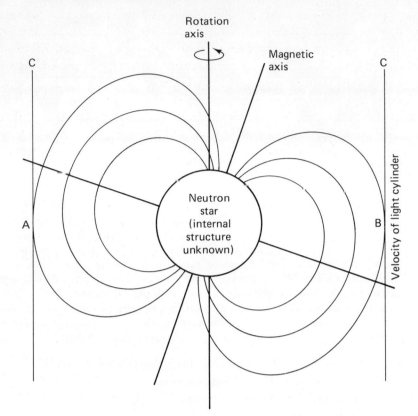

Fig. 12.11. A schematic model of a neutron star. The rapidly spinning neutron star with a diameter of 10–15 kilometers has a mass of 1.5 *m*(sun). There is a smooth, solid crust composed mostly of neutron-rich nuclei and iron. The magnetic field ~ 10^{13} gauss) deforms the atoms into needle-like structures. At a depth of about 1 km, the density is about 4×10^{11} g/cm³ and neutrons begin to be squeezed out of the nuclei, until possibly we get a medium of free neutrons. There is no way of knowing the exact properties of matter under the extreme densities found in neutron stars. Some possible neutron star structures are discussed in George Greenstein's *Frozen Star* (see bibliography for Chapter 11).

The magnetic axis may be inclined with respect to the axis of rotation. The lines of force act as though welded to the star, spinning around with it like a solid body out to the point where the rotational velocity approaches that of light. *C* marks the velocity of light cylinder. Beyond this the lines of force trail away. The rapid spin of the magnetized pulsar (30 revolutions per second for the Crab neutron star) induces a powerful electric field, $E \sim 10^{12}$ volt/cm, capable of accelerating charged particles to cosmic-ray energies. Charged particles must spiral along magnetic lines of force. At points A and B they can move parallel to the sides of the light cylinder. Escaping from this vicinity they supply the pulsar emission and pump energy into the nebula.

lines of force are squeezed together which means that the intensity of the field is enhanced. If the sun were to shrink from a radius of 700 000 km to 5 km, the initial magnetic field would be enhanced about 2×10^{10} times, so a 50-gauss field would become one of 10^{12} gauss. Since angular momentum must be conserved, the solar rate of rotation would speed up to 0.01 second or less!

Various models have been proposed to account for the way in which a spinning neutron star converts rotational energy to accelerate particles up to cosmic-ray energies. Perhaps the most obvious model is one in which particles are accelerated from the neighborhood of the magnetic pole along the magnetic axis which may be inclined to the axis of rotation as it is on earth. Contrary to observation this model predicts that the pulsar width should depend on the period.

Another (and more promising) possibility, suggested by T. Gold and amplified and extended by others, is that since the magnetized region or magnetosphere rotates like a solid body out nearly to the point where the velocity of rotation approaches that of light, particles could be trapped just inside this critical distance (light cylinder) and escape outward along it (see Fig. 12.11). Graham Smith noted that as a particle approaches the light cylinder it must pick up enough speed (~ 0.95 the velocity of light c) to break away from the magnetic field lines. His model predicted the correct pulse width. Models invoking particle escape along the inside of the light cylinder also can explain the rotation of the plane of polarization during a pulse. The synchrotron radiation from the pulsar itself indicates that the particles are escaping from a region where the magnetic field $\sim 100\,000\ H$ gauss, which is much smaller than the magnetic field on the surface of the pulsar itself, viz. 10^{12} gauss.

When we translate the slow-down in the rotation rate of the Crab pulsar into a rate of loss of mechanical energy, we find that it roughly matches the energy output of the pulsar as required by the emission of the supernova remnant. We might expect the energy output of the Crab Nebula to be slowly declining, reflecting the slowing down of the central flywheel. Indeed, it is. Steve Reynolds and Hugh Aller measured a decay rate of 0.16 ± 0.02 percent per year in the 8 gigahertz flux, a rate that agrees well with the predictions by S. Reynolds and R. Chevalier for a pulsar-driven supernova remnant evolution.

At last the energy source of the Crab Nebula and perhaps also a source of cosmic rays appears to have been found, and in an object that represents an entirely new, superdense, state of matter. Most stars expire as white dwarfs. In our time and neighborhood, only one star, Sanduleak's star in the Large Magellanic Cloud, has been observed to explode as a supernova. Supernova remnants such as the Network Nebula may be observable only for a few millenia, a pulsar may be optically observable only for a few decades or centuries, while persisting as a radio-frequency source for tens or hundreds of thousands of years. Neutron stars are manufactured in supernovae and only in supernovae where the core is subject to a huge compression; a 'normally' expiring star can produce only a white dwarf.

Epilogue

There is much in the domain of present-day astrophysics that we have not mentioned. The whole problem of the Big Bang, the inflationary universe, and the early history of matter and radiation lies outside the scope of this book. We deal with a universe in which stars and galaxies are already formed and have focussed our attention on what is going on at the present time. The many puzzles posed by the generation and maintenance of cosmic magnetic fields, high-energy particles, and intense energy sources will keep us busy for many years to come.

Probably the greatest mystery of all, though is the 'missing mass'. As we saw in our discussion of binary stars, mass can be measured by its gravitational effects. The mass and mass distribution of a galaxy can be ascertained from its rotation, or more specifically from how rotation rate varies with distance from the center. If most of the mass of a galaxy is concentrated toward its center, as the light distribution often is, the rotation curve will recall that of the planets about the sun, that is, the velocity will fall off markedly with distance. If there is a lot of mass in the outer regions, the rotation curve will decline more gently or may even continue to rise as far out as we can observe it. When Horace Babcock measured the rotation curve of the Andromeda Galaxy, M31, he found much of the mass to be concentrated in the outer regions, i.e. the mass-to-light ratio rose steadily outwards from the nucleus. A similar, although less striking effect, was also found in the Triangulum Spiral, M33. Stars of low mass, planets, or small dark bodies like snowballs were then invoked to explain the discordance. At about the same time, Sinclair Smith, who measured the velocities of galaxies in the Virgo cluster of galaxies, found them moving too rapidly, if one assumed that the mass-to-light ratio was the same as in our Galaxy. The mass discrepancy was about a factor of ten.

Subsequent work has only exacerbated the problem. Some of the discordance can be taken care of by dark bodies, and some by high-temperature gas, but there are theoretical reasons (based on cosmology and elementary particle theory) for believing that much of this mass is 'non-baryonic', that is, not like ordinary material at all. It does not interact with ordinary electromagnetic waves at a presently detectable level. Its presence is revealed only by its gravitational attraction. What is this stuff? Neutrinos or superneutrinos or some exotic particle? At this moment we cannot say. Ninety percent of the mass of the universe may be presently undetectable!

314

For some questions we can give answers that are reasonably secure. We have seen how the chemical composition of a stellar atmosphere or a gaseous nebula may be deduced from its spectrum, and that the same chemical elements in about the same proportions as in the sun are found in galaxies thousands of millions of light-years distant. In broad outlines, we understand the history of a star from the time of its formation to its demise as a white dwarf. The earliest steps in building a star or solar system are obscured in mystery. The interaction of gas and dust under the influence of radiation pressure, magnetic fields, and gravity is going to be complex and, although some promising scenarios have been proposed, particularly by Frank Shu, the picture is far from satisfactory. Once gravity can play an active role, the star contracts quickly until its internal temperature rises high enough for nuclear reactions to occur. It then shines by converting hydrogen into helium – first as a main-sequence star and later as a giant or supergiant when the core hydrogen is gone. Then, as the nuclear fuels become exhausted, the outer parts of the star escape back into space. With ordinary stars like the sun, the core settles down as a white dwarf. More massive stars may produce supernovae, neutron stars, and even black holes.

The evolution of a star is not really all that complicated, even though a high-speed computer may be needed to track its changes through the most interesting phases of its evolution. The brightest jewel in the sky does not have the complexity of a single lotus flower, or, for that matter, a single virus.

The earth – the solid ground under our feet, the mountains that glisten in the sunset, the steel of our industrial civilization, and the salt of our oceans – is but the ash of long-dead stars that adorned the canopy of heaven for millions of millenia before the creation of the earth.

What, then, of the future? The author of the book of Ecclesiastes had no doubts for he asserts that one generation passes away and another takes its place but the earth abides for ever. Lao Tse in the Chinese classic, the *Tao Teh King*, remarks that: 'The earth is everlasting; heaven is eternal'

Another point of view, though, is expressed in the Buddhist text of the *Dhammapada*: 'It is better to understand for one day the fleeting nature of ALL things than to live for a hundred years in the ignorance of the same.'

The fiery end of the earth can be described with greater certainty than the circumstances of its birth. Over the coming thousands of millions of years, the sun will gradually brighten, more and more rapidly as time goes on. The temperature of the earth will rise and all life will perish as oceans boil. Finally, the outermost envelope of the sun, possibly by then a Mira variable, will envelop the denuded cinder that had been the earth and drag it inwards. The sun will experience only a relatively brief life as a giant star before its outermost layers escape into space and the core shrinks to a white dwarf with a density between a hundred thousand and a million times that of water.

Each electron and each atomic nucleus is assigned a place in the huge, incredibly dense, crystal-like structure that had once been the busy hub of a star. No particle can move without another taking its place. Everywhere there exists complete harmony and total organization; there is no disorder and there are no dissenters. No deviations are permitted anywhere at any time. This is the ultimate death of matter from which there is no resurrection, for material that gets locked up in such a state stays there until the end of time. One possible exception may occur in a white dwarf binary, the sum of whose component masses is less than 1.4 solar masses. These may merge together to give a type I supernova, but such circumstances are extremely rare.

What happens to the material that escapes from the dying sun and drifts out into the clouds of interstellar smog and gas? We will never know, but perhaps the vision of St. John the Divine was as good as any other: 'For I beheld a new heaven and a new earth for the old heaven and the old earth had passed away '

Appendix A: Designations of stars and nebulae

The brighter stars in each constellation are denoted by letters of the Greek alphabet (see below) and the name of the constellation (in the genitive case); for example Betelgeuse $= \alpha$ Orionis, Rigel $= \beta$ Orionis. Usually the letters are assigned to the stars in order of brightness, but sometimes in order of position, as for the Big Dipper. For fainter naked-eye stars, Flamsteed numbers are sometimes used, for example, 53 Tauri. Most bright stars are listed in the *Bright Star Catalogue* of Yale University Observatory, where they are designated by their numbers in the *Harvard Revised Photometry* (HR numbers). This catalogue has been revised and updated by D. Hoffleit and gives positions, galactic coordinates, magnitudes, colors, spectrum and luminosity classes, and data on parallaxes and motions. *Sky Catalogue 2000* (A. Hershfeld and R. Sinnett, 1982 Cambridge University Press and Sky Publishing Corp.), gives data for stars down to the eighth magnitude.

Stars between the naked-eye limit, near the sixth magnitude, and the ninth or tenth magnitude are listed in the *Bonner Durchmusterung*, or BD, catalogue, or in the Henry Draper, or HD, catalogue. In the BD catalogue, which has been extended to far southern skies in the *Cordoba Durchmusterung*, stars are listed in zones of declination. Thus BD $+ 30°$ 3639 means star number 3639 in the zone at $+ 30°$ declination (it would pass directly overhead for an observer at 30° N latitude). The HD catalogue lists magnitudes, positions, and spectral classes according to the equatorial coordinate system of 1900, in order of right ascension. Both catalogues include bright naked-eye stars and one star may be given several designations; for example ι Herculis $=$ HD 160762.

The older catalogues have been replaced by the SAO $=$ *Smithsonian Astrophysical Observatory* catalogue of 258 997 stars (1965). A revision of the HD catalogue that gives luminosity as well as Draper classes is being published by Nancy Houk and her associates at the University of Michigan.

Variable stars have their own special designations by letters preceding constellation names, e.g. R Andromedae (R And), AX Persei (AX Pers), RY Tauri (RY Tau), etc. When possible alphabetical designations are exhausted, as in the constelllation of Cygnus, the letter 'V' followed by a number is employed, e.g. V444 Cyg. A given star may have several designations.

Special objects, such as radio, X-ray, or gamma-ray sources, supernova remnants, quasars and pulsars have their own peculiar systems of nomenclature. For example, radio sources are often designated by their number in a catalogue of such objects, thus 3C 273 is number 273 in the third Cambridge catalogue. Pulsars are indicated by notations such as CP 1919, which denotes a 'Cambridge pulsar'; here the number gives information on its position. Other objects are often described by their positions with respect to some well-identified stars.

A few of the brightest nebulae, galaxies, and star clusters were listed in a catalogue compiled by C. Messier, and are given M numbers, e.g. the Andromeda Spiral, M31, but for most such objects we employ J.L.E. Dreyer's careful compilation in his *New General Catalogue*, NGC, or the two supplementary *Index Catalogues* (IC or NGCI, NGCII). These catalogues give brief descriptions of each object, together with its position for 1860 (right ascension and north polar distance), plus precession constants for updating the position. A revised version was prepared by J. Sulentic and W.G. Tifft (University of Arizona Press, 1973).

Fast, wide-field cameras with narrow band-pass filters have contributed much to the study of diffuse nebulae and H II regions of low surface brightness (see *Atlas of Galactic Nebulae*, Th. Neckel and H. Vehrenberg (1985), which gives both positions and direct photographs of each object). L. Perek and L. Kohoutek's *Catalogue of Galactic Planetary Nebulae* (1967) is being supplanted by a more comprehensive catalogue by A. Acker and her associates at Strasbourg Observatory (France).

For the general purposes of amateurs, perhaps the single most suitable reference is Antonin Becvar's *Skalnate Pleso Atlas* and accompanying catalogue (Sky Publishing Corp.) which lists naked-eye stars, and also important binaries, variables, galactic and globular clusters, bright diffuse nebulae, planetary nebulae, galaxies, and radio-frequency sources.

The Greek alphabet is as follows:

α	alpha	ν	nu
β	beta	ξ	xi
γ	gamma	o	omicron
δ	delta	π	pi
ε	epsilon	ρ	rho
ζ	zeta	σ	sigma
η	eta	τ	tau
θ	theta	υ	upsilon
ι	iota	φ	phi
κ	kappa	χ	chi
λ	lambda	ψ	psi
μ	mu	ω	omega

Appendix B: Some physical quantities and relations useful in astronomy

1. Concerning units

In this book I have generally used c.g.s. metric units – the *centimeter* (cm) is the unit of length, the *gram* (g) is the unit of mass, and the *second* (s) is the unit of time. The popular practice in science and engineering now is to use m.k.s. or SI (*Système International*) units – the *meter* (m) is the unit of length, the *kilogram* (kg) is the unit of mass, and the *second* (s) is the unit of time. Conversion from one set of metric units to the other is usually straightforward.

Some conversions between the more frequently used English units and their metric analogues are given below, since many of us find it appropriate to use these old-fashioned units in popular talks.

1 meter = 39.37 inches	1 inch = 2.54 cm = 25.4 mm
1 kilometer = 0.621 mile	1 mile = 1.609 km
1 liter = 1000 cm³ = 1.06 quarts	1 cubic inch = 16.387 cm³ (milliliters)
1 kilogram = 2.20 pounds (avoird.)	1 ounce = 28.35 g = 0.02835 kg

Astronomers generally use the Kelvin or absolute temperature scale. Zero on the Kelvin scale is $-273\,°$Celsius; hence 273 K is $0\,°$Celsius or $32\,°$Fahrenheit. We convert Celsius to Fahrenheit degrees (as for popular illustrations) by the formula $F = 1.8C + 32°$.

In all systems of units where very large or very small numbers are involved powers of ten are used. For example, 30 000 000 000 is written as 3×10^{10} (which means 3 followed by ten zeros). For a very small number like 0.000 000 0001 we write $1 \times 10^{-10} = 1/10^{10}$; thus 1.71×10^{-10} means 1.71 divided by 10^{10}.

2. Definitions and units of force, energy, power, and magnetic and electric fields

Acceleration is the rate of change of velocity. If velocity is expressed in centimeters per second, acceleration is expressed as centimeters per second per second. The acceleration due to gravity at the earth's surface is 980 (cm/s)/s or 980 cm/s², which means that the velocity of a freely falling body is increased by 980 cm/s every second of its fall.

Force is given by *mass times acceleration*; the unit of force in c.g.s. units is the *dyne*. A force of 1 dyne will give a mass of 1 gram an acceleration of 1 cm/s². The force of gravity upon objects at the earth's surface is 980 dynes per gram of mass. The unit of force in the m.k.s. system is the *newton*[N]. It is the force required to give a mass of 1 kg an acceleration of 1 meter/s², i.e. 100 000 dynes.

Pressure is force per unit area and is often measured in dynes per square centimeter, or in atmospheres. One atmosphere of pressure is 1 013 250 dynes/cm². It is equivalent to the pressure exerted by the weight of a column of mercury 760 mm high at 0 °C. The m.k.s. unit of pressure, 1 newton/meter² (10 dynes/cm²) is called the *pascal*.

A force of 1 dyne acting over a distance of 1 centimeter will do 1 *erg* of work. Since the erg is a very small unit, the m.k.s. unit of energy, which is called the *joule*, is often used. A force of 1 newton acting over a distance of 1 meter will do 1 joule of work. Thus 1 joule = 10^7 ergs. The ability of a system to do work involves *energy*, which is measured by the work that is done. Energy appears in various forms, e.g. as mechanical, electrical, thermal, and radiation.

A frequently used unit of heat energy is the *calorie*, which is the amount of heat needed to raise the temperature of 1 gram of water 1 °C. The mechanical equivalent of heat is the ratio of a quantity of work to the quantity of heat into which that work can be converted (with the work measured in joules and the heat measured in calories). The mechanical equivalent of heat is 4.1854 joules/calorie.

The *rate of doing work* is called *power* and is expressed in *ergs/second, watts*, or *kilowatts*. In astrophysics, power is often encountered in the context of energy received from a celestial source, or the rate of energy output by a star. One watt of power is equivalent to the deliverance of 1 joule or 10^7 ergs/second. A kilowatt is 1000 watts and amounts to a work rate of 1000 joules/second or 10^{10} ergs/second. The power output of the sun is 4×10^{23} kilowatts. Sometimes reference is made to an archaic unit called the '*horsepower*'; this is equivalent to 0.746 kilowatts.

In c.g.s. units, magnetic fields are expressed in *gauss*. By way of orientation, the earth's magnetic field is about 0.5 gauss. The m.k.s. unit, 1 *tesla*, equals 10 000 gauss.

In the electrostatic system of units, often employed in atomic and molecular calculations, the unit of electrical charge (or esu) may be defined on the basis of the following considerations. Similarly charged bodies repel, oppositely charged bodies attract each other. Suppose two small insulated bodies (pith balls are often used in such experiments) are similarly charged and placed a distance r apart in vacuum. They will repel each other with a force F given by Coulomb's law:

$$F = \frac{q_1 q_2}{r^2},$$

where q_1 and q_2 are the charges on the two bodies. If $F = 1$ dyne, $q_1 = q_2$, and $r = 1$ cm. the amount of charge (q_1 or q_2) is 1 electrostatic unit (esu).

In the world of practical affairs, electrical charge is usually measured in *coulombs*;

1 coulomb $= 3 \times 10^9$ esu. For example, the charge on the electron is 4.80×10^{-10} electrostatic units or 1.602×10^{-19} coulombs.

3. Some useful physical constants

Velocity of light, $c = 2.99793 \times 10^{10}$ cm/s ($\times 10^8$ m/s)

Constant of gravitation, $G = 6.670 \times 10^{-8}$ dynes cm^2/g^2 ($\times 10^{-11}$ N m^2/kg^2)

Volume of 1 mole $= 22.4136 \times 10^3$ cm$^3 = 22.4136$ litres

Standard atmospheric pressure $= 1.013250 \times 10^6$ dynes/cm^2 ($\times 10^5$ N/m^2 or pascals)

Melting point of ice 273.16 K (absolute scale)

Acceleration of gravity, $g = 980.665$ cm/s$^2 = 9.80665$ m/s^2

Avogadro's constant, $N = 6.02217 \times 10^{23}$ molecules or atoms/mole

Loschmidt number (number of atoms or molecules at 0 °C and 1 atmosphere pressure), $n = 2.6868 \times 10^{19}$/cm^3

Charge of electron, $\epsilon = 4.80325 \times 10^{-10}$ esu $= 1.602192 \times 10^{-19}$ coulombs

Mass of electron, $m = 9.10956 \times 10^{-28}$ g ($\times 10^{-31}$ kg)

Mass of proton, $M_p = 1.67261 \times 10^{-24}$ g ($\times 10^{-27}$ kg)

Mass of hydrogen atom, $M_H = 1.67352 \times 10^{-24}$ g ($\times 10^{-27}$ kg)

Radius of first Bohr orbit, $a_0 = 5.29177 \times 10^{-8}$ cm ($\times 10^{-10}$ m)

Proton mass/electron mass, $M_p/m = 1836.11$

Gas constant/mole, $R_0 = 8.3143 \times 10^7$ erg/K mole (joules/K mole)

Boltzmann's constant, $k = R_0/N = 1.38062 \times 10^{-16}$ erg/K ($\times 10^{-23}$ joules/K)

Planck's constant, $h = 6.6262 \times 10^{-27}$ erg s ($\times 10^{-34}$ joule s)

Rydberg constant for hydrogen, $R_H = 109677.58$ cm^{-1}

Energy of 1 electron volt (eV), $E_0 = 1.602192 \times 10^{-12}$ ergs ($\times 10^{-19}$ joules)

Wave number associated with 1 eV, $\tilde{\nu} = 8065.46$ cm^{-1}

4. Some relations concerning gases

The pressure of a gas, which is the force it exerts per unit area on the walls of its container, is related to the mass of the gas, the volume in which it is enclosed, and the temperature.

Standard conditions of temperature and pressure 0 °C or 273 K and 1 standard atmosphere or 760 mm-of-mercury pressure. Under standard conditions, 22.414 liters of a gas will weigh μ grams, where μ is the molecular weight of the gas – 28.02 for molecular nitrogen, N_2, 32.00 for O_2, and so on.

The *gas law* is

$$PV = RT,$$

where P (dynes/cm^2) is the pressure of the gas, V (cm^3) is the volume occupied by 1 mole, or μ gram of the gas, R (the gas constant) is 8.314×10^7 ergs/K mole, and $T(K)$ is the temperature of the gas. The gas law is frequently written in the form

$$p = nkT,$$

where p (dynes/cm^2) is the pressure, n is the number of atoms or molecules per cubic centimeter, and k is Boltzmann's constant, 1.380×10^{-16} erg/K.

Example. If the electron pressure in the atmosphere of the sun is 10 dynes/cm^2 and the temperature is taken as 5800 K, what is the number of electrons per cubic centimeter? We have

$$10 = n \times 1.380 \times 10^{-16} \times 5800,$$

whence

$$n = 1.25 \times 10^{13} \text{ electrons/cm}^3.$$

The most probable velocity (v) of a particle of mass M, in a gas at temperature T in kelvins is

$$v = \sqrt{2kT/M},$$

while the velocity corresponding to the average energy (u) is a bit higher, viz.

$$u = \sqrt{3kT/M}.$$

5. The radiation laws

In Chapter 4 we saw how the astronomer measures the temperature of a star by studying:

(a) The distribution of radiation intensity with respect to color or wavelength (application of Planck's law).

(b) The wavelength of the position of maximum intensity (application of Wien's law).

(c) The amount of energy radiated per unit area of the surface (application of Stefan's law).

These laws refer to the emission of energy by perfect radiators. In 1859, G. Kirchhoff showed that for any temperature *the ratio of the emissive power of a body to its absorptivity is a constant for all objects and equals the emissive power of a black body*, that is, one that absorbs all the radiation that falls upon it. It is a matter of familiar experience that dark-colored objects are much better absorbers of heat than light-colored ones, and they are also much better emitters of energy. No perfectly black surface has been produced but it is possible to realize experimentally the essential conditions of a black body, insofar as we wish to study radiation.

Planck's law gives the relation between the intensity (I) in a frequency interval Δv, at frequency v, *for a temperature T*, in unit solid angle (there are 4π unit solid angles in a whole sphere), as

$$I_v \Delta v = \frac{2hv^3}{c^2} \frac{1}{e^{hv/kT} - 1} \Delta v.$$

where h is Planck's constant, k is Boltzmann's constant, and c is the velocity of light. If we want the amount of radiation flowing over all directions we multiply this expression by 4π; to compute the energy density we multiply $I_v \Delta v$ by $4\pi/c$. If we wish

Planck's law in wavelength units instead of in frequency units, we make use of the relations

$$v = \frac{c}{\lambda} \qquad \text{and} \qquad \Delta v = \frac{c}{\lambda^2} \Delta \lambda,$$

and obtain

$$I_\lambda \Delta \lambda = \frac{2hc^2}{\lambda^5} \frac{1}{e^{hc/\lambda kT} - 1} \Delta \lambda,$$

which is the form in which this radiation law is most often written.

Stefan's law gives the relation between the total amount of radiation emitted by a black body and the temperature of the body. It is

$$E = \sigma T^4,$$

where E (erg cm^{-2} s^{-1}) is the rate of emission of energy, the Stefan–Boltzmann constant $\sigma = 5.66956 \times 10^{-5}$ erg cm^{-2} K^{-4} s^{-1} (or 5.67×10^{-8} watts m^{-2} K^{-4}), and T (K) is the temperature.

Example. What would be the amount of energy radiated per square centimeter per second by the surface of a star whose temperature is 5700 K?

$$E = 5.67 \times 10^{-5} (5700)^4 = 5.99 \times 10^{10} \text{ ergs/cm}^2 \text{ s}$$
$$= 5.99 \text{ kilowatts/cm}^2.$$

Wien's law gives the wavelength at which the intensity of the radiated energy is a maximum. The relation is

$$\lambda_{max} T = 0.28978 \text{ cm K}.$$

Example. At what wavelength in the spectrum is the intensity a maximum for a star whose temperature is 5000 K?

$$\lambda_{max} = \frac{0.2898}{5000} \text{ cm} = 5.796 \times 10^{-5} \text{ cm} = 5796 \text{ Å},$$

since 1 Å $= 10^{-8}$ cm. Hence the maximum intensity occurs in the yellow near 5800 Å.

Stefan's law and Wien's law may be derived from Planck's law.

Appendix C: The ionization and excitation formulas, and curves of growth

As we saw in Chapter 4, the theory of ionization explains the great changes exhibited by the spectra of the stars as we proceed along the spectral sequence from a hot O star to a cool M dwarf or giant. In the hotter stars the metals are ionized and no longer absorb radiation in the spectral ranges where we can observe them; the spectra of cooler stars are jammed full of metallic lines. We shall devote some attention to both the excitation and the ionization formulas.

1. The meaning of thermal equilibrium

Returning to the discussion of Chapter 4, let us fix our attention upon the excitations and ionizations of a group of atoms in our hypothetical box whose walls are maintained at some temperature T. If the temperature is sufficiently high, say 4000 or 5000 K, atoms will dash wildly about, absorb and emit energy, collide with one another, and lose and regain electrons.

In Fig. C.1, let A be the ground level, and B, C, D, . . . excited levels. A fast electron may hit an atom in level A, lift it to level B, and then go away with less energy. Similarly, another electron may hit an atom in level B, de-excite it to the ground level A, and bounce off with increased energy. In an enclosure, these two processes will exactly balance. Likewise, the collisional excitations of level C will exactly equal the collisional de-excitations. A similar situation obtains for the emission and absorption of radiant energy:

number of absorptions A→B = number of emissions B→A;

number of absorptions A→C = number of emissions C→A.

The number of ionizations from the level B will exactly equal the number of recaptures by the ion of electrons upon level B, and so on. Thus every process is exactly balanced by its inverse process. Under such conditions, the assemblage of atoms is said to be in *thermodynamic equilibrium*.

2. The excitation equation

Under conditions of thermodynamic equilibrium, the relative number (N) of atoms in two levels A and B is given by the Boltzmann equation:

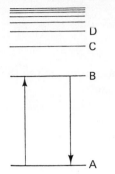

Fig. C.1. Schematic energy-level diagram.

$$\frac{N_B}{N_A} = \frac{g_B}{g_A} e^{-\chi_{AB}/kT},$$

where k is Boltzman's constant, 1.380×10^{-16} erg/K, e is the base of natural logarithms, 2.718, T is the absolute temperature, and χ_{AB} is the energy necessary to excite the atom from level A to level B. The factors g_A and g_B are constants depending on the level involved; they are called statistical weights. In each instance they are numerically equal to the number of Zeeman states into which the level is split when the atom is placed in a magnetic field, and may easily be computed from atomic theory. If ν_{AB} is the frequency of the line emitted in the transition from B to A, then

$$\chi_{AB} = h\nu_{AB}.$$

Generally, it is more convenient for numerical computations to have this formula in another form. If we take logarithms to the base 10 and express χ_{AB} in electron volts, then

$$\log\frac{N_B}{N_A} = -\frac{5040}{T}\chi_{AB} + \log\frac{g_B}{g_A}.$$

For a derivation of the Boltzmann equation and the ionization equation (next section) see, for example, L.H. Aller, 1963 *Astrophysics I: Atmospheres of the Sun and Stars*, Ronald Press, NY, chapter 3.

In many cases g_A and g_B are small numbers about the same size, and for a qualitative notion of the ratio N_B/N_A we may omit them and write simply:

$$\log\frac{N_B}{N_A} \sim -\frac{5040}{T}\chi_{AB}.$$

Example. If A is the ground level of the O III ion (in this case A is actually a group of three levels close together, but we may treat them as one level for the present problem; see Fig. 10.7) and the excitation potential of level B is 2.48 volts, $g_A = 9$, $g_B = 5$, what will be the relative numbers of atoms in level B in thermodynamic equilibrium at a temperature of 10,000°?

$$\log\frac{N_B}{N_A} = -1.25 - 0.25 = -1.50,$$
$$N_B = 0.032 N_A.$$

Example. What would be the fraction of hydrogren atoms excited to the second energy level in the sun, if the excitation temperature is 5800°?

$$\chi_{AB} = 10.16 \text{ volts}, \qquad g_B = 8, \qquad g_A = 2;$$
$$\log\frac{N_B}{N_A} = -\frac{5040 \times 10.16}{5800} + \log 4 = -8.83 + 0.60 = -8.23,$$
$$N_B = 5.9 \times 10^{-9} N_A,$$

that is, under these conditions, about six atoms in every thousand million are excited to the second level and thus become capable of absorbing the Balmer lines.

3. The ionization equation

Under conditions of thermodynamic equilibrium, the relative numbers of ionized and neutral atoms are given by the Saha equation:

$$\log\frac{N_1}{N_0}P_\epsilon = -\frac{5040}{T}I + 2.5\log T - 6.48 + \log\frac{2B_1(T)}{B_0(T)}$$

where N_1 is the number of ionized atoms, N_0 the number of neutral atoms, P_ϵ (atmos) the electron pressure, I (volts) the ionization potential, and T (K) the temperature. The correction term $\log[2B_1(T)/B_0(T)]$ is a function of the temperature for any given atom. It depends on the number and kind of energy states and may be computed from atomic theory. For some elements, such as He, C, N, O, Ne, Ar, and their ions, it is often accurate enough to replace this term by $\log 2\,(b_1/b_0)$, where b_1 and b_0 are atomic constants depending on the kind of ground energy levels in each ion.

Table C.1 gives the ionization potentials for a number of the most abundant elements of astrophysical interest, together with values of the term $\log(2B_{r+1}/B_r)$ for $r=0$ at $T=5800$ K and for $r=1$ at 10 000 K. These numbers are chosen to illustrate the effects in the neighborhood of the solar temperature, where neutral atoms are becoming singly ionized, and near a Class A0 star, where singly ionized atoms are tending to become doubly ionized. For most elements the ratio changes slowly with temperature.

Example. What are the relative proportions of neutral and ionized sodium in the sun if $T=5800$ K and $P_\epsilon = 10^{-5}$ atmosphere? The ionization potential of sodium is 5.14 electron volts, $\log(2B_1/B_0) = -0.08$, $2.5\log T = 9.41$, and $5040I/T = 4.46$. Hence we find that

$$\log(N_1/N_0) = -4.46 + 9.41 - 6.48 - 0.08 + 5.0 = 3.39.$$

that is, $N_1/N_0 = 2460$, and only 0.041 percent of the sodium in the atmosphere of the sun would be neutral.

Example. What is the relative amount of singly ionized iron (Fe II) in the

Table C.1 *Data for calculation of ionization equilibrium*

Atom		Atomic no.	χ_0	χ_1	χ_2	$\log 2B_{r+1}/B_r$ $r=0$ $T=5800$	$r=1$ $T=10\,000$
Hydrogen	H	1	13.60	—	—	0.00	
Helium	He	2	24.58	54.40	—	0.60	0.00
Carbon	C	6	11.26	24.38	47.87	0.11	−0.48
Nitrogen	N	7	14.53	29.59	47.43	0.65	0.07
Oxygen	O	8	13.61	35.11	54.89	−0.05	0.63
Neon	Ne	10	21.56	41.07	63.50	1.08	0.48
Sodium	Na	11	5.14	47.29	71.65	−0.08	1.08
Magnesium	Mg	12	7.64	15.03	80.12	0.58	0.00
Aluminum	Al	13	5.98	18.82	28.44	−0.47	0.58
Silicon	Si	14	8.15	16.34	33.46	0.08	−0.47
Sulfur	S	16	10.36	23.40	35.00	−0.03	0.50
Argon	Ar	18	15.76	27.62	40.90	1.03	0.51
Potassium	K	19	4.34	31.81	46.00	−0.14	1.05
Calcium	Ca	20	6.11	11.87	51.21	0.54	−0.25
Titanium	Ti	22	6.82	13.57	27.47	0.48	−0.20:
Vanadium	V	23	6.74	14.65	29.31	0.27	0.15:
Chromium	Cr	24	6.76	16.49	30.95	0.16	0.50:
Manganese	Mn	25	7.43	15.64	33.69	0.37	0.23:
Iron	Fe	26	7.87	16.18	30.64	0.49	−0.07
Cobalt	Co	27	7.86	17.05	33.49	0.25	0.08
Nickel	Ni	28	7.63	18.15	35.16	−0.12	0.31
Strontium	Sr	38	5.69	11.03	—	0.50	−0.24
Barium	Ba	56	5.21	10.00	—	0.43	−0.55

Note:
: indicates approximate values

atmosphere of Sirius, for which we take $T=10\,000$ K and $P_e = 3 \times 10^{-4}$ atmosphere (see Table C.2)? The first ionization potential of Fe 1 is 7.87 eV; we take $\log\ (2B_1/B_0)=0.49$. Then $5040I/T=3.96$, $2.5\ \log T=10.000$, $\log(N_1/N_0)= -3.96+10-6.48+0.49+3.52=3.57$, or $N_1/N_0=3720$, that is, iron is almost completely singly ionized. Is it appreciably doubly ionized? Apply the ionization equation again with the following data: second ionization potential of iron is 16.18 eV, $\log\ (2B_2/B_1)= -0.07$. Then we find $\log\ (N_2/N_1)= -1.18$, $N_2/N_1=0.066$, $N_2/(N_1+N_2)=0.062$. Hence about 6 percent of the iron is already doubly ionized and 94 percent is singly ionized.

In applications of the ionization formula we must know the electron pressure as well as the temperature. For most types of calculation it suffices to use a mean value of the electron pressure as well as the temperature. Table C.2 gives representative mean values of the electron pressure for giant and supergiant stars as well as data from main-sequence (dwarf) objects.

Finally, to compare the numbers of, say, neutral atoms above the photosphere of

Table C.2 *Electron pressure and mass above photosphere, representative values*

T(K)	$\frac{5040}{T}$	Electron pressure (10^{-6} atm)			Mass 'above photosphere' (g/cm²)		
		Dwarfs	Giants	Supergiants	Dwarfs	Giants	Supergiants
10080	0.5	320	—	33	0.071	—	0.37
8400	0.6	190	—	23	0.25	—	0.71
7200	0.7	70	—	13.6	0.76	—	2.5
6300	0.8	23	10	4.0	1.7	3.0	10
5600	0.9	7.9	2.7	1.0	3.0	10.0	26
5040	1.0	4.4	0.76	0.22	3.7	20	60
4582	1.1	2.8	0.35	0.10	3.9	28	76
4200	1.2	1.8	0.19	0.052	3.9	40	89
3800	1.3	1.2	0.10	0.025	4.0	50	112
3600	1.4	0.81	0.04	0.012	4.1	76	162
3360	1.5	0.57	0.018	0.004	4.1	100	200

the sun with those above the photosphere of some other star, we must have some knowledge of the relative transparencies of the two atmospheres, that is, the opacity variation. If N_S is the number of atoms above the photosphere of a star of temperature T, and $N_\odot$ is the number above the photosphere of the sun, we may tabulate the ratio of N_S to $N_\odot$ for different values of T for dwarfs, giants, and supergiants. Alternatively, we may give the number of grams per square centimeter above the photosphere for different temperatures and types of stars (Table C.2).

4. A simple illustration of the curve of growth (see Chapter 5)

In order to use the curve of growth, it is necessary to combine line intensity data from different spectral regions. The exact theory shows that we should use $\log(W/\lambda)(c/v)$ rather than W as the ordinate. Here c is the velocity of light and

$$v = \sqrt{2kT/M + \xi^2},$$

where k is Boltzmann's constant, T is the gas kinetic temperature, and M the mass of the atom; ξ is the most probable value of random vertical velocities in the atmosphere. Further, we should use as abscissa $\log Nf(\lambda/5000)p$, where λ is measured in Å, rather than $\log Nf$. N is the number of absorbing atoms and f is the oscillator strength. Here, p allows for the variation of the continuous absorptivity or opacity of the solar atmosphere with wavelength. For the sun, we know (c/v) so we can simply shift the theoretical curve vertically so as to use $\log(W/\lambda)$ as ordinate. In this application, we will neglect the p factor. The solid curve in Fig. C.2 is the theoretical curve of growth.

Our empirical curve of growth consists of a graph of $\log(W/\lambda)$ against $\log f(\lambda/5000)$, which is then fitted to the theoretical curve, Fig. C.2. The scale at the bottom is that of the theoretical curve; the scale at the top is that of the empirical curve. From a comparison of the two scales, we find that

$$\log N = 14.98,$$

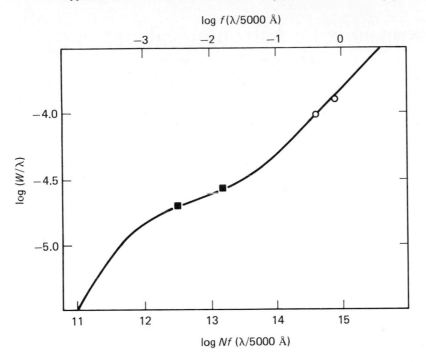

Fig. C.2. Curve of growth for sodium in the sun. The solid curve is the theoretical curve of growth for the sun. Here W is the equivalent width and λ the wavelength (both in ångström units) for a given spectral line. N is the number of atoms 'above the photosphere' in the lower level of the transition, capable of absorbing the line in question, and f is the oscillator strength. Squares denote the observed points for the 3300 Å lines and circles denote the 'D' lines, 5889 and 5890 Å. The empirical plot of $\log (W/\lambda)$ versus $\log f(\lambda/5000)$ is fitted to the theoretical curve by a horizontal shift, whose magnitude, 14.98, corresponds to $\log N$, N being the number of neutral sodium (Na I) atoms above the photosphere.

that is, the number of neutral sodium atoms in the ground level above the photosphere of the sun is 0.95×10^{15}. An application of Boltzmann's formula will show that nearly all sodium atoms are in the ground level. With sufficient accuracy we can take the number of neutral sodium atoms above the solar photosphere to be 1×10^{15}. If $P_\epsilon = 10^{-5}$ atmosphere and $T = 5800$ K, N(neutral Na)/N(total Na) $= 4.1 \times 10^{-4}$, so the total number of sodium atoms above the photosphere is 2.46×10^{18}. Since the mass of the sodium atom is $23 \times 1.66 \times 10^{-24} = 3.8 \times 10^{-23}$ g, the total mass of sodium above the photosphere is 9.4×10^{-5} g $- 0.094$ mg. Similar analyses can be carried out for other metals so their relative abundances can be found. Hydrogen produces most of the continuous absorption and its abundance controls the 'depth of the photosphere'.

Appendix D: Astronomical constants

One astronomical unit	149 597 892 km = 92 955 635 miles
One light-year	9.4605×10^{17} cm
One parsec	3.085678×10^{18} cm = 3.26 light-years
Mass of the sun	1.989×10^{33} g
Radius of the sun	6.960×10^{10} cm
Mean density of the sun	1.410 g/cm^3
Surface gravity of the sun	2.740×10^4 cm/s^2
Energy radiated by the sun	3.83×10^{33} erg/s
Absolute bolometric magnitude of the sun	+4.69 (Popper)
Absolute B magnitude of sun	+5.47 (Popper)
Absolute photovisual magnitude of the sun	+4.83 (Popper)
Temperature (T_{eff}) of the sun	5776 K (Labs and Neckel)
Mass of the earth	5.976×10^{27} g
Mean radius of the earth	6.3710×10^8 cm
Mean density of the earth	5.517 g/cm^3
Surface gravity of the earth	980.665 cm/s^2
Number of seconds in one sidereal year	3.1558×10^7

Appendix E: Stellar magnitudes and colors

1. The relation between apparent magnitude, absolute magnitude, and distance

Magnitude differences are related to ratios of brightness by the basic expression

$$0.4(m_2 - m_1) = \log l_1/l_2,$$

where m_1 and m_2 are the apparent magnitudes of two stars and l_1 and l_2 are their corresponding apparent brightnesses. One magnitude difference corresponds to a brightness difference of 4 decibels.

To obtain the absolute magnitude, which is defined as the magnitude the star would have if it were at a distance of 10 parsecs, we note that brightness varies inversely as the square of the distance. Hence if l is the apparent brightness of a star at the distance r and L is the brightness it would have at the distance R.

$$l/L = (R/r)^2.$$

Taking logarithms and setting $R = 10$ parsecs,

$$\log l/L = 2 \log R - 2 \log r = 2 - 2 \log r$$

since $\log 10 = 1$.

If M is the absolute magnitude of a star of apparent magnitude m, the relation between brightness and magnitude becomes

$$2.5 \log l/L = M - m,$$

or

$$M = m + 5 - 5 \log r.$$

If we use parallax, p, instead of distance, then, since r (parsecs) $= 1/p$ (arcsecs),

$$M = m + 5 + 5 \log p.$$

Example. γ Geminorum, visual magnitude 1.93, has a parallax of 0.040 arcsec according to the Yale *Catalogue*. What are its distance, absolute magnitude, and absolute brightness, and how do they compare with those of the sun? The distance of the star is $r_s = 1/0.040 = 25$ parsecs, and $\log r_s = 1.40$. Hence the star's absolute magnitude is

$$M_S = 1.93 + 5 - 5 \times 1.40 = -0.07.$$

The star lies above the main sequence, from which it has evolved.

The absolute visual magnitude of the sun $M_\odot = +4.83$. Hence γ Geminorum is $4.83 - (-0.07) = 4.90$ magnitudes brighter than the sun.

The basic magnitude relation holds for absolute as well as apparent magnitudes. Hence

$$0.4 \ (M_\odot - M_S) = \log L_S/L_\odot,$$

where L_S is the absolute brightness of the star and $L_\odot$ that of the sun. Hence, to compare γ Geminorum and the sun.

$$0.4 \times 4.90 = 1.960 = \log L_S/L_\odot,$$

whence $L_S/L_\odot = 91$.

Thus γ Geminorum is nearly 100 times as bright as the sun.

If there is absorption of light in space by obscuring matter (see Chapter 10), the equation relating apparent and absolute magnitudes must be modified. If the light of a distant star is dimmed P magnitudes by interstellar clouds, then

$$m_{obs} = m_{true} + P;$$

hence

$$M = m + 5 - 5 \log r - P.$$

The amount of interstellar absorption may be inferred from its effect on the colors of distant stars, since the interstellar cloud reddens as well as dims starlight.

2. The kinds of magnitudes

The magnitude of a star depends on the color sensitivity of the detector employed. With a photocell and a violet filter, a red star may appear as a faint object, although it may yet be quite bright to the eye. Visually, a star such as χ Cygni may be of the tenth magnitude, yet a thermocouple, which measures the total energy that reaches it, may show that this star is emitting as much radiation as a normal star of the fifth magnitude. Therefore, in expressing the magnitude of a star, we must specify the color sensitivity of the detection system employed.

Suppose we observe a star with a photoelectric photometer equipped with a blue filter. The red, yellow, most of the green and violet, and the ultraviolet light will be cut out. Only blue light over a range of a few hundred ångström units will affect the sensitive surface of the photocell. Experiments have shown that, although radiations covering a span of several hundred ångströms may fall on the photocell, the magnitudes determined will be just the same, to a fair degree of approximation, as though all the light were concentrated at one mean wavelength, which is called the *effective wavelength*, λ_{eff}.

The modern U, B, V system of photometry, due to Harold Johnson, employs filter–photocell combinations with the following characteristics (Fig. E.1):

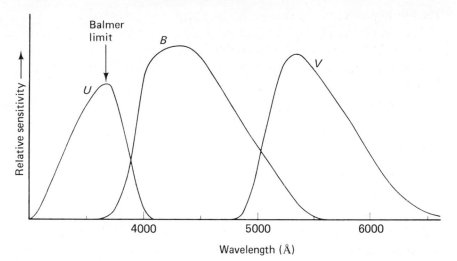

Fig. E.1. The wavelength dependence of the response of the U, B, V color system. These curves give the response for a U, B, V photometer as employed in a Cassegrain or Newtonian telescope, with reflection from two aluminized mirrors and atmospheric extinction for a star observed in the zenith at Mount Wilson. (After Johnson, Code, and Melbourne, as quoted by H.C. Arp, 1961, *Astrophysical Journal*, **133** p. 875.)

Color	λ_{eff} (Å)	$(\Delta\lambda)_{1/2}$ (Å)	$F_\lambda (0.0)$ $(10^{-7}\ erg/cm^{-2}(100\ Å)^{-1}\ s^{-1})$
U	3650^{-100}_{+200}	530	4.35
B	4400^{-70}_{+100}	1000	7.20
V	5470^{-10}_{+30}	850	3.92

The effective wavelengths are given for a star of temperature 10 000 K.

The corrections to the effective wavelength of -100, -70, and -10 Å apply for a star of temperature 20 000 K, whose energy distribution is richer in ultraviolet light, and the corrections $+200$, $+100$, and $+30$ refer to a cool star of 4000 K, whose radiation is mostly concentrated in the red; see C.W. Allen, 1973, 3rd edition *Astrophysical Quantities*, Oxford University Press, NY, p. 195. The 'half-band' width $(\Delta\lambda)_{1/2}$ is the width of the wavelength range over which the sensitivity exceeds half its maximum value. The last column gives the flux received (for band-pass of 100 Å) from a star of magnitude 0. The zero points are so adjusted that a main-sequence star of spectral class A0 has exactly the same magnitude in each system, that is $U - B = B - V = 0$.

Other combinations of photocells and filters are also employed. Thus for work in the red and infrared Harold Johnson has introduced the systems R ($\lambda_{eff} \sim 7000$ Å), I (9000 Å), J (12 500 Å), K (22 000 Å), L (34 000 Å), M (50 000 Å), and N (102 000 Å $= 10.2\mu m$). It is possible to employ narrow band-passes and to select wavelengths

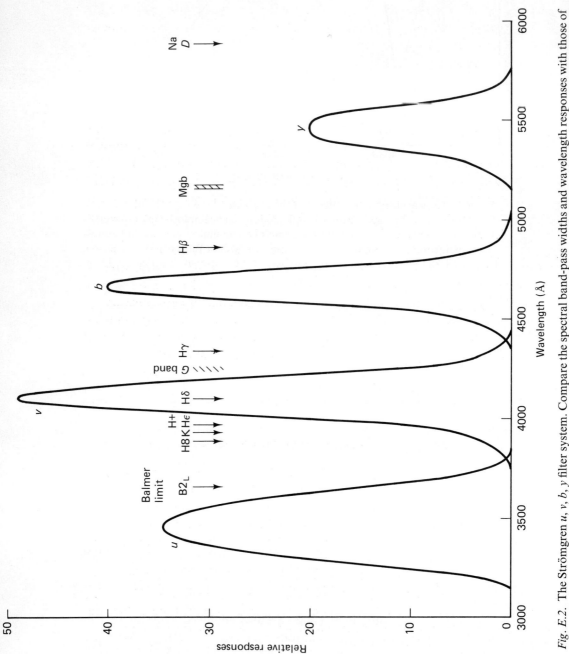

Fig. E.2. The Strömgren u, v, b, y filter system. Compare the spectral band-pass widths and wavelength responses with those of the Johnson U, B, V system (Fig. E.1). Notice that the u-filter transmits radiation only beyond the Balmer limit; there is no overlap as with the U filter. The positions of some important solar spectrum lines are indicated.

to emphasize certain spectral features, such as those depending on luminosity or metal abundance, for example.

The modern U, B, V, system as revised by Harold Johnson and W.W. Morgan, is a powerful tool for stellar photometry, but it suffers from the shortcoming that for some purposes the band-passes are too broad and the U filter straddles the Balmer jump. Strömgren developed a photometric system that employs narrower band-pass filters and four colours as follows:

u	v	b	y
3480	4010	4620	5450

See Fig. E.2. He also devised a useful scheme for evaluating absolute luminosities for B, A, and F stars by means of the β index. One compares the flux in a 60 Å band-pass centred on Hβ with that in a 300 Å band-pass similarly centered. A dwarf A star with a wide Hβ line would have a much smaller flux through the 60 Å filter than would a supergiant of the same visual magnitude and Draper spectral class. Fluxes through the 300 Å band-pass filter would be nearly the same. Thus the ratio F(Hβ, 60 Å)/F(Hβ 300 Å) would depend strongly on luminosity.

The $(b-y)$ colour index is analogous to Johnson's $(B-V)$ which is explained in the next section. The Balmer discontinuity index is $c=(u-v)-(v-b)$, while a 'metallicity index' is defined by $m=(v-b)-(b-y)$. These indices can be used to estimate T_{eff}, M_v and surface gravity, g, for a star and even establish its evolutionary status! That is, by narrow band-pass photoelectric photometry, it is possible to obtain spectral and luminosity classes. Strömgren showed that the effects of interstellar extinction can be eliminated by calculating the combinations

$$[c] = c - 0.20(b-y), \quad [m] = m + 0.18(b-y).$$

The β-index is not affected by interstellar extinction.

3. Color indices, spectral classes, and bolometric corrections

The difference between the magnitudes of a star as measured in two-color systems is called a color index. The most commonly employed color indices are $B-V$ and $U-B$. Consider, first of all, stars that are unaffected by space absorption.

The $B-V$ color index then depends simply on the temperature. It is negative for very blue stars, such as the nuclei of planetary nebulae, because such objects are brighter in the blue than in the visual spectral regions. For cool stars, however, the $B-V$ index is positive and may become very large. Thus it serves as an index of temperature; in a Hertzsprung-Russell diagram V is plotted against $B-V$.

What happens if we plot $U-B$ against $B-V$? Consider first stars that are unaffected by space absorption, and suppose that stars radiate as black bodies. Then a graph of $U-B$ against $B-V$ would simply compare the slopes at two points in the Planckian curve and we would expect to get very nearly a straight line. The actual two-color graph for main-sequence stars resembles a 'lazy S' (Fig. E.3). Although

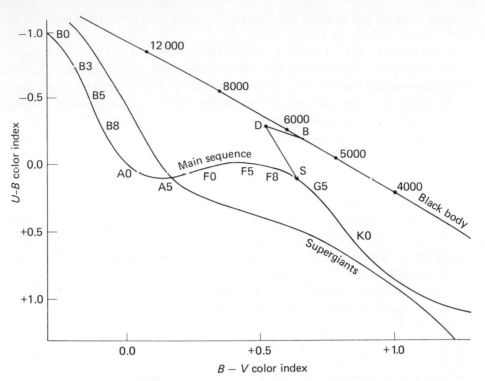

Fig. E.3. The relation between $U-B$ and $B-V$ color indices. The mean relation derived by Harold Johnson between $U-B$ and $B-V$ color indices is plotted for main-sequence stars and supergiants. The corresponding curve as calculated by H.C. Arp for a sequence of black bodies is also included. Compare the smooth slope of the black-body relation with the S-shaped curve for the main-sequence stars. The supergiant curve shows a less marked curvature because the Balmer jump is less prominent in supergiant F, A, and B stars than in dwarfs. A sequence of pure hydrogen stars would show a considerable departure from a black-body curve but absorption by strong metallic lines produces marked effects also. S denotes the position of the sun, B the position of a black body of the same effective temperature as the sun, and D the position of a metal-deficient star of the same temperature as the sun. It is displaced from B by the amount DB because the negative hydrogen ion is a non-gray absorbing agent (see Fig. 5.6). The displacement DS corresponds to the influence of strong metallic absorption in the sun. (Adapted from H.C. Arp, 1961, *Astrophysical Journal*, **133**, pp. 878, 880, copyright University of Chicago Press.)

$B-V$ changes smoothly with temperature, $U-B$ does not, primarily because of the pronounced distortion produced by the absorption at the head of the Balmer series (Fig. 5.5) and because further distortions are produced by the influence of strong absorption lines. The two-color graphs for metal-deficient stars differ appreciably from those for normal stars. Supergiant stars show a less-pronounced kink because the Balmer jump is less important for stars near Class A0.

Another very important use of the two-color $(U-B)-(B-V)$ graph is to determine space absorption. Interstellar reddening affects the two-color indices by different amounts, so by comparing the observed $(U-B)-(B-V)$ graph for a star cluster with the standard graph one can determine the extra coloring or color excess $\Delta(B-V)$ produced by space absorption. Magnitudes and colors corrected for space absorption are usually denoted by V_0, $(B-V)_0$, $(U-V)_0$. See Appendix E.4.

The various kinds of magnitudes that we have been discussing, U, B, V, and infrared magnitudes, utilize radiation over a limited wavelength range. For many problems, such as those of stellar evolution, we want to compare the luminosities of two stars with reference to the radiation summed over all wavelengths.

We express the total luminosity of a star in terms of its so-called bolometric magnitude. The difference between the bolometric and V (essentially visual) magnitudes is called the *bolometric correction*:

$$BC = M_{bol} - M_v.$$

The system of bolometric magnitudes is so adjusted that the corrections vanish for a Class F0 star and are small for the sun. They become very large for hot stars, where most of the energy is in the far ultraviolet, and for very cool stars, which radiate most of their energy in the infrared.

The corrections may be established reasonably well for stars similar to the sun; they are uncertain for very hot stars, where we have to rely on predictions of model-atmosphere theory with occasional checks from rocket and satellite observations. They are likewise uncertain for cool stars whose spectra are strongly distorted by molecular-band absorption and most of whose energy is extinguished by absorption by water vapor in the earth's atmosphere.

Table E.1 summarizes our data on temperatures, intrinsic, that is, $(B-V)_0$, colors, and bolometric corrections for various types of stars. We discussed the temperature scale in Chapter 4. The $B-V$ colors are taken primarily from the work of Harold Johnson. The bolometric corrections and T_{eff} are derived from a number of sources; for early-type stars we rely on discussions by A.D. Code, J. Davis, R.C. Bless, R. Hanbury-Brown, and R. Shobbrock (1976); for a wide range of spectral classes and temperatures there are discussions by D. Hayes (1978), D.M. Popper (1980), and T. Schmidt-Kaler (1982). Colour indices and bolometric corrections for giants and supergiants are very similar. Notice that except near F0, the bolometric corrections are always negative so that the star is always brighter bolometrically than visually.

4. Some uses of color and luminosity measurements: Relation between absolute magnitude, temperature, and radius of a star

If we know the absolute brightness of a star and its size we can find its temperature. Alternatively, if we know the temperature and the true brightness of a star we can find its size.

The relation between the absolute magnitude of a star M_λ (as measured in a magnitude system of effective wavelength λ_{eff}), its radius R, and its temperature T (assuming that it radiates approximately like a black body of temperature T) is

$$M_\lambda = C_\lambda - 5 \log R + \frac{1.561}{\lambda_{eff} T} + X_\lambda;$$

Table E.1 *Temperatures, $(B-V)_0$ colors, and bolometric corrections (BC) for normal stars*

(a) *Luminosity type* V *(main sequence)*

Spectral class	$(B-V)_0$	T_{eff}	BC	Zero age main sequence M_v
O5	−0.33	45000	−4.3	−5.7
O6	−0.32	41000	−3.9	−5.5
O7	−0.31	38000	−3.64	−5.2
O8	−0.305	35700	−3.45	−4.9
O9	−0.30	33100	−3.27	−4.5
O9.5	−0.295	31400	−3.1	−4.2
B0	−0.200	29900	−2.90	−4.0
B0.5	−0.280	28500	−2.83	−3.6
B1	−0.26	25800	−2.59	−3.2
B2	−0.24	23000	−2.36	−2.4
B3	−0.20	18900	−1.94	−1.6
B5	−0.165	15450	−1.44	−1.2
B7	−0.13	10970	−0.95	−0.6
B8	−0.10	10700	−0.61	−0.2
B9	−0.06	10420	−0.31	+0.2
A0	0.00	9470	−0.15	+0.6
A1	+0.02	9150	−0.11	+1.0
A2	+0.06	8870	−0.08	+1.3
A3	+0.08	8700	−0.06	+1.5
A5	+0.14	8170	−0.02	+1.9
A7	+0.19	7810	−0.01	+2.2
F0	+0.31	7100	−0.01	+2.7
F2	+0.35	6800	−0.02	+3.0
F5	+0.43	6500	−0.03	+3.5
F8	+0.53	6150	−0.08	+4.1
G0	+0.59	5950	−0.10	+4.4
G2	+0.63	5800	−0.13	+4.7
G5	+0.66	5710	−0.15	+5.1
G8	+0.74	5520	−0.18	+5.5
K0	+0.82	5190	−0.24	+5.9
K2	+0.92	4900	−0.35	+6.4
K5	+1.15	4320	−0.66	+7.4
K7	+1.32	4040	−0.93	+8.1
M0	+1.41	3870	−1.22	+8.8
M1	+1.47	3720	−1.50	+9.3
M3	+1.51	3470	−2.0	+10.4
M4	+1.54	3370	−2.28	+11.3
M5	+1.64	3250	−2.54	+12.3
M6	+1.73	3220	−2.93	+13.5
M7	+1.80	2880	−3.46	+14.3
M8	+1.93	2620	−4.0	+16.0

(b) *Luminosity class III stars (giants)*

Spectral class	$(B-V)_0$	T_{eff}	BC	Mean M_v
G0	0.65	5800	−0.15	+1.0
G2	0.77	5300	−0.25	+0.9
G5	0.89	4800	−0.31	+0.9
G8	0.95	4700	−0.36	+0.8
K0	1.01	4530	−0.42	+0.7
K2	1.16	4370	−0.53	+0.5
K3	1.26	4100	−0.70	+0.3
K5	1.51	3950	−1 19	−0.2
M0	1.57	3750	−1.28	−0.4
M1	1.58	3700	−1.36	−0.6
M2	1.60	3640	−1.51	−0.6
M3	1.61	3320	−1.91	−0.5
M4	1.62	3200	−2.55	−0.5
M5	1.63	3100	−2.7	−0.2
M6	1.64	3000	−3.0	−0.2

(c) *Supergiants*

Spectral class	$(B-V)_0$	T_{eff}	BC
O5	−0.31	40000	−3.84
O9	−0.27	38000	−3.12
B0	−0.23	26400	−2.53
B1	0.20	21000	−1.90
B2	−0.17	18600	−1.59
B3	−0.13	16500	−1.30
B5	−0.09	14000	−1.00
B8	−0.03	11500	−0.70
A0	0.00	9700	−0.40
A2	+0.03	9050	−0.29
A5	+0.09	8480	−0.13
F0	+0.18	7600	−0.01
F2	+0.23	7300	−0.01
F5	+0.32	6850	−0.03
F8	+0.57	6100	−0.09
G0	+0.76	5500	−0.16
G2	+0.87	5200	−0.21
G5	+1.02	4800	−0.35
G8	+1.15	4600	−0.42
K0	+1.23	4400	−0.51
K2	+1.35	4260	−0.61
K5	+1.60	3860	−1.00
M0	+1.65	3660	−1.27
M1	+1.69	3560	−1.37
M2	+1.70	3450	−1.62
M5	+1.80	2800	−3.5:

Note:
: indicates the value is uncertain.

(see Russell, Dugan, and Stewart, 1938, *Astronomy*, 2nd edition, Ginn, Boston, vol.2, p. 733), where C_λ is a constant depending on wavelength and X_λ is a small correction factor (see below) which may be important at high temperatures:

$\dfrac{1.561}{\lambda_{\text{eff}} T}$	5.0	4.0	3.0	2.0	1.0
X_λ	-0.01	-0.03	-0.07	-0.19	-0.55

For *V*-magnitudes, let us adopt

$$\lambda_{\text{eff}} = 5480 \text{ Å} = 5.48 \times 10^{-5} \text{ cm}.$$

To evaluate C_λ we note that a black body of the same size and effective temperature as the sun would have approximately the absolute *V*-magnitude 4.83. Since R is measured in terms of the solar radius, $R = 1.0$, $\log R = 0.00$. Hence, if the effective temperature of the sun is 5780 K,

$$4.83 = C_\lambda + 4.93 - 0.01$$

or

$$C_\lambda = -0.09.$$

The relation between visual magnitude, radius, and effective temperature is

$$M_V = -0.09 - 5 \log R + \frac{28\,500}{T}.$$

For *B* magnitudes, we might take $\lambda_{\text{eff}} = 4400$. The absolute *B*-magnitude of the sun is $M_B = 5.47$. Hence a similar calculation give $C_B = -0.66$, so

$$M_B = -0.66 - 5 \log R + \frac{35\,500}{T}.$$

Eliminating R between these two equations and noting that $M_B - M_V = B - V$, we find

$$T = \frac{7000}{B - V + 0.57},$$

which yields a crude conversion from $B - V$ color indices to temperature for stars whose temperatures do not differ very much from that of the sun.

Example. The *V*-magnitude of Wolf 359 is 13.66. Its parallax is 0.425 arcsecs. The corresponding absolute magnitude is 16.80. The temperature corresponding to its spectral class is 2960 K. What is its radius in terms of that of the sun? The equation for M_V gives

$$16.80 = -0.09 - 5 \log R + 9.63,$$

or

$$\log R = -1.45,$$

whence

$$R = 0.035.$$

Example. The radius of a B3V component of an eclipsing binary system is 4.23 in terms of the sun. What is the absolute magnitude of the star? The temperature of a B3V star (Table E.1) is 18 900 K. Then

$$M = -0.09 - 5 \times 0.626 + 1.51 = -1.71.$$

The assumption that stars radiate like black bodies is only a crude approximation, particularly rough for Class A stars, which show a strong continuous absorption due to hydrogen, and for cool stars, which show strong molecular bands.

Although effective temperatures may be estimated from the $B - V$ index corrected for interstellar extinction, i.e. $(B - V)_0$, it is better to use the Strömgren narrow band-pass color system for this purpose. Of course, if bolometric magnitudes can be accurately estimated from the observational data and if effective temperatures are known, we can determine stellar radii readily in view of the relationship

$$L = 4\pi R^2 \sigma T_{\text{eff}}^4$$

If we measure L and R in terms of the corresponding solar values, $L_\odot$ and $R_\odot$, we obtain

$$\frac{L}{L_\odot} = \left(\frac{R_*}{R_\odot}\right)^2 \left(\frac{T_*}{T_\odot}\right)^4,$$

or in terms of bolometric magnitudes

$$M_\odot - M_* = 5 \log R + 20 \log(T/5800).$$

In both these equations the asterisk indicates the stellar values.

5. Estimates of interstellar extinction using U, B, and V colors

Since the interstellar material reddens the light of the stars it dims, it acts to increase their color indices. Suppose one observes the main-sequence stars of a distant cluster on the U, B, V system. If one plots the $(U - B)$ color indices against the $(B - V)$ color indices, the standard S-shaped curve will be displaced (Fig. E.4). To fit the standard $(U - B)_0 - (B - V)_0$ curve to the observations, one shifts it along a line whose slope is given by

$$\frac{(U - B)_{\text{color excess}}}{(B - V)_{\text{color excess}}} \sim 0.7$$

until coincidence is obtained. In the example illustrated, $\Delta(B - V) = 0.20$, $\Delta(U - B) = 0.14$. The total space absorption is then

$$A_V = 3.0\Delta(B - V) = 0.6 \text{ magnitude},$$

although some observers, notably Harold Johnson, have concluded that the coefficient may vary from point to point in the Galaxy. It is also different in the

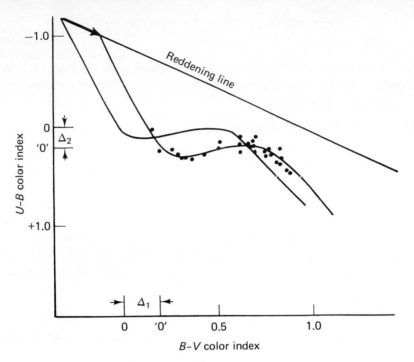

Fig. E.4. Determination of space reddening for a star cluster. The observed main sequence is indicated by the points to which the standard $(U-B)$ vs. $(B-V)$ curve is fitted by a shift, whose amount is indicated by the heavy arrow. The displacements $\Delta(B-V) = \Delta_1$ and $\Delta(U-B) = \Delta_2$ determine the color excess. The ratio $\Delta_2/\Delta_1 = 0.7$ defines the slope of the 'reddening line' along which the curve is to be shifted.

Magellanic Clouds. This method is strictly applicable for stars of 'normal' chemical composition and could not be applied for a cluster of metal-deficient stars; supplementary spectroscopic data would then be needed.

6. Analysis of a color-magnitude diagram for a star cluster

The detailed procedures differ for galactic and globular clusters. For the former, one often can obtain spectroscopic as well as photometric observations. Galactic-cluster main sequences usually extend to early types, F, A, or even B or O, whereas main sequences of globular clusters do not reach much earlier than F8 or G0. Spectroscopic observations of individual stars in globular clusters require very large telescopes. Although we can now observe many stars simultaneously with a modern multi-image-fiber-optic-equipped spectrograph, we can still observe only the brightest stars. For the most part, we must rely on color–magnitude diagrams.

The steps in the procedure are as follows:

(1)   Construct the best possible color-magnitude diagram using charge-coupled device (CCD) observations. Photoelectric photometry is sometimes used to set up basic standards of color and brightness.

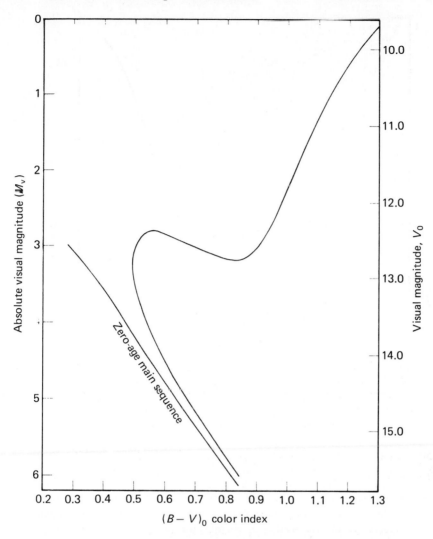

Fig. E.5(a). The color–luminosity relations for the galactic cluster M67. 'Visual'
magnitudes V_0 are plotted against $(B-V)_0$ colors for mean points for this cluster.
The left-hand ordinates give the absolute magnitude M_v obtained by fitting the
curve to a standard Hertzsprung–Russell diagram. The zero-age main sequence is
included. (After A.R. Sandage, 1962, *Astrophysical Journal,* **135**, p. 349, copyright
University of Chicago Press.)

(2)   Evaluate interstellar extinction by the method described in the previous
      section or some variation thereof. Then convert V-magnitudes and $(B-V)$
      colors to extinction-corrected V-magnitudes and true $(B-V)_0$ colors.

(3)   Next compare the V_0 versus $(B-V)_0$ diagram thus obtained with the
      standard Hertzsprung–Russell diagram, which gives $M_V - (B-V)_0$, to
      obtain the distance modulus $y = V - M_V = 5 \log r - 5$, and hence the
      distance r of the cluster and the absolute magnitude of its members.

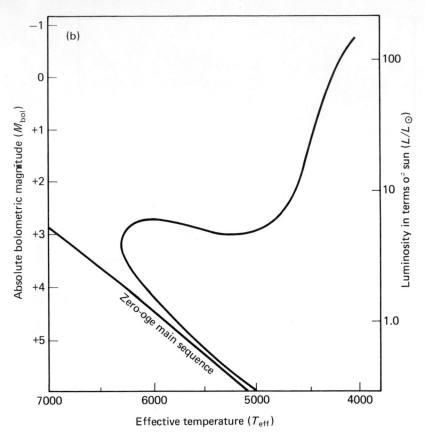

Fig. E.5(b). The luminosity–effective-temperature relation for M67. By use of the data of Table E.1, Fig. E.5 has been transformed from a graph of M_v vs. $(B-V)_0$ to one giving bolometric magnitude vs. effective temperature. The right-hand scale gives the luminosity in terms of that of the sun. This curve differs from that given by Sandage, 1962, *Astrophysical Journal*, **135**, p. 349, because a different color–temperature and bolometric-correction–temperature relation has been used.

The resulting HR diagram can be compared with published data to decide whether the cluster is a normal metal-rich one or a metal-deficient one.

We may also transform the color–magnitude diagram to a bolometric-magnitude–effective-temperature diagram for comparison with predictions of stellar evolution theory for a star of appropriate metallicity.

Consider the color-magnitude diagram for the old galactic cluster M67 (Fig. E.5a), which is taken from the work of A. Sandage. Apparent magnitude V is given on the right-hand side and absolute magnitude M_V on the left-hand side. The quantities V, M_V, and $B-V$ are all corrected for space absorption. Proceeding from fainter to brighter stars, the main sequence departs increasingly from the zero-age main sequence and finally breaks off to run continuously up into the giant region.

Using the data of Table E.1, we may transform M_V versus $(B-V)_0$ measurements to M_{bol} and T_{eff} as follows. At the upper right-hand corner of the diagram Fig. E.5a

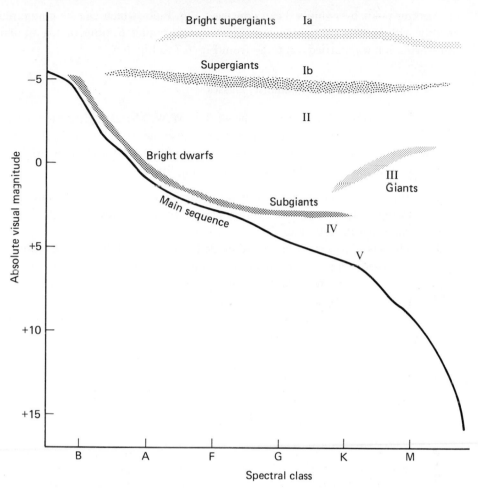

Fig. E.6. The relation between luminosity class, absolute magnitude, and spectrum. From the appearance of the spectrum, one can estimate not only spectral class but also a luminosity class, Ia, Ib, II, III, IV, or V. These have to be calibrated in terms of absolute magnitude, as has been done by W.W. Morgan and others. For supergiants, and even for giants, there is a considerable spread in intrinsic brightness for each luminosity class. Conventional giants constitute luminosity class III. Notice that in the later spectral classes luminosity class IV refers to subgiants, whereas in earlier classes it refers to bright dwarfs – in each instance stars that have evolved away from the main sequence, but by differing amounts.

we extrapolate the curve slightly to $M_V = 0$ at $(B-V)_0 = 1.31$. These stars are obviously giants so we go to Table E.1, luminosity class III, and interpolate between K3 and K5 to get $BC = -0.75$ and $T_{eff} = 4050$. Thus we obtain the first entry in a table:

M_V	$(B-V)_0$	BC	M_{bol}	T_{eff}
0.0	1.31	-0.75	-0.75	4050

Proceeding point by point, we can construct Fig. E.5b, which can be compared directly with theoretical predictions. Turning to Chapter 6, note that a similar transformation was carried out to go from Fig. 6.7 to Fig. 6.8.

7. Interpretation of luminosity classes

Fig. E.6, which is based primarily on the work of W.W. Morgan, gives the relation between luminosity and spectral classes and absolute visual magnitudes. The calibration is accurate for the main sequence (except perhaps for the very brightest stars), but the supergiant data are more uncertain, for two reasons:

(a)   Since these stars are very remote, determinations of their distances are intrinsically difficult because they depend on association with star clusters and other uncertainties.

(b)   There is a large intrinsic spread in the brightnesses of these stars; although main-sequence stars, particularly young main-sequence stars, tend to hug the dwarf sequence, giant and supergiant stars both show a huge range in intrinsic brightness, depending on the mass and chemical composition of the main-sequence star from which they evolved. Accordingly, giant and supergiant luminosity classes have been indicated by broad bands rather than by a narrow line as for main-sequence dwarfs.

Appendix F: Interstellar molecules

Except for robust structures, such as the durable CO (carbon monoxide) molecule, which is widespread throughout the Galaxy, most species are concentrated in dense molecular clouds, such as Sagittarius B-2, where they are protected from the ultraviolet radiation that tends to destroy them. Many molecules are found in the envelopes of highly evolved stars such as IRC + 10216 (see Chapters 9 and 10).

Table F.1, compiled from lists by A. Hjalmarson and by M.R. Morris, gives many of the molecules that have been identified so far. Most of the vibrational and rotational lines fall in the range from 77 μm (far infrared) to 43 cm. The majority are found in the millimeter range, but identifications are often difficult. Frequencies of individual transitions have to be measured very accurately and one must determine the Doppler shift of the cloud. One must identify and eliminate the rotational and vibrational transitions of known molecules. Often, this is not an easy task. Methyl formate, $HCOOCH_3$, an asymmetrical molecule with restricted modes of rotation, contributes about 200 lines. But sometimes, the interstellar medium data help the organic chemist to construct accurate structural models, while many molecules and radicals found in the interstellar medium are of an unstable form which would react quickly in a terrestrial setting.

Fig. F.1 shows the structures of some well-known interstellar molecules. Diagrams for many more may be found in *Larousse Astronomy*; editors P. de la Cotardiere and M.R. Morris (English edition), 1986, New York, Facts on File, pp. 270–1. A schematic diagram for a polycyclic aromatic hydrocarbon (PAH) is also shown. It has been suggested that these may produce some of the diffuse features such as the 4430 Å broad absorption line.

A number of isotopes are clearly detected in the interstellar medium. They are often readily revealed. For example the substitution of ^{13}C for ^{12}C changes the vibrational and rotational frequencies of any molecule, no matter how complex its structure. See Table F.2.

Table F.1 *Interstellar and circumstellar molecules*

H_2 HD CO	NH	OCS	CH_3NC
H_2O	CN	SO	HC_2N
CH_2OH	HCN	SO_2	HNC
H_2CO	CH_2OCHO	SiO	HC_3N
CN	CH_3C_2H	H_2S	C_3N
C_2H	CH_3CN	H_2CS	C_4H
$(CH_3)_2O$	HCO	CS	HC_7N
OH	CH	HCS	CH_2CHCN
C_7	CH_3OH	NO	SiS
C_3H	CH_3CH_3OH	NS	SiH_4
C_4H	CH_3CHO	HNO	HNCS
C_5H	$HCOOCH_3$	NH_2CHO	CH_3C_3N
C_6H	CH_3C_4H	CH_2NH	HC_3N
C_3H_2	CH_2CO	HNCO	CH_3SH
C_3O	HCOOH	NH_2CH	SiC_2
C_2S	HC_2H	CH_3NH_2	HC_9N
C_3S	$H_2C_2H_2$	CH_2CHCN	$HC_{11}N$
CH_3C_5N	CH_3COOH	CH_3CH_2CH	HCl
S C	C_4Si	H_2CCO	CH_2CN
NaCl	AlCl	KCl	AlF

N.B. We have written formulae to show atoms involved in each molecule, but in organic chemistry, formulae are written in a way to show actual structures involved. For example propynal (HC_2CHO) is depicted thus

$$HC \equiv C - C \underset{\displaystyle H}{\overset{\displaystyle O}{\diagup\kern-0.6em\diagdown}}$$

Molecular ions

CH^+	HCO^+	$HOCO^+$	H_3^+	N_2H^+	$HCOOH^+$
NH_2^+	HCS^+	HOC^+	$HOH^+(?)$	CO^+	$HCNH^+$

Table F.2 *Isotopic molecular structures*[a]

Hydrogen	H_2	HD			
Hydroxyl	OH	^{17}OH	^{18}OH		
Carbon monoxide	CO	^{13}CO	$C^{17}O$	$^{13}C^{18}O$	$C^{18}O$
Water	HOH	HDO	$H^{18}OH$		
Carbon monosulfide	CS	^{13}CS	$C^{33}S$	$C^{34}S$	
Hydrocyanic acid	HCN	DCN	$H^{13}CN$	$HC^{15}N$	
Hydrogen isocyanide	HNC	DNC			
Formaldehyde	H_2CO	$H_2{}^{13}CO$	$HC^{18}O$		
Cyanoacetylene	HC_3N	$H^{13}CC_2N$			
Silicon monoxide	SiO	^{29}SiO	^{30}SiO		
Ammonia	NH_3	NH_2D			
	NH_2CHO	$NH^{13}CHO$			

Isotopic molecular ions

CH^+ $^{13}CH^+$ HCO^+ DCO^+ $H^{13}CO^+$ NH^+ NO^+

Note:

[a] The notation used is that the most abundant isotopic species are not designated. Thus, HCN indicates $^1H^{12}C^{14}N$ while, if ^{13}C is substituted for ^{12}C, we write $H^{13}CN$, etc. For a detailed account of molecular isotopes in the interstellar medium see P.G. Wannier, 1980, *Annual Reviews of Astronomy and Astrophysics*, **18**, p. 399.

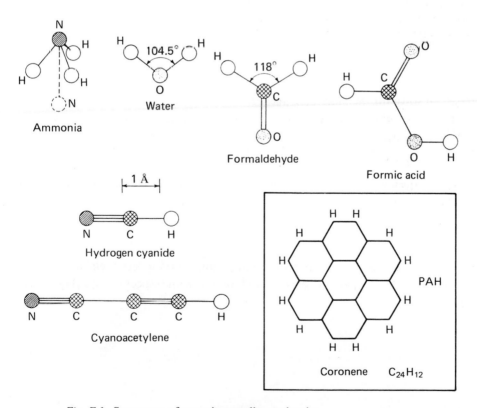

Fig. F.1. Structures of some interstellar molecules.

Appendix G: The determination of stellar masses

If two stars in a binary system are separated by an average distance a and move about one another with a period P, Kepler's third law says that the sum of the masses, $m_1 + m_2$, is given by

$$m_1 + m_2 = \frac{a^3}{P^2}.$$

Here P is measured in years, a in astronomical units (mean distance from earth to sun), and m_1 and m_2 in units of the sun's mass.

Suppose we know the velocities of the two stars and that they move in circular obits. Then

$$2\pi a_1 = V_1 P, \qquad 2\pi a_2 = V_2 P,$$

and

$$m_1 a_1 = m_2 a_2,$$

so we can solve for the individual masses.

For a visual binary where the velocities are not directly measured but where the distance is known,

$$a = a'' r,$$

where a'' is the mean separation in arcsecs and r is the distance in parsecs. In terms of parallax $p \, (= 1/r)$,

$$m_1 + m_2 = \frac{a''^3}{P^2 p^3}.$$

We cannot determine the masses of the individual stars unless their motions are determined with respect to the background, that is, with respect to a fixed reference frame.

A selected bibliography

A very large number of good to excellent books and articles on the topics covered in this book have appeared in recent years. The following selection is intended as a representative sample of much of this material. First some general texts are mentioned, then other references are grouped by chapter or related chapters. In addition to popular books, I list a number of more advanced texts and compendia. These are denoted with a (*) and require a knowledge of calculus and physics often beyond the first-year college course level. A book denoted with (**) requires at least an undergraduate degree in physics. To those with such a background attention is drawn to specialized symposia reports published by Kluwer Publishers in Dordrecht, the Netherlands, and to the summarizing articles in *Annual Reviews of Astronomy and Astrophysics*.

Some general books and references

Three good, general astronomy references are:
Field Guide to the Stars and Planets, D.H. Menzel and J.M. Pasachoff, 1983, Houghton Mifflin (popular, informative).
Cambridge Atlas of Astronomy, Jean Audouze and Guy Israel (eds.), 1985, Cambridge University Press (excellent).
Larousse Astronomy, Philippe de la Cotadiere, translated by M. Morris, 1986, Facts on File Inc., NY (covers all topics of astronomy and is beautifully illustrated).
A very readable account at the elementary/intermediate level of selected important topics in modern astrophysics is available in:
The New Astronomy, N. Henbest and M. Marten, 1985, Cambridge University Press.
A good source for astronomical facts is:
Mcgraw-Hill Encyclopedia of Astronomy, Sybil Park (ed.), McGraw-Hill, NY.
At a more advanced level try:
(**) *The New Physics*, Paul Davies (ed.) 1989, Cambridge University Press (see especially chapter 6, 'New Astrophysics', M. Longair).
(*) *Astrophysics: I Stars; II Interstellar Matter and Galaxies*, R. Bowers and T. Deeming, 1984, Jones and Barlett (an intermediate-level account of some topics).

Chapter 1

There are many elementary textbooks at the first year college level, requiring only high-school mathematics. (Although the most recent dates of publication are given, they soon become obsolete as new editions of popular textbooks tend to appear every two or three years.)

Astronomy, the Cosmic Perspective, M. Zeilik, 1983, Harper and Row.
Dynamic Universe, T.P. Snow, 1985, West Publishers.
The Universe, W.J. Kaufmann III, 1988, W.H. Freeman, NY.
Contemporary Astronomy, J.M. Pasachoff, 1989, Saunders, NY.
Exploration of the Universe, G. Abell, D. Morrison, and S.C. Wolff, 1989, Holt, Rinehart and Winston, NY.
At a somewhat more advanced level, requiring some college mathematics, I recommend:
The Physical Universe, Frank Shu, 1982, University Science Books.
Introductory Astronomy and Astrophysics, M. Zeilik and E.V.P. Smith, 1987, Saunders College Publishing, NY.
A concise summary of basic mathematics and physics relevant to astrophysics is given by:
(*) *Astrophysical Concepts*, M. Harwit, 1988, Springer-Verlag, NY.
Among books emphasizing techniques used and discoveries in different spectral ranges are:
The Invisible Universe, G.B. Field and E.J. Chaisson, 1985, Birkhauser, Boston (discusses how new instrumentation has opened additional windows through which we may explore the universe).
Astronomical Observations, an Optical Perspective, G. Walker, 1987, Cambridge University Press (the techniques of optical astronomy are compared with those in other spectral ranges).
The Invisible Universe, G.L. Verschuur, 1987, Springer-Verlag (how radio astronomy led to the discovery of giant molecular clouds, cosmic masers, the microwave background, pulsars, etc.).
(*) *Radio Astronomy*, J.D. Kraus, 1986, Cygnus-Quasar Books, Powell, Ohio (explains basic methods and concepts of radio astronomy).
(*) *Galactic and Extragalactic Radio Astronomy*, G.L. Verschuur and K.I. Kellerman (eds.), 1988, Springer-Verlag (research papers).
(*) *Exploring the Universe with the IUE Satellite*, Y. Kondo (ed.), 1987 (this collection of papers describes the fundamental contributions made with a modest-sized space telescope – stars, pp. 119–463, interstellar medium and gaseous nebulae, pp. 467–602).

Chapters 2 and 3

The material in these chapters is covered in many physics texts at the elementary and intermediate levels. Such books are ubiquitous in good libraries and bookstores and I make no attempt to provide a full list. One 'classic' is:
(*) *Atomic Spectra*, H.E. White, 1934, McGraw-Hill.
For optics, consider:
Light, M.I. Sobel, 1987, University of Chicago Press.
(*) *Optics*, E. Hecht, 1987, Addison Wesley.
Popular accounts of the quantum theory showing how it plays a role in astronomy and even in daily life are:
Other Worlds, Paul Davies, 1980, Simon and Schuster.
Quantum Universe, Tony Hey and P. Walter, 1987, Cambridge University Press.

Chapters 4, 5 and 6

An historical account of the advance of stellar spectroscopy from 1815 to 1965 may be found in:
Analysis of Starlight, J.B. Hearnshaw, 1986, Cambridge University Press.

A very well written elementary account of stellar spectra is:

Stars and their Spectra, J.B. Kaler, 1989, Cambridge University Press.

For accounts of the analysis of stellar atmospheres and the flow of radiation through matter, see:

(*) *Observational Analysis of Stellar Photospheres*, David Gray, 1976, Wiley, NY also discusses observational techniques).

(*) *Spectroscopy of Astrophysical Plasmas*, A. Dalgarno and D. Layzer (eds.), 1987, Cambridge University Press (this 'Festschrift' volume for Leo Goldberg contains articles not only on stellar atmospheres, but also on gaseous nebulae, the interstellar medium, the sun, and laboratory astrophysics).

The mathematical theory of radiative processes in astrophysics, including the theory of radiative transfer, synchrotron radiation, bremsstrahlung, Compton scattering, and plasma effects, is given in:

(**) *Radiative Processes in Astrophysics*, G. Rybicki and Alan P. Lightman, 1979, Wiley, NY.

The best and most comprehensive book on the modern theory of stellar atmospheres is:

(**) *Stellar Atmospheres*, Dimitri Mihalas, 1978, W.H. Freeman, San Francisco.

Recommended texts at the intermediate level are:

(*) *Introduction to Stellar Astrophysics, I Basic Stellar Observations and Data, II Stellar Atmospheres*, Erika Bohm-Vitense, 1989, Cambridge University Press.

(*) *The Fundamentals of Stellar Astrophysics*, G.W. Collins, 1989, Freeman.

Chapters 7, 8, and 9

Excellent popular accounts of stellar evolution are found in:

Stars and Clusters, C. Payne-Gaposchkin, 1979, Harvard University Press.

A Hundred Billion Suns, R. Kippenhahn, 1983, Basic Books, NY (An account of life histories of stars from the point of view of one of the leading contributors to the field).

Readers with a good undergraduate background in physics, especially nuclear physics, should find much of interest in:

(*) *Principles of Stellar Evolution and Nuclear Synthesis*, Donald Clayton, 1968, McGraw-Hill, NY.

(*) *Essays on Nuclear Astrophysics*, C.A. Barnes, D.D. Clayton and D.N. Schramm (eds.), 1982, Cambridge University Press.

(*) *Nucleosynthesis, Challenges and New Developments*, W.D. Arnett and J.W. Truran, 1985, University of Chicago Press.

The interesting evolution of binary stars is explored in: *Binary and Multiple Systems of Stars*, A.H. Batten, 1973, Pergamon Press (a good intermediate-level introduction).

(*) *Interacting Binary Stars*, J. Sahade and F.B. Wood, 1978, Pergamon Press (a somewhat more advanced treatment than that of Batten).

(*) *Close Binary Stars*, M.J. Plavec, D.M. Popper and R.K. Ulrich (eds.), 1980, Reidel (these papers discuss observational and theoretical studies involving optical, infrared, ultraviolet, X-ray, and polarimetric data).

The role of accretion disks, not only in binary systems but also in quasistellar objects and active galactic nuclei, is given in:

(*) *Accretion Power in Astrophysics*, J. Frank, A.R. King and D.J. Raine, 1985, Cambridge University Press.

The advanced mathematical theory of objects produced at the end of stellar evolution is given in:

(**) *Black Holes, White Dwarfs, and Neutron Stars*, S.L. Shapiro and S.A. Teukolsky, 1982, Wiley, NY.

Chapter 10

Two general references include *In Darkness Born*, Martin Cohen, 1988, Cambridge University Press (formation of stars in the interstellar medium and pre-main-sequence evolution), and *Searching Between the Stars*, Lyman Spitzer Jr, 1972, Yale University Press (a semi-popular account of discoveries made with the Copernicus Satellite).

Among advanced monographs and summaries are:

(*) *Physical Processes in the Interstellar Medium*, L. Spitzer, 1978, Wiley, NY.

(*) *The Orion Complex, A Case Study of the Interstellar Medium*, G. Goudis, 1982, Reidel (a summary of Orion Nebula research).

(*) *Planetary Nebulae*, S.R. Pottasch, 1983, Reidel.

(*) *Physics of Thermal Gaseous Nebulae*, L.H. Aller, 1984, Reidel.

(*) *Astrophysics of Gaseous Nebulae and Active Galactic Nuclei*, D.E. Osterbrock, 1989, University Science Books, Mill Valley, California.

(*) *Interstellar Processes*, D.J. Hollenbach and H.A. Thronson (eds.), 1987, Reidel (papers presented at a symposium on the interstellar medium).

(*) *The Center of the Galaxy*, Mark Morris (ed.), 1989, Kluwer.

(*) *Interstellar Matter*, G.L. Verschuur, 1989, Springer.

Chapter 11

A number of popular essays have been written on special aspects of late stages of stellar evolution, including supernovae. Three very well written books are:

The Crab Nebula, Simon Mitton, 1979, Charles Scribner's and Sons.

The Quest for SS 433, David H. Clark, 1985, Penguin Books, NY.

Frozen Star, George Greenstein, 1983, Freundlich, NY (stresses neutron stars, black holes, and relativity effects).

Observational characteristics of novae are described in *Galactic Novae*, C.H. Payne-Gaposchkin, 1964, Dover, NY.

A good account of η Carinae, as well as supernovae, supernova remnants, pulsars, black holes, and the missing mass is given in:

Monsters in the Sky, Paolo Maffei, 1980, MIT Press.

Specialized accounts of topics in this chapter are found in:

(*) *Observational Tests of Stellar Evolution Theory*, A. Maeder and A. Renzini (eds.), 1984, Reidel.

(*) *Classical Novae*, N. Evans and M. Bode (eds.), 1988, Wiley, NY.

Popular accounts of supernovae and supernova remnants are found in:

Superstars, David Clarke, 1984, McGraw-Hill (historical emphasis).

Supernovae, Paul and Lesley Murdin, 1985, Cambridge University Press.

There is a very extensive technical literature on supernovae. The earlier work is nicely summarized by Virginia Trimble in *Reviews of Modern Physics*, **54**, 1183 (1982); **55**, 4511 (1983), while an update and bibliography are given by W. Weiler and R.A. Sramek, *Annual Reviews, Astronomy and Astrophysics*, **26**, 295 (1988).

Chapter 12

Earlier work on X-ray astronomy (much of it pertinent to close binaries, cf. Chapter 11) is summarized in:

Glimpsing an Invisible Universe, R.F. Hirsch, 1983, Cambridge University Press.

The X-ray Universe, W. Tucker and R. Giaconni, 1985, Harvard University Press.

For a yet higher energy range of radiation see:
(*) *Gamma-Ray Astronomy*, P.V. Ramana Murthy and R.F. Hirsch, 1983, Cambridge University Press.
The standard advanced text on the subjects of this chapter is
(**) *High Energy Astrophysics*, M.S. Longair, 1981, Cambridge University Press. (Attention is also called to his chapter in *The New Physics*, which is listed under the general books.)

Name index

Abell, G., 264
Acker, A., 348
Adams, W.S., 114, 116, 223
Alfven, H., 252
Allen, C.W., 333
Allen, L.R., 64
Aller, H.D., 313
Allwright, J., 122
Alpher, H., 169
Ambarzumian, V., 144, 206
Anders, E., 101, 102
Anderson, C., 136
Angel, R., 53, 202
Anglo-Australian Telescope, 94
Ångstrom, A.J., 19
Arnett, W.D., 273
Arp, H.C., 333, 336
Avrett, E.A., 63, 64

Baade, W., 121, 270, 282, 307, 310
Babcock, H.W., 53
Bahcall, J., 153, 155
Balick, B., 211–214
Balmer, J.J., 41
Baluja, K.L., 220
Barnard, E.E., 255
Barnes, T.G., 64
Batten, A.H., 32
Bayer, J., 267
Becklin, E., 14
Becvar, A., 318
Bell, Jocelyn, 308
Bessel, F.W., 201
Bethe, H., 139, 141, 273
Blaauw, A., 144
Bless, R.C., 337
Bohr, N., 41–43, 50
Bonner Durchmusterung, 317
Bowen, G., 192–196
Bowen, I.S., 216, 226, 227
Brackett, F.S., 41
Brault, J., 96
Bright Star Catalogue, 317
Bunsen, R., 20, 50
Burbidge, G., 169–171
Burbidge, M., 169–171
Burger, H., 96

Cacciani, A., 200
Cameron, A.G.W., 170, 171
Cannon, A.J., 26
Carilla, C.L., 300
Caswell, J.I., 263
Cerro Tololo Observatory, 269
Chandrasekhar, S., 81, 82, 169, 201, 223
Chevalier, R., 171
Christensen-Dalsgaard, 198
Christy, R.F., 172, 189, 190
Chu, Y.H., 211
Clark, D.B., 263
Clark, D.H., 263, 264
Clark, G.W., 262
Clayton, D., 170
Code, A., 64, 337
Cohen, M., 181
Cooke, W.J., 310
Copernicus satellite, 210, 233
Cordoba Durchmusterung, 317
Cox, A., 181
Cox, J., 189
Crampton, D., 264
Critchfield, C.L., 139
Curtiss, R.H., 70, 98
Czyzak, S.J., 178, 220
Dalgarno, A., 102
Danielson, G.F., 211
Davidson, K., 269
Davis, J., 337
Davis, R., 140
De Greve, J.P., 161
De Loore, C., 161
De Marque, P., 153, 157, 158
Deubner, F.L., 198
Dirac, P.A.M., 136
Disney, M.J., 310
Dombrovsky, V.A., 282, 307
Dominion Astrophysical Observatory, 77
Dreyer, J.L.E., 318
Duncan, J.C., 301
Dunham, T., 233, 269

Eddington, A.S., 187–189, 265
Edlen, B., 292
Einstein, A., 128, 237, 286, 302
Einstein (X-ray) Observatory, 292

Elvey, C.T., 78
Evans, D.S., 64
Ewen, H., 232

Fabian, A.C., 263, 264
Faraday, Michael, 51, 248
Fath, E.A., 189
Feifenstein, R.C., 310
Ferland, G.J., 260
Fermi, E., 298
Fitzgerald, M., 176
Flammarion, C., 269
Fleming, W., 26
Ford, H., 262
Foster, J.S., 76
Fowler, W.A., 169, 170, 171
Fraunhofer, J., 22
Fuller, Buchminster, 131

Gamow, G., 169, 171, 272
Garmire, G.P., 282
Gatley, I., 214
Gehrz, R. 268
Giacconi, R., 282, 292
Gold, T., 373
Gough, D., 198, 199
Grandi, S., 263
Gratton, L., 208, 269
Greenstein, G. 312
Greenstein, J.L., 14, 71, 201, 202
Grevesse, N., 101, 102
Grindlay, J., 297
Gull, T.R., 269
Gum, C. 231
Gursky, R., 282, 292

Hale, G.E., 52, 53
Halley, E., 267
Hanbury-Brown, R., 64, 109, 337
Haro, G., 144, 145
Harris, W.E., 122
Hart, M., 156
Harvard College Observatory, 26
Harvard Revised Photometry, 30
Hawkins, I., 225
Hayashi, C., 143, 147
High Energy Astrophysical Observatory, HEAO, 292
Helmholz, H., 128
Henry Draper catalogue, 26, 317
Herbig, G., 144, 145
Herschel, Sir John, 142
Hertzsprung, E., 11, 104, 105
Hess, V.F., 281
Hesser, J.E., 122
Heubner, W.F., 155
Hewish, A., 308, 309
Hinkle, K., 191
Hjalmarson, A., 347
Hjellming, M.E., 263
Hoffleit, D., 271, 317
Hofmeister, E., 189
Holweger, H., 63
Houck, Nancy, 26, 27, 115, 317

Hoyle, F, 169, 170, 171
Huggins, W., 22, 24, 25
Humphrey, C.J., 44

Iben, I., 147, 149, 161, 172, 174, 275
Index Catalogue of Nebulae and Clusters, IC, 318
Infrared Astronomical satellite, IRAS, 252
Innes, R.T.A., 10
International Ultraviolet Explorer, IUE, 59, 179, 210, 212, 220, 221, 233, 269
Iowa State University, 192

Jacoby, G., 96, 211
Jenkins, E.B., 234
Jewitt, D.C., 211
Johnson, Harold, 242, 332, 334, 337, 341, 342
Joy, A.H., 144, 191
Jugaku, J., 93
Jura, M., 206, 234, 235, 254

Kallman, T., 227
Kamiokande II, 217
Kelvin, Lord, 128
Kemp, J.C., 53, 202
Kingston, E., 222
Kippenhahn, R., 149, 160, 161
Kirchhoff, G.K., 20–23, 50
Kirshner, R., 171
Kohlschutter, A., 114, 116
Kohoutek, L., 318
Kraft, R., 189
Kraushaar, W.L., 282
Krisz, S., 163
Kroto, H., 181
Kudritzki, R., 99, 215
Kupferman, D.N., 211
Kurucz, R.L., 93, 94
Kwok, S., 176, 211

Labs, D., 60, 330
Lambert, D.L., 260
Landstreet, J.D., 53, 202
Large, M.L., 310, 311
Larson, R.B., 143, 153
Lawrence Radiation Laboratory, 137, 138
Layzer, D., 102
Leavitt, H.S., 185
Ledoux, P., 192
Leeds, University of, 297
Leibacher, J.W., 197, 198
Leighton, R., 197
Libby, W.F., 39
Lick Observatory, 4, 21, 24, 25, 32, 224, 225, 247, 304, 306, 310
Liller, W., 215
Lockyer, N., 24
Loeser, R., 63, 64
Lubow, S.H., 155
Lyman, T., 41

McCray, R., 227
McKellar, A., 233
McLaughlin, D.B., 257
McMath–Hulbert Observatory, 197

McNamara, D., 197
Maeder, A., 266, 267
Margon, B., 362–365
Masson, C.R., 215
Mathis, J., 241
Maury, A.C., 26
Mayall, N.U., 304, 306
Mendoza, C., 220
Menzel, D.H., 22
Messier, C., 3, 318
Meyer, J.P., 298, 299
Meyer, W.F., 197
Michaud, G., 203
Michigan, University of, 26, 27, 71, 197
Milgrom, M., 263, 264
Mihalas, D., 99
Miller, J.S., 310
Mills, B.Y., 223, 247, 249, 310
Minkowski, R., 270
Moffett, T., 64
Mohler, O.C., 71
Molonglo Radio Telescope, 311
Moore, J.H., 32
Morgan, W.W., 115, 345, 346
Morris, M., 248, 249, 347
Morton, D., 64
Mount Stromlo Observatory, 3, 230
Mount Wilson Observatory, 23, 46, 175, 200
Muller, E., 63
Munch, G., 304, 305

Neckel, H., 60, 330
Neckel, T., 318
Nelson, G., 249
Neubauer, F.J., 32, 115
New General Catalogue of Nebulae and Clusters,
    NGC, 318
Newberry, M.V., 26, 27
Newton, Sir Isaac, 15
Ney, E., 181, 268
Nicholl-Bohlin, J., 179
Nomoto, K., 273, 275
Norris, Ray, 238, 239
Nussbaumer, H., 220, 227

O'Connell, R., 202
Oke, B., 64
O'Mara, B.J., 94
Oppenheimer, J.R., 110
Orbiting Astronomical Observatory, OAO, 59

Pacsynski, B., 147, 161
Palomar Observatory, 304, 305
Panek, R.J., 179
Paoline, F., 292
Parker, E., 180
Parker, P.D., 155
Paschen, F., 41
Payne-Gaposchkin, C., 70
Pease, F.G., 109, 112
Peimbert, M., 171, 223
Penzias, A., 282
Perek, L., 318
Peterson, B., 94

Pfund, A.H., 41
Pickering, E.C., 26, 32, 50
Plaskett, H.H., 50
Plavec, M., 161, 163, 165
Popper, D.M., 64, 107–111, 124, 337
Pouillet, C.S., 127
Pradhan, A.K., 220
Purcell, E., 232
Purton, C., 176

Rees, M., 213
Renzini, A., 147, 172
Reynolds, S., 313
Rhodes, E., 198
Ritter, A., 187
Ross, J.E., 94–96
Rossi, B.R., 282, 292
Russell, H.N., 11, 97–99, 105

Sackmann, I.J., 172
Saha, M.N., 67
Salpeter, E., 170
Sandage, A., 151, 254, 343, 344
Sanduleak, N., 263, 313
Saraph, H.E., 227
Savage, R., 241
Scalo, J.M., 172
Schmidt-Kaler, T., 337
Schwarzschild, M., 147, 188
Seaton, M.J., 220, 227
Secchi, P.A., 25, 26
Shapley, H., 187
Shields, G., 260
Shklovsky, I.S., 307
Shobbrock, R., 337
Shott, P., 122
Simon, G., 198
Skalnate-Pleso Sky Atlas, 318
Sky and Telescope, 214, 239
Slee, O.B., 249
Smith, Harlan, 196
Smithsonian Astrophysical Observatory, 317
Sneden, C., 260
Spitzer, L., 234
Staelin, D.C., 310
Stein, R.F., 197
Stephenson, C.B., 263
Stretson, P., 122
Stone, R., 262
Strand, K., 11
Stromgren, B., 231, 234, 335
Struve, O., 78, 196, 197
Sulentic, J., 318
Swedlund, J.B., 53, 202
Sweigart, A.V., 147
Sydney University, 311, 347

Tayler, A.J., 310
Testerman, L., 96
Thackeray, A.D., 269
Thomas, H., 149, 160
Tifft, W.G., 318
Tomczyk, S., 200
Toomre, J., 198

Tousey, R., 242
Treffers, R.R., 9
Trimble, V., 306
Trumpler, R.J., 115
Tutukov, A.V., 161, 275
Twiss, R.W., 109

Ulrich, R.K., 153, 155, 197, 198, 200
Uppsala Schmidt Camera, 3

van de Hulst, H.C., 232
Vandenberg, D.A., 122
Van den Berg, S., 171
Vaughan, A.E., 310
Vehrenberg, H., 318
Vernazza, J., 63, 64
Very Large Array, VLA, 9, 211, 300
Very Long Baseline Interferometer, VLBI, 7, 238
von Weizacker, C.F., 141

Walborn, N.R., 269
Walker, M.F., 145
Walraven, T., 196
Wampler, E., 310
Wannier, P.G., 349

Wasson, J., 102
Weaver, T., 273
Webbink, R.F., 161
Weigert, A., 149, 160, 161
Westerlund, B., 3
Weymann, R., 153, 155, 227
Wheeler, C., 270
White, H.E., 49, 76
Wildt, R., 80
Williams, R.E., 227
Willson, Lee Ann, 192–195
Wilson, J.R., 273
Wilson, R., 282
Wollaston, W., 21
Woolf, N., 181
Woosley, S.E., 273, 275, 277–279
Wunner, G., 202

Yusef-Zadei, M., 249

Zanstra, H., 223
Zeeman, P., 51, 52
Zeippen, C.J., 220
Zuckerman, B., 14,. 214
Zwicky, F., 270, 271, 310

Object index

Abell 30, 213
Achernar, 116
Aldebaran, 8, 65, 195, 112
Algol, 13, 113, 163, 207
Altair, 8, 61, 65
Andromeda Galaxy (M31), 123, 229, 270
Antares, 29, 65, 111, 113, 116
CY Aquarii, 196
Nova Aquilae 1918, 255, 257
Arcturus, 9, 29, 65, 104, 106, 112, 116

BD (*Bonner Durchmusterung*) + 30° 3639, 266
Becklin–Neugebauer object, 253
Betelgeuse, 29, 65, 106, 113, 116, 179
Big Dipper, 32

VZ Cancri, 196
ζ Cancri, 11
Canopus, 29, 65, 106, 206, 255
α_2 Canum Venaticorum, 90
Capella, 29, 65, 104, 106, 112
η Carina nebula, 204, 205, 209, 230, 231
η Carina (star), 266–270
Cassiopeia A (supernova remnant), 171, 279
Castor, 11
Castor C, 77
α Centauri, 10, 106
β Centauri, 106
Centaurus X-1, 311
δ Cephei 196
DQ Cephei, 196
VV Cephei, 105, 113
μ Cephei, 105
Coma Berenices cluster, 119
T Coronae Borealis, 260
Crab Nebula (M1), 279, 282, 287, 304–313
Nova Cygni, 255
P Cygni, 79, 99
Y Cygni, 109, 143
31 Cygni, 78
61 Cygni, 104, 106, 116
V1500 Cygni, 259, 260
Cygnus A, 298, 300, 302
Cygnus X-1, 294, 261, 297
Cygnus X-3, 229, 263
3C 273 (quasi-stellar object, QSO), 301

Deneb, 106
Denebola, 65
30 Doradus, frontispiece, 209, 254

σ_2 Eridani B, 111

Galaxy (Milky Way System), 2, 3, 5, 8, 105, 123,
    169, 204, 205, 209, 210, 232, 234, 247, 282
center of, 236, 249, 251, 284, 296
halo of, 123, 247
rotation of, 123
γ Geminorum, 116

HD (Henry Draper) 30353, 167
HD 215441, 53
Hercules X-1, 294, 311
α Herculis, 113
AM Herculis, 294
BL Herculis, 185, 261
DQ Herculis (Nova Herculis 1934), 256, 257, 260
HZ Herculis, 261
Hyades, 119–121, 157
VZ Hydrae, 159

IC (Index catalogue) 443, 278
IC 2165, 225, 228
IC 3568, 225
IC 4846, 225
ε Indi, 104
IRC 2, 238

Kepler's supernova, 270, 279
Kleinman–Low object, 238, 251
Kochab, 104
Krueger 60, 106
Krueger 60B, 112

10 Lacertae, 106
Lagoon Nebula (M8), 231, 250
RR Lyrae, 196
β Lyrae, 207

Magellanic Clouds, 4, 210, 229, 279, 280, 342
    Large, 3, 14, 95, 123, 251, 261
    Small, 185
M (Messier) 8, 119

M11, 119, 121, 157
M13 (globular cluster in Hercules), 158
M15, 158
M31, *see* Andromeda Galaxy
M33, *see* Triangulum Galaxy
M67, 120, 121, 157, 158
M87 (giant elliptical galaxy in Virgo), 302
M92, 158
M101, 169, 339
M (Minkowski) 2–9, 213
Mira, 190–195, 238, 245
Mizar, 32

Network Nebula, 279
NGC (New General Catalogue) 40, 224, 226
NGC 188, 120, 121, 157, 158
NGC 604, 254
NGC 2244, 119
NGC 2264, 144, 145
NGC 2362, 119–121, 157
NGC 2440, 178
NGC 3242, 213
NGC 6058, 178
NGC 6445, 178
NGC 6537, 228
NGC 6543, 211
NGC 7008, 178, 211
NGC 7027, 228, 245

ρ Ophiuchi complex, 250
Orion Nebula, 21, 24, 25, 143, 208–210, 223, 228, 231, 247, 251, 282
Orion region, 1, 2, 29

β Pegasi, 104
Nova GK Persei (1901), 257
h and χ Perseus cluster, 119–121, 157
SX Phoenicis, 196
β Pictoris, 144, 252
ZZ Piscium, 14
Pleiades, 119–121, 157

Pollux, 65
Praesepe cluster, 119
Procyon, 10, 65, 106, 116, 157
Procyon B, 110
ζ Puppis, 50, 106, 179, 206

Regulus, 65
Rigel, 9, 65, 106, 116, 206

FG Sagittae, 168
υ Sagittarii, 71, 164
Sagittarius B2, 236, 251, 347
Sanduleak $-69°$ 202 (precursor of supernova 1987A) 277
τ Scorpii, 46, 177
Scorpio–Centaurus Association, 144
Scorpio X-1, 295
δ Scuti, 189, 196
Sirius, 7, 8, 65, 109, 157, 255, 327
Sirius B, 107, 110, 198
Southern Cross, 230
Spica, 46, 177
SS 433, 262–265, 299
Supernova 1987A, 273, 276–278

T Tauri, 252
Taurus interstellar clouds, 143, 251
Triangulum spiral galaxy (M33), 3, 4, 5, 143, 169, 209, 251, 279, 286
Trifid Nebula, 231
47 Tucanae, 123, 158
Tycho's supernova, 270, 279

van Biesbroeck's star, 112
Vega, 65, 109, 116, 144, 152
Vela Supernova Remnant, 279, 310
AI Velorum, 196
γ Velorum, 206
W Virginis, 185
Z Vulpeculae, 163

Subject index

absolute magnitude effects, 113–117
absorption of radiation
    by earth's atmosphere, 18, 19
    by interstellar medium, *see* interstellar
        medium, extinction . . .
accretion disk, 144, 160, 164, 165, 261
acoustical waves, in sun, 197–199
active galactic nuclei, AGN, 292, 295
Alfven waves, 252
alpha particle, 139, 140
amorphous silicate, 242
ångstrom unit, 19
anti-neutrino, 136, 285
apsidal motion in binary stars, 166
astronomical constants, 330
astronomical unit, 7, 330
asymptotic giant branch, AGB, *see under* stellar
    evolution
atomic weight, 37–39
atoms
    constituents of, 36, 37
    complex, 45
    energy level diagrams of, 44
    ionization of, 42, 66, 326–328
auroral type transitions defined, 217
Avogadro number, 321

background ($3/K$) radiation, 169, 282
Balmer series, 21, 40–45, 74, 191, 222
beta decay, 136
big bang, 103, 169, 171, 254
black hole, 14, 117, 206, 207, 251, 273, 292, 296,
    299, 303
Bohr model of hydrogen atom, 41–45
bolometric magnitude and corrections, 62, 107,
    118, 337–339
Boltzmann constant, 321, 322, 325
Boltzmann's law, 237, 238, 324–326
bound–free atomic processes, 81–83
bouyancy or gravity waves in the sun, 197
Bowen fluorescent mechanism, 226–227
Brackett series, 41, 44, 50
Bremsstrahlung (free–free emission), 222, 246,
    286, 295, 301
brown dwarfs, 112, 146

bubble chamber, 136, 137
bursters, 292, 295

β Canis Majoris variables, 182, 184
carbon burning, 170
carbon monoxide, 66, 235, 236
carbonaceous chondrites, 101, 102, 167
catastrophic variables (*see also* novae), 206, 207,
    260–262, 294
CCD, Charge Coupled Device, 8, 117, 211
Cepheid variables, 133, 195, 185–190
    classical type, 182, 184, 185
    dwarf type, 184, 186
    overtone pulsations of, 189
    period–luminosity relation, 185
    population type II variety, 184, 185
    pulsation theory for, 187–190
Cerenkov radiation, 284, 296, 297
chemical compositions
    of cosmic rays, 299
    of gaseous nebulae 227–229
    of interstellar medium, 234, 235
chemical compositions, stellar, 72–103
    curve of growth method for, 83–91
    differences in, 70, 167–169
    influence on stellar structure, 132
    refined methods, 92–95
chromosphere
    solar, 40, 83
    stellar, 95, 145, 179, 191
circumstellar envelope, 176, 180–182
CNO (carbon–nitrogen–oxygen) cycle, 140, 141,
    228
collisonal broadening of spectral lines, 78, 79,
    85–88
color index, 117, 335–337
combination variables, 294
Compton effect, 287, 288, 295
convection
    in stellar interiors, 132, 133, 146, 147, 166, 171,
    198
    turbulent, 265
cores, stellar, 148
corona, 99, 179, 191, 292
cosmic rays, 5, 51, 137, 138, 173, 210, 236, 247,

362

283–285, 290, 291
  origin of, 297–298
Coulomb's law, 320
curve of growth
  in interstellar medium, 234
  stellar, 84–91, 328, 329
cyanogen radical, 54, 56, 68

definitions and units of force, energy, power,
    etc., 319–321
deformation oscillations of stars, 196–200
degenerate gas, 148, 149, 201, 310
deuterium, 39, 135, 136, 169
diffraction grating, 15
diffuse gaseous nebulae, *see* H II regions
Doppler effect, 8, 30, 32
  on spectral lines, 73–75, 78, 85, 86–88, 234,
    256
double stars, *see* stars, binary, and eclipsing
    binaries
dredge-up cycle, 174
dust (circumstellar and interstellar), *see* grain
    formation and occurrence
dwarf (main-sequence stars) 14, 105, 106

eclipsing binaries, 12, 13, 109–113, 158–163
effective temperatures of stars, 62, 63, 338, 339
effective wavelength, 232, 233
electric fields in stellar atmospheres, 53
electromagnetic radiation, 5, 18, 32–34
electron
  charge and mass of, 321
  spin of, 47
electron scattering, 132
electron temperatures in gaseous nebulae, 219,
    224, 227
equivalence of mass and energy, $E = mc^2$, 128,
    286, 362
equivalent width of a spectral line, defined, 83
excitation of atomic levels, 324–328

Faraday effect, 248, 309
Fermi mechanism for acceleration of cosmic
    rays, 296
flares, 78, 144
fluorescence
  Bowen mechanism of, 226
  in gaseous nebulae, 220–227
  in long-period variables, 491
forbidden lines, 216–220, 223, 256
free–free absorption and emission processes (*see*
    also Bremsstrahlung), 81, 82, 246, 286
f-value, 84, 91

galaxy (Milky Way System), *see* Object index
gamma rays, 5, 14, 19, 210, 272, 278, 282, 283,
    286, 287, 295–297
  as tracers of cosmic rays, 295
gamma-ray bursters, 284
gas laws, 131, 132, 148, 321, 322
giant molecular clouds, *see* molecular clouds,
    giant
glitches, 309
globules, 204, 250
grain formation and occurrence

  in interstellar medium, 240–243
  in stellar atmospheres, 173, 180–182
graphite, 242
gravitation, constant of, 321
Greek alphabet, 318
greenhouse effect, 153
gyromagnetic radiation, 293–294

H II regions, 231, 232
helium flash, 149
Hertzsprung gap, 106, 149, 157
Hertzsprung–Russell diagram, 103, 355
  for galactic clusters, 120, 121, 124–126, 343,
    344
  for globular clusters, 122, 123 126
  theoretical interpretation via stellar evolution,
    127, 148, 149, 156–158
high-latitude cirrus, 250
Hubble–Sandage variables, 266, 270
hydrogen
  atomic model of, 41–45
  molecular, 244, 245
  21-cm radiation of, 48, 209, 232, 233
  two-photon emission of, 222
hydroxyl radical, 66
hyperfine structure, 79, 89, 90, 232

induced emission (negative absorption), 237
interferometer, stellar, 58, 61, 108, 109, 113, 179,
    182
interstellar medium, 5, 177, 204–254
  absorption lines produced by, 233–235
  chemical composition of, 234, 235
  cycling of material in, 207
  extinction by dust grains in (*see also under*
    grain formation and occurrence), 62, 117,
    124, 143, 145, 209, 240–243, 336, 337, 341,
    342
  McKee–Ostriker model of, 208, 209
  magnetic fields in, 209, 238, 243–249
  molecular clouds in, *see* molecular clouds
  reddening by dust grains in, 181, 240–242,
    326–342
inverse Compton effect, 287, 288, 301, 302
ionization of atoms, 42, 66, 324–328
ionization potential, 43, 326, 327
ionized atoms, spectra of, 50
iron peak, 171
isotopes, 39, 100, 135, 235
  effects in molecules, 55, 56, 347, 349

jansky, defined, 238
Johnson–Morgan system of photometry, 116, *see*
    also UBV photometry

Kelvin or absolute temperature scale, 57, 319
Kirchhoff's laws
  of spectroscopy, 21, 207
  of radiation, 58

Lagrangian point, 161, 258
Larmore frequency, 293, 294
lasers 236, 237
lepton, 36

limb darkening in sun, 63, 82
long-period variables, 176, *see also* Mira
    variables
    complex spectra of, 190–192
    interpretation via pulsation theory, 193–195
    relation to OH–IR stars, 195
    as suppliers of material to interstellar medium,
        207
Loschmidt number, 301
Lyman series, 43, 44, 84

Mach number, 243
magnetic fields, 34, 283
    detection by Zeeman effect, 51–54
    effects on charged particles' motions, 280–290,
        293, 294
    effects on spectral line broadening, 74, 78
    in interstellar medium, 209, 238, 243–250, 284,
        298
    in neutron stars, 302, 309–313
    in sunspots, 52
    units of, 320
magnetic stars, 53, 90, 202, 261
magnitude, stellar
    absolute, M, 9, 119, 330, 337, 340, 341
    apparent, m, 8
    bolometric, 8, 62, 119, 337
    photoelectric, 8, 116–119
    relation between, M, m, and distance, 331
    systems of, 116–119, 332–335
main sequence, 106, 108, 119, 120, 144, 145, 157,
    158
masers, 56, 175, 236–240
mass loss in close binaries, 162–166
    in winds, *see* winds, stellar
mass–luminosity correlation, 107, 108, 110, 127,
    134
mesons, 36, 137, 285, 295, 296
metallicity index, 335
metastable levels, 217, 218
meteorites, 101, 167
Michelson stellar interferometer, 58, 112
Mira variables, 176, 180, *see also* long-period
    variables
model stellar atmospheres, 62–64, 92
molecular clouds, 204, 208, 238, 249–253, 347
    giant, 143, 243, 250, 253
molecules
    dissociation of, 66, 97
    in interstellar medium, 56, 210, 236–240, 250,
        251, 347–349
    in stellar atmospheres, 96–98, 180
    structure and spectra of, 54–56
muon, 285

natural line broadening, 73, 74, 85
nebulae, diffuse, *see* H II regions and Orion, 30
        Doradus, and η Carina (in Object index)
    planetary, *see* planetary nebulae
nebular type line, defined, 217
negative absorption (induced emission), 237
negative hydrogen ion, 80–82
neutrino, 136, 140, 141, 272, 273, 285

neutron, 36, 135, 136
neutron capture, 169–173
neutron stars (*see also* pulsars), 14, 177, 201, 206,
        254, 279, 284, 292, 294, 295–311
Novae, 79, 106, 177, 210, 255–260
    light curves of, 257
    runaway nuclear reactions in, 259–260
    spectra of 256, 260

Oddo and Harkins' rule, 103
OH–IR objects, 59, 175, 176, 184, 235, 238
opacity
    of solar atmosphere, 63
    of stellar atmospheres, 79–83
    of stellar interiors, 132
oscillations of stars, *see* stellar pulsations
ozone, 18, 19

P Cygni
    line profile types, 177, 179, 223
    phenomenon, 177
PAH, polycyclic aromatic hydrocarbons, 181,
        236, 242, 347, 349
pair production, 286
parallax, stellar
    defined, 6
    spectroscopic, 113–116, 345
    trigonometric, 7, 8, 11
parsec, defined, 7
Paschen series, 41, 44, 222
photoelectric cell, 19
photoelectric photometer, 116
photometry
    of stars, 116–119, 332–337
    of star clusters, 121–124
photon correlation interferometers, 58, 109
photosphere, 22, 63, 94
physical constants, tables of, 321
pion, 285
Planck's constant, 43
Planck's radiation law, 58, 59, 237
planetary nebulae, 178, 206, 207, 210–229
    appearance and structure of, 210–214
    catalogues of, 318
    central stars of, 223, 224, 261, 266, 211–214
    distances of, 215
    fluorescence in, 220–223, 225–227
    relation to late stages of stellar evolution, 176,
        177
    shock waves in, 213, 214
    spectra of, 215, 228
plerion, 279
polarization of light, 32–34
    by circumstellar grains, 182
    by interstellar grains, 209, 240, 243
    in gyromagnetic radiation, 294
    in synchrotron radiation, 288–290
    in Zeeman effect, 51, 52, 238, 248
population types, stellar, 123
principal component of a binary system, 159
profiles (or shapes) of spectral lines, 86–88, 94,
    95
proper motion, 6, 31

proton, 36, 135, 321
proton–proton reaction, 139, 148
proto-star, 143–146, 252
pulsars (*see also* neutron stars), 51, 248, 279, 308–311
pulsations of stars, *see* stellar pulsations

quantum, 43
quantum mechanics, 35, 49
quark, 36, 135
quarter-wave plates, 34
quasar, QSO, or quasi-stellar object, 210, 301

radiation laws, 57–62, 237, 322, 323
radiation pressure, 180, 195, 265
radiative transport of energy, 132, 133
radio frequency waves, 5, 18, 19, 143, 248
    from Crab nebula, 307–310
    from extended sources, 299–303
    polarization of, 248, 249, 290, 291
radio galaxies, 210
radioactivity, 129
Raman spectra, 242
Rayleigh scattering of light, 181
reddening by grains, 181, 241, 242, 336–342
relativity
    general, 275
    special, 283, 286–289
reversing layer 22
Roche surface (or lobe), 158–163, 258, 273
r-process (rapid neutron capture), 171, 274
RR Lyrae stars, 124, 182, 184, 190
RV Tauri star, 183, 184
Rydberg constant, R, 41, 43, 321

Saha theory, 67–71, 91, 98
Schwarzschild radius, 296
semi-regular variables, 182–184
Seyfert-type galaxies, 227
shock waves, 76, 143, 176, 179, 180, 293, 298
    in interstellar medium, 210, 236, 240, 243–245, 250, 253
    in long-period variables, 193–195
    in planetary nebulae, 211
    from supernovae, 229, 245, 251, 280, 291
    within supernovae, 273
spallation, 170, 285, 299
speckle interferometer, 61, 109
spectra, 20–21
    absorption type, 21
    atomic, 39–54
    continuous, 28, 45, 79–83
    emission, 21, 222
    of gaseous nebulae, 215–228
    of ionized atoms, 50
    Kirchhoff's law of, 21
    molecular, 54
    of sunspots, 52–53
spectral classification (stellar)
    Henry Draper system, 25–30
    Morgan–Keenan system, 115, 116
spectral line formation (stellar), 73–91
spectral sequence, 25–30

interpretation of, 67–71
    splitting of, 70, 167, 174
spectroheliograph, 82
spectroscope, 15, 30
spectrum–luminosity relation (absolute magnitude effects), 113–116
spectrum scanner, 30
spectrum synthesis, 95–97
spontaneous emission, 237
s-process (slow neutron capture), 171–173
star clusters
    color–magnitude diagrams for, 120–126, 342–345
    galactic clusters, 119–121, 124–126, 144, 157
    globular clusters, 119, 121–126, 157, 158, 168, 185
    interpretation of, 151, 156–158
stars
    binary, 10–13, 107–113, 158–166
    catalogues of, 317
    chemical compositions of, *see* chemical compositions, stellar
    colors of, 116–119, 332–337, 338–339
    densities of, 13, 108–111
    diameters of, 13, 107–113, 337
    distances of, 6–8
    element building in, 169–174, 273–276
    energy generation in, 127–148
    evolution of, *see* stellar evolution
    formation of from interstellar medium, 143–146
    of high luminosity, 265–267
stark effect in stellar spectra, 74–78, 89
Stefan's law, 57–62
Stefan–Boltzmann constant, 57
stellar associations (*see also* star clusters), 114, 206
stellar evolution 13, 142–165
    in asymptotic giant branch, AGB, 173–176, 206
    for binary stars, 158–166
    dependence on mass, 143, 157
    observational checks on, 156–158
stellar pulsations, 123, 167, 182–199
Strömgren photometric system, 334, 335
Strömgren sphere, 231, 232
subdwarfs, 106, 125
subgiants, 26, 106, 143
sun
    age of, 128
    atmospheric model for, 62–64
    chemical composition of, 99–103
    chromosphere of, 83, 99
    effective temperature of 63
    energy distribution in spectrum, 60, 63
    evolution and internal structure of, 152–156
    luminosity of, 127, 128, 330
    oscillations of, 197–200
    spectrum of, 22, 23, 79–83
sunspots, 52, 71
supergiant stars, 14, 106, 143, 158
supernovae, 142, 171, 172, 177, 206, 207, 251, 253, 270–280, 306, 313

supernovae (*cont.*)
  characteristics of, 272
  as galaxy distance indicators, 276
  heavy element building in, 273, 274, 277, 278
  of type I, 274, 276
  of type II, 270, 271
supernova remnants, 208–210, 245, 278–280,
    291–293, 298
synchrotron radiation, 248, 279, 282, 287–291,
    295, 301, 307, 310, 317

temperature, stellar
  color, defined, 61
  dependence of radiation output on, 57–59
  effective, 62, 63, 338, 339
  interiors, 134–143
  relation to spectrum, 64–67
  scale of, 57, 319
  of solar corona, 292
  in supernova remnants, 280, 291
  in X-ray sources, 295
thermocouple, 19
thermodynamic equilibrium, 324
thermostat action of nebular line excitation,
    223–225
three-degree 'background cosmic' radiation, 169,
    282
tidal effects in close binary stars, 166
transmutation of elements, *see* stars, element
    building in
triple alpha process, 139, 150, 152, 266, 272
T Tauri variables, 144, 180
turbulence in stellar atmospheres, 78, 89, 90, 116
two-photon emission, 272

UBV photometry, 116–119, 132, 133, 332–335,
    341–345

units of physical forces, energy, power, etc.,
    319–321
URCA process, 272

variable stars, 183
  designations of, 317
  see also Cepheids, long-period variables, RR
    Lyrae stars, novae, and supernove

wave mechanics, 35
wave model of atom, 49, 50
white dwarfs, 10, 11, 14, 106, 110, 147, 175, 199–
    203, 206, 207, 211, 254, 275
  accretion disks in, 164, 292
  in evolved binary stars, 164, 165, 258–263, 275
  magnetic fields in, 53, 202, 294
  pulsating, 202
  spectra of, 202
Wien's law, 322, 323
winds
  solar 99
  stellar, 167, 175–180, 204, 205, 211–214,
    266–269
Wolf–Rayet stars, 71, 79, 168, 206, 223, 224, 226,
    267, 270, 276

X-rays, 5, 19, 78, 134, 205, 210, 244, 278, 280,
    283, 286, 291–295, 302
  absorption of, 284
  diffuse background of, 295
X-ray binaries, 227, 261–263
X-ray bursters, 292, 295

Zeeman effect, 51–54, 71, 89, 238
zero-age main sequence, ZAMS, 120, 132, 147,
    149, 151, 156, 176, 343, 344
ZZ Ceti variables, 202